W9-CDT-726

Renaissance and Revolution is a collection of fifteen essays on some of the problems presently associated with the Scientific Revolution of the sixteenth and seventeenth centuries. The topics treated include the dissemination of Greek science, medical empiricism, natural history, the relations of scholars and craftsmen from the fifteenth to the seventeenth century, the so-called 'mechanical philosophy' in France and England, the work of Isaac Newton, and the difficulties encountered by Newtonianism in Italy in the early eighteenth century. Figures discussed include Leonardo Fioravanti, Jan Swammerdam, Piero della Francesca, Johannes Hevelius, Jonas Moore, Robert Boyle, Isaac Newton, Christiaan Huygens, Francesco Algarotti and Luigi Ferdinando Marsigli. There is an introduction by the editors and an afterword by A. Rupert Hall. The authorship is international, including scholars with established reputations as historians of science.

Renaissance and Revolution

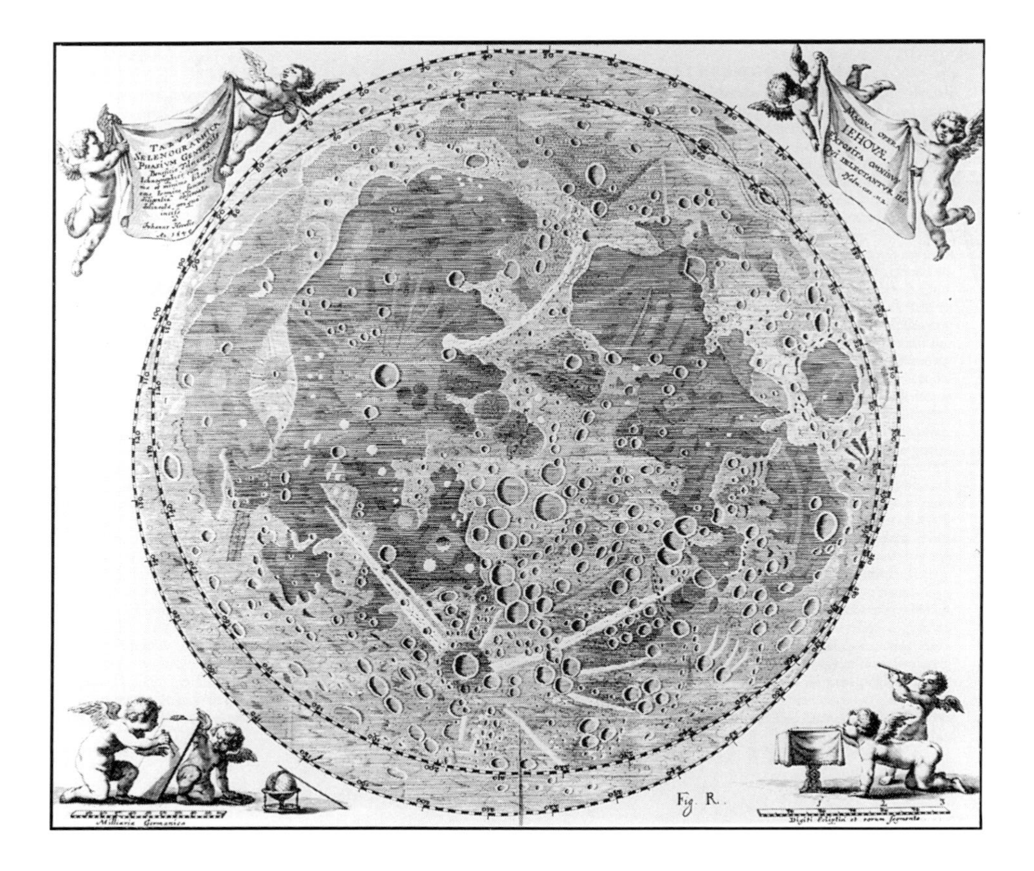

Full Moon engraved by Hevelius. Putti probably added by Adolph Boij. Hevelius, *Seleno-graphia*, (Gdańsk, 1647) pp. 262–3.

Renaissance and Revolution

Humanists, scholars, craftsmen and natural philosophers in early modern Europe

Edited and introduced by

J. V. FIELD
Birkbeck College, University of London

and

FRANK A. J. L. JAMES
Royal Institution Centre for the History of Science and Technology

PUBLISHED BY THE PRESS SYNDICATE OF THE UNIVERSITY OF CAMBRIDGE
The Pitt Building, Trumpington Street, Cambridge CB2 1RP, United Kingdom

CAMBRIDGE UNIVERSITY PRESS
The Edinburgh Building, Cambridge CB2 2RU, United Kingdom
40 West 20th Street, New York, NY 10011–4211, USA
10 Stamford Road, Oakleigh, Melbourne 3166, Australia

First published 1993
Reprinted 1997

Printed in the United Kingdom at the University Press, Cambridge

A catalogue record for this book is available from the British Library

ISBN 0 521 43427 0 hardback
ISBN 0 521 62754 0 paperback

UP

Contents

Preface

The papers collected in this volume originate from the British Society for the History of Science Summer Meeting of 1990, held in Keble College, Oxford, with lectures in the University's Engineering Department. In honour of the seventieth birthday of Professor Rupert Hall, the title of the meeting was 'The Scientific Revolution'. Its subject matter ranged widely, by no means being confined to topics considered in Hall's similarly titled book of 1954 (second edition 1962) or its slightly less similarly titled third edition of 1983.

The conference was attended by about 160 people. Despite the heat – which strained the credulity of foreign visitors and the air-conditioning of the lecture theatre used for the plenary sessions – discussions, both formal and informal, were lively and friendly. References to some of them are to be found in footnotes of papers in this collection.

That everyone at the meeting seemed happy to talk shop with everyone else suggested a community of historical interests that was not immediately apparent on scanning the titles of the papers presented. In selecting papers for publication, we strove to preserve something of the diversity that enlivened the conference while at the same time bringing out some of the unifying elements that made the conference so friendly. Such elements have helped to give coherence to this book.

Like most of the members of the Organizing Committee of the Oxford conference, the editors were students in the Department of History of Science and Technology at Imperial College, where they both wrote their Ph.D. theses (J. V. Field's research being supervised by Rupert Hall, F. A. J. L. James's by Marie Boas Hall). In drawing up the programme of the conference there was general agreement that the guiding principle should be not to look for personal connections with the Halls but rather to ask for papers on subjects with which they had been concerned. This use of the Halls' scholarly work as an organizing principle was never intended to lead, and in the event did not lead, to a conference that was a homage to the Halls in any sense other than that of showing that some of the intellectual hares they had started were still up and running. We consider that this, together with being read, is the healthiest form of homage to a historian. It has been repeated, in a more durable mode, in this book.

The two editors were co-secretaries to the Organizing Committee for the conference, which developed from an idea put forward by John Hendry. The other members of the Committee were Janet Browne, Robert Fox, Willem Hackmann, Graham Hollister-Short, Malcolm Oster and Mari E. W. Williams.

We should like to record our gratitude to John Pickstone (Programme Secretary, BSHS) and Wing Commander Geoffrey Bennett (Executive Secretary, BSHS) for their help; and to the British Academy, the International Union for the History and Philosophy of Science, the Royal Society of London and the Wellcome Trust for generous financial support for the conference.

We are grateful to our contributors for their patience in responding to editorial queries and for allowing the emendations required to give the volume an acceptable degree of stylistic coherence. In particular, we are grateful to Michael Hunter for guidance based on his own experience as an editor, and, most particularly of all, to Rupert Hall, who kindly read and commented on an earlier draft of our Introduction. We owe thanks also to John Henry, Anita McConnell and Simon Schaffer, who provided us with bibliographic references and other items of aid.

During most of the time that work was being done on this book, JVF was an academic visitor in the Mathematics Department of Imperial College of Science, Technology and Medicine (University of London). We should like to express our gratitude to the College and the Department for facilitating our work in this way.

Much of the editorial work involved using FAJLJ's word processor. We are grateful to the other members of the James household – most particularly Joasia James – for their forbearance in tolerating the consequent disruption to their normal lives.

JVF FAJLJ *London, August 1992*

Illustrations

Notes on contributors

PAOLO CASINI earned his *venia docendi* in history of philosophy at the University of Rome, 'La Sapienza', where he currently teaches. He has also given courses in the Universities of Lecce, Trieste and Bologna. Among his publications are: *L'Universo-Macchina. Origini della Filosofia Newtoniana* (Bari, 1969); *Introduzione all' Illuminismo. Da Newton a Rousseau* (Bari, 1973); *Newton e la coscienza europea* (Bologna, 1983); a critical edition of the *Classical Scholia* by Newton, *Hist. Sci.*, 1984, **22**: 1–58. He is preparing a book concerning various aspects of the Pythagorean tradition from the sixteenth to the nineteenth century.
Address: Istituto di Filosofia, Università di Roma 'La Sapienza', Via Nomentana 118, 00161 Roma, Italy.

HAROLD J. COOK, Associate Professor of the Departments of the History of Medicine and History of Science at the University of Wisconsin-Madison, has mainly written on seventeenth-century English medicine. His book *The Decline of the Old Medical Regime in Stuart London* was published by Cornell University Press in 1986. More recently, he has been working on comparative studies of Dutch and English science and natural history, and has contributed an article titled 'The new philosophy in the Low Countries' for *The Scientific Revolution in National Context*, edited by R. Porter and M. Teich.
Address: Department of the History of Medicine, University of Wisconsin Medical School, 1415 Medical Sciences Center, 1300 University Avenue, Madison, WI 53706, USA.

GIORGIO DRAGONI is Professor of Physics at the University of Bologna. He has written many articles on the history of science in Bologna and elsewhere.
Address: Dipartimento di Fisica, Università degli Studi di Bologna, Via Irnerio 46, 40126 Bologna, Italy.

WILLIAM EAMON is Associate Professor and Head of the History Department at New Mexico State University. He has published numerous articles on science and popular culture and on the 'books of secrets' tradition in late medieval and early modern Europe. His book *Science and the Secrets of Nature* is forthcoming.

Address: Department of History, New Mexico State University, Box 3H/Las Cruces, NM 88003, USA.

J. V. FIELD, who has research interests in mathematics and the mathematical sciences in the period from about 1400 to 1650, is the author of *Kepler's Geometrical Cosmology* (1988) as well as co-author, with J. J. Gray, of *The Geometrical Work of Girard Desargues* (1987) and, with E. J. Aiton and A. M. Duncan, of a translation of Kepler's *Harmonices mundi libri V* (Linz, 1619) now in press.
Address: Department of History of Art, Birkbeck College, University of London, 43 Gordon Square, London WC1H 0PD.

KARIN FIGALA studied pharmacy at the Universities of Bern, Bonn and Hamburg. Following industrial employment, she wrote her doctoral thesis on the history of pharmacy (1969) and joined the Technical University, Munich, (now Zentralinstitut für Geschichte der Technik) as assistant professor. She gained her *Habilitation* (1977) with a study of Newtonian alchemy and theory of matter. Further topics of her studies are the history of sixteenth- and seventeenth-century chemistry and alchemy in general, and the life and work of the German alchemist and Rosicrucian Michael Maier.
Address: Institut für Geschichte der Technik, Deutsches Museum, 8000 Munich 26, Germany.

GAD FREUDENTHAL is *Chargé de recherche* at the *Centre National de la Recherche Scientifique (CNRS)*, Paris. His main research interests are the history of theories of matter before the seventeenth century and the history of science within medieval Jewish communities. He has recently edited *Studies on Gersonides – A Fourteenth-Century Jewish Philosopher-Scientist* (Leiden: Brill, 1992).
Address: 156 rue d'Aulnay, F – 92290 Châtenay-Malabry, France.

ALAN GABBEY teaches philosophy at Barnard College, Columbia University, New York, and is former Reader in History and Philosophy of Science, Queen's University Belfast. He is a *membre effectif* of the Académie Internationale d'Histoire des Sciences. His interests lie mainly in Renaissance and seventeenth-century philosophy and science, especially natural philosophy and the philosophy of the mind. Currently he is working on a commentated translation of the More–Descartes letters, an inventory of the manuscripts and papers of Roberval, and a study of the mechanical philosophy.
Address: Department of Philosophy, 326 Milbank Hall, Barnard College, 3009 Broadway, New York, NY 10027–6598, USA.

A. RUPERT HALL first opened the Pandora's box of Newton's Cambridge manuscripts in 1948. His book on the Scientific Revolution was first published in 1954. He has taught in universities in Britain and in the United States of America.
Address: 14 Ball Lane, Tackley, via Kidlington, Oxfordshire, OX5 3AG, England.

R. W. HOME studied physics and then history and philosophy of science at the University of Melbourne before completing a Ph.D. in history and philosophy of science at Indiana University. He has been Professor of History and Philosophy of

Science at the University of Melbourne since 1975. He has published extensively on the history of eighteenth-century physics and on the history of science in Australia.
Address: Department of History and Philosophy of Science, University of Melbourne, Parkville, Victoria 3052, Australia.

MICHAEL HUNTER is Professor of History at Birkbeck College, University of London. He is author of various studies of the early Royal Society and its context, and is currently working on the life and thought of Robert Boyle.
Address: Department of History, Birkbeck College, Malet Street, London, WC1E 7HX, England.

FRANK A. J. L. JAMES is Lecturer in History of Science at the Royal Institution Centre for the History of Science and Technology. He has written, from a variety of perspectives, on the history of the physical sciences during the first two-thirds of the nineteenth century. He is currently editing the correspondence of Michael Faraday (1791–1867), volume one of which was published in 1991.
Address: Royal Institution Centre for the History of Science and Technology, Royal Institution, 21 Albemarle Street, London, W1X 4BS, England.

ROBERTO DE ANDRADE MARTINS is a Brazilian historian of science. After a first degree in physics, he took his Ph.D. in logic and philosophy of science in 1987 at UNICAMP (State University of Campinas). He has published two books (on Copernicus and on Pascal) and several papers on history of science, philosophy of science and physics (relativity theory).
Address: Group of History and Theory of Science, DRCC/IFGW/UNICAMP, Caixa Postal: 6165, 13081 Campinas – SP, Brazil.

VIVIAN NUTTON is Senior Lecturer in the History of Medicine at the Wellcome Institute for the History of Medicine, London.
Address: Wellcome Institute for the History of Medicine, 183 Euston Road, London, NW1 2BP, England.

ULRICH PETZOLD studied chemistry at the Universities of Konstanz and Zurich and has worked as a freelance translator and editor. He cooperated in Karin Figala's studies on Newtonian alchemy and is preparing a thesis on J. F. Vigini and chemical education at Cambridge University. He is also interested in the history of sixteenth- and seventeenth-century alchemy and the publishing history of alchemy.
Address: Institut für Geschichte der Technik, Deutsches Museum, 8000 Munich 26, Germany.

ALBERT VAN HELDEN is Lynette S. Autrey Professor of History at Rice University, where he teaches the history of science and technology. His research area is the history of telescopic astronomy.
Address: Department of History, Rice University, PO Box 1892, Houston, TX 77251, USA.

RICHARD S. WESTFALL is Professor Emeritus of the History of Science at Indiana University. He has made the Scientific Revolution his special field of study and is the author of a biography of Isaac Newton. At present he is working on a social history of the scientific community of that era.
Address: Department of History and Philosophy of Science, Indiana University, Bloomington, IN 47405, USA.

FRANCES WILLMOTH studied history at the University of Birmingham and completed an archive traineeship at the Bodleian Library, Oxford. She prepared the present catalogue of John Flamsteed's papers at the Royal Greenwich Observatory before moving on to work elsewhere as an archivist; her subsequent Ph.D. thesis, a biography of Flamsteed's patron Sir Jonas Moore, was published in 1993. She has undertaken the task of finishing the late Professor Eric Forbes's edition of Flamsteed's correspondence.
Address: 74 Canterbury Street, Cambridge, CB4 3QE, England.

MARY G. WINKLER, an art historian, is Assistant Professor of Medical Humanities at the Institute for the Medical Humanities at the University of Texas Medical Branch in Galveston. Her research area is the artistic tradition of the Northern Renaissance.
Address: Institute for the Medical Humanities, The University of Texas Medical Branch at Galveston, Galveston, TX 77550–2778, USA.

Abbreviations

BWN *Biographisch Woordenboek der Nederlanden.*

DBI *Dizionario Biografico degli Italiani.*

DNB *Dictionary of National Biography* [British].

DSB *Dictionary of Scientific Biography.*

KGW Kepler, Johannes, *Johannes Kepler gesammelte Werke*, ed. M. Caspar, *et al.*, Munich, 1938–.

OCCH Huygens, Christiaan, *Œuvres complètes de Christiaan Huygens*, ed. D. Bierans de Haan, J. Bosscha, D. J. Kortweg and J. A. Vollgraff, 22 volumes, The Hague, 1888–1950.

OGG Galilei, Galileo, *Opere di Galileo Galilei*, ed. Antonio Favaro, 20 volumes, Florence, 1890–1909, 1929–1939, 1964–1966.

NNBW *Nieuw Nederlandsch Biografisch Woordenboek.*

TBDOO Brahe, Tycho, *Tychonis Brahe Dani Opera Omnia*, ed. J. L. E. Dreyer and J. Ræder, 15 volumes, Copenhagen, 1913–1929.

Introduction

J. V. FIELD AND FRANK A. J. L. JAMES

The essays in this volume are concerned with the history of science in Europe between, roughly, 1400 and 1750. Consequently, the adjective 'scientific' hovers behind the nouns 'Renaissance' and 'Revolution' in the title.

The terms 'Renaissance' and 'Scientific Revolution' are both recommended to be used with some degree of circumspection by the most direct heirs of the intellectual communities that gave them birth – that is, cultural historians and historians of science. Both terms, indeed, tend to be applied to periods defined in terms of their intellectual products, somewhat in the manner in which the Bronze Age is defined by the nature of its material artefacts. This form of definition has the advantage of not imposing a spurious unity on the products of a particular time or a particular place.

In combining two such fluid terms in the title of this book we are asking for trouble. We are not, however, asking for lingering reflections upon the meaning of the terms themselves. The trouble takes the form of stirring up the two terms together to see whether the result looks historically interesting. The emphasis is thus on continuing processes rather than on abrupt and tongue-twisting '*gestalt*-switches'. We believe that the terms 'Renaissance' and 'Scientific Revolution' both denote real historical phenomena, but not short, sharp ones well-defined in number, measure and weight. 'Revolution' in particular has shown itself to be sufficiently complex to form the subject of a recent book, whose author, I. B. Cohen, points out that the word itself was not used by the natural philosophers historians now most readily identify as having done revolutionary work.[1] The Renaissance is, of course, by now well accepted as

[1] I. B. Cohen, *Revolution in Science* (Cambridge MA, 1985). The comment about 'revolution' not being an actors' category in the sixteenth and seventeenth centuries is on page 6. G. E. R. Lloyd, *The Revolutions of Wisdom: Studies in the Claims and Practice of Ancient Greek Science* (Berkeley, 1987), which deals with the Ancient World, uses the word 'revolution' in its title but does not include it in the index, or discuss its meaning in connection with references to Kuhn (all of which are very brief). In fact, Lloyd appears to regard the use of 'revolution' as unproblematic for the period with which he is concerned. There is, of course, no question of its being mistaken for an actors' category in this context.

A discussion of the use of the term 'Scientific Revolution' is given in D. C. Lindberg, 'Conceptions of the Scientific Revolution from Bacon to Butterfield: A preliminary sketch', in D. C. Lindberg and R. S. Westman (eds.), *Reappraisals of the Scientific Revolution* (Cambridge, 1990), 1–26. For Rupert Hall's reflections on the term, see below, pp. 239–49.

having been a rebirth so slow and ill-defined as to allow adequate freedom to the most incompetent astrologer in drawing up its horoscope.[2] Both 'Renaissance' and 'Scientific Revolution' are, moreover, misnomers to the extent that 'What was reborn in the Renaissance?' makes an ideal question for university entrants but is conceivable for an older scholar only on a highly festive occasion, while the Scientific Revolution, if seen as culminating and essentially completed in the work of Newton, resulted in something better described as a mathematized natural philosophy than as 'Science' in today's sense of the word. If one borrows the logical style of Lewis Carroll's Alice and demands that the product of the Scientific Revolution shall be Science, then this book may well need extending to the year 2001.[3] For our present purposes, we are taking the period of the Scientific Revolution as ending in the mid eighteenth century, while bearing in mind that the changes that characterize the Scientific Revolution were not all completed by that time.

The contributors vary widely in the extent to which they engage explicitly with questions of historiography. However, in view of the Carrollian problem mentioned above, we exercised editorial jurisdiction – in no case extending beyond *territio* – over the use of the words 'science' and 'scientist'. In all cases where any uncertainty might be supposed possible, we prescribed the use of 'actors' categories' in the limited sense that the words used should be those that might have been used by participants in the events concerned.[4] The one apparent exception confirms our rule: Westfall's subject made it inevitable that he should use 'science' and 'scientist' in the way he does, and we are sure his usage will not give rise to confusion.

Westfall's is a general study of the scientific community, within wide bounds of space and time. Nutton and Gabbey have written similarly general studies. Most other contributors have dealt with more 'local' subjects, that is, they have concentrated on one particular time and place or one particular figure. For instance, Eamon on Leonardo Fioravanti, Hunter on Robert Boyle, Figala and Petzold on Newton and Yworth, Home on Newton. As will be clear, our use of the term 'local' is not meant to indicate that such studies have no implications beyond the subject with which they are primarily concerned. Such a position would, in any case, be untenable in regard to a

[2] Something of the character of the term is apparent even in the title of Erwin Panofsky's classic *Renaissance and Renascences in Western Art*, first published in 1960. For a historiographic assessment see A. Chastel (ed.), *The Renaissance: Essays in Interpretation* (London, 1982, first published in Italian, 1979). Our use of the term 'Renaissance' in the context of science is, of course, intended as a reference to the title of Marie Boas Hall, *The Scientific Renaissance, 1450–1630* (Cambridge, 1962).

[3] Elements that might be construed as hinting at the possibility of such a conception are to be found in A. Cunningham, 'How the *Principia* got its name; or, taking natural philosophy seriously', *Hist. Sci.*, 1991, **29**: 377–92. Our choice of the year 2001 refers to Arthur C. Clarke, *2001: A Space Odyssey* (London, 1968), a novel based on the screenplay by Clarke and Stanley Kubrick for the latter's film of the same name, which was hugely successful at the time – partly due to topicality (appearing in a period of fierce USA–USSR rivalry in interplanetary travel), but mainly due to slick special effects and loud music by more than one Strauss. The plot, which has no obvious connection with Homer, is partly a development of Clarke's short story 'The Sentinel' (1951). This story is the true subject of our reference since it depends upon mankind's being considered to have reached intellectual maturity in science when it succeeds in tampering with a 'sentinel' left on the Moon by representatives of a higher civilization capable of interstellar travel.

[4] Sociologists have used the term 'actors' categories', in a wider sense than that intended here, since the 1930s. The term became familiar to historians of science through its use by Shapin and Barnes in the late 1970s.

figure of such acknowledged importance as Newton, and all our authors have in fact taken pains to point out the wider implications of their work.

The clearest example of the usefulness of such 'local' history is perhaps provided by Hunter. His discussion of Boyle's concern with religious questions of conscience is a reminder that religion bulked much larger in the lives of most of the people historians write about than it does in the lives of their counterparts today. Moreover, in providing a believable intellectual portrait of Boyle, Hunter shows up crucial weaknesses in earlier versions of some events given by historians for whom Boyle was essentially only one diagrammatic figure in a story not his own. A banal moral, which Hunter does not draw, is that this is, of course, a peril for every historian who works on a figure whose personal papers survive in quantity, particularly if they are largely unpublished.[5] What Hunter does do is to show that taking a new look at one particular area of Boyle's religious life can provide alternative explanations of matters relevant to Boyle's work as a natural philosopher.

Hunter's conclusions in regard to Boyle are notable for relying on 'external' factors in an argument directed in part against the neglect of them in work done by some historians who, at the time, appeared to regard themselves as more externalist than thou. A fair degree of externalism is, however, by now normal in history of science. Any debate between internalism and externalism appears to be dead – though one might, perhaps, by a suitable choice of battlements, arrange to encounter its ghost. There are no references to it among the historiographic remarks made by the contributors to this volume. Readers will see that we have chosen our words with care in referring to 'historiographic remarks'. Apart from Gabbey, whose subject is historiographic, and Cook, whose main conclusions are historiographic, none of our contributors is greatly concerned with historiographic issues, and even Cook's example tends to prove the rule in that he is directing attention to an imbalance in historians' choice of particular fields of scientific enquiry, rather than discussing a particular style of historiography. There seems to be general agreement that we are all cultural historians now – at least in regard to the period under consideration here. We are tempted to suggest that historians of the early modern period may have been brought to such a recognition by making a comparison between their work and that of the increasing number of their colleagues who are working on the history of science in much later periods, such as the nineteenth and twentieth centuries, for which much larger quantities of general information are available. In any case, the apparent community of outlook between our contributors has not been imposed either by the editors' choice of papers or by suggesting revisions. Similarly, these collected papers are notable for an absence of direct prescription and an absence of buzzwords. We are not in search of a label to apply to such a state.

This absence of policing of boundaries may be more apparent than real, but it is none the less accompanied by some of our contributors being willing to use their skills in areas which might be considered proper to other kinds of historian (other, that is,

[5] Hunter is currently working on Boyle's papers, and has published a guide to them. See Michael Hunter, *Letters and Papers of Robert Boyle: A Guide to the Manuscripts and Microfilm* (Bethseda, 1992).

than a historian of science). For instance, Nutton's classical scholarship is turned to the task of examining the activities of his humanist predecessors, whose work he has, of course, been using in his own studies of their period and in his reading of classical texts; Hunter is contributing to, as well as making use of, the history of religious thought; Field, and Winkler and Van Helden, are effectively making contributions to the history of art as well as that of science; and, for instance, Dragoni's paper might well be seen as describing an episode of social history, though its relevance to the history of science is at once apparent if it is considered in relation to Casini's paper, in which several of the same actors appear. It seems to us that some, but not all, of this boundary crossing represents a commitment to using actors' categories. That buzzword is surely so old as to have become merely a technical term,[6] but behind it there stands the less well-observed but equally important notion of attending to the fact that disciplinary boundaries should also be seen in actors' terms. Westfall's paper can be read as, in a sense, a contribution to a better understanding of the position and nature of these disciplinary boundaries. Gabbey's paper gives a sharply focused view of the historiographic effects of working with an unfocused picture of them.

The problem of nomenclature in 'mechanics' to which Gabbey draws attention is of historians' own making – some of the historians in question being distinguished amateurs, such as Ernst Mach. Eamon's chapter, however, shows that such problems may sometimes be intrinsic: he makes it clear that Fioravanti's use of the word 'scientia', which one inevitably translates as 'science', is both loose and flexible. Neither of these forms of vagueness is likely to commend itself to the historian, except pragmatically, and in the short term. None the less, we all know that they are characteristic of the use of the same word in our own time, by the general public at least. Indeed, Fioravanti's slapdash, and polemical, usage of 'scientia' sometimes as meaning true understanding, and sometimes as indicating a collection of abstruse theory-encumbered prejudices, is both apparently 'modern' and apparently Paracelsian. Eamon's account of Fioravanti's work indicates, however, that his preoccupations were, if not conventional, at least closely bound up with issues that engaged the attention of his contemporaries, and that his general philosophy was not very close to that of Paracelsus. What Eamon calls Fioravanti's 'medical primitivism' involved relying on direct experience of nature. In making out his own case, however, Eamon may perhaps also be seen as suggesting, by his example, the usefulness of a form of historiographic 'primitivism', namely relying on a careful reading of primary sources.

Typified, perhaps, by Paolo Rossi's book about Francis Bacon (first published in Italian in 1957),[7] one tendency in modern historiography has been to show that many of the first generation of Heroes of the Scientific Revolution, figures such as Tycho (b. 1546), Galileo (b. 1564), Kepler (b. 1571) and Harvey (b.1578), had their feet firmly planted in a culture best described as Renaissance. For example, they all made creative use of newly established good texts or translations of ancient Greek

[6] See note 4 above.

[7] Paolo Rossi, *Francesco Bacone: Dalla magia alla scienza* (Bari, 1957); English translation *Francis Bacon: from Magic to Science* (London, 1968).

scientific works. Copernicus (b. 1473) is conspicuous by his absence from this list because he clearly cannot be described otherwise than as a Renaissance humanist figure. Instead of describing figures of this first generation as 'transitional', it seems better to admit that historians themselves have supplied the Rubicon whose crossing they then proceed to celebrate. Looking at intellectual maps drawn up for other kinds of history suggests that we do not need to copy Ancient Roman tidy-mindedness – except perhaps for administrative purposes – and might find it useful to forget about drawing lines between a certainly ill-defined Renaissance, as Transalpine Gaul, and the Cisalpine Gaul of an at least somewhat ill-defined Scientific Revolution that is either the first phase of Modernity or the prelude to it.

An example of the usefulness of regarding the Scientific Revolution as a continuation of the Renaissance can be seen by comparing Eamon's chapter with that of Gabbey. Eamon is concerned with sixteenth-century Italy. Gabbey deals mainly with work done North of the Alps in the following century, but his chapter can be seen as importing into the study of what has been regarded as a characteristically seventeenth-century problem – perhaps reflecting the relationship of scholars and craftsmen – the intellectual tools which the work of the late Charles Schmitt has shown to be so useful in studying the Renaissance. Indeed, one of the authors to whose work Gabbey refers, namely Tycho Brahe, is in many ways a typically 'Renaissance' figure, as the late Victor Thoren's biography of him, aptly titled *The Lord of Uraniborg*, makes abundantly clear.[8] Furthermore, in allowing Kepler access to his books of astronomical observations, in a manner Kepler found infuriatingly restrictive, it would seem that Tycho was according him the privilege of making use of an astronomical version of a cabinet of natural curiosities put together in a spirit not unlike that which animated the seventeenth-century students of natural history discussed in Cook's paper.

Abolishing this metaphorical Rubicon is a suggestion from the editors. It will, however, be noted that the contributors seem inclined to eliminate Caesar. Their attention is, for the most part, turned away from the standard heroes or heroic episodes in the Scientific Revolution, and towards figures and events that tended to be regarded as minor or marginal by earlier historians. This attitude seems to be part of an established trend in modern historiography. An example is provided by the British Society for the History of Science Summer Meeting of 1987, a year much bedecked with tercentenary garlands for Newton's *Principia*. The meeting concerned itself with the life and work of Newton's contemporary, and sometime adversary, Robert Hooke. No bones were made about his being, compared with Newton, a relatively 'minor' figure, though there was the entirely justifiable claim that historians had paid him disproportionately little attention.[9] In the event, historians voted with their feet: the meeting was very well attended and the volume of selected papers edited by the

[8] V. E. Thoren, *The Lord of Uraniborg: A Biography of Tycho Brahe* (Cambridge, 1990). Thoren chaired one of the sessions at the Oxford conference.

[9] Michael Hunter and Simon Schaffer (eds.), *Robert Hooke: New Studies* (Woodbridge, 1989), Introduction.

organizers has been equally well received.[10] Hooke is, of course, somewhat exceptional among 'supporting actors' in being relatively well documented – for instance by the survival of his private diary for certain years, which tells us a great deal about his day-to-day personal contacts. As Rupert Hall points out in his Afterword to this volume, on the whole 'major' figures tend to have left us with more information to go on. Presumably today's trend towards studying 'minor' figures and 'minor' episodes goes with the habit of setting them in their wider cultural context, since understanding this context can compensate for the relative paucity of direct evidence. In a way, such studies show the context becoming the content. However, it must be admitted that they also ignore the principle of looking for the lost coin under the lamp post because that is the only place one has a hope of finding it.

One previously marginalized area which becomes more central in this volume is that of work done in Italy. In view of the industrial scale of research on Galileo, it may seem unreasonable to claim that Italian natural philosophy has been neglected, but a closer inspection of the literature shows that, apart from much excellent work on Renaissance Aristotelianism, historical investigation has in fact been remarkably closely concentrated on Galileo and his immediate contacts. Relatively little attention has been paid to subjects in which he did not make what Whig historians would recognize as revolutionary advances. Moreover, possibly due to long-term influence by historians engaged in covert Protestant polemic, such as David Brewster, it seems too readily to have been assumed that the condemnation of Galileo by the Catholic Church led to an inexorable decay of the scientific elements of Italian civilization, whose later story could thus present little of interest. (Thus the rich field of nineteenth-century Italian science has been largely neglected.) Casini's account of the reception of Newton's optical work in Italy is a distinguished contribution to proving such a view mistaken. He uncovers a lively world of intellectual activity and academic and social careerism in which the background threat of the Inquisition, though real, could nevertheless be avoided if one took sufficient care.[11] Intellectual, religious and political threat are seen to go together. Thus the situation Casini describes integrates history of science into a wider history. In this case, good internal history of science and good external history of science have simply become the same thing. Perhaps a hidden assumption of their necessarily being antithetical has helped to perpetuate the Brewsterian neglect of events in Italy.

Protestant polemic having faded out of historians' overt agenda, Brewster's portrayal of Galileo as a Martyr to Catholicism (1841)[12] has become transmuted and generalized into the perception of a possibly antagonistic interaction between Religion and Science. The case of the reception of Darwin's theory of evolution has apparently

[10] *Ibid.* Reviewed by M. B. Hall in *Arch. Int. Hist. Sci.*, 1990, **40**: 400–1, which endorses the editors' opinions about the earlier neglect of Hooke and his work and points out that much still remains to be done.

[11] The place of the Inquisition in this world is rather like an attenuated version of that of the Austrian system of political repression in the world of Stendhal's novel *La Chartreuse de Parme* (1839) – which is based on historical characters who lived in Italy in the sixteenth century, so the similarity is probably not entirely fortuitous.

[12] See J. R. R. Christie, 'Sir David Brewster as an historian of science', in A. D. Morrison-Low and J. R. R. Christie (eds.), *'Martyr of Science': Sir David Brewster 1781–1868* (Edinburgh, 1984), 53–6.

tended to harden today's scientists', and sometimes also today's historians', belief that the possibility of antagonism is important. This would seem to be the most natural reading of John Brooke's recent *Science and Religion* (1992). There appears, however, to be a tacit assumption that, while Science is a continually changing system – indeed changing so much that the use of the word 'science' is dubious for its earlier forms – Religion, or rather religious belief, has no history. Given the Darwinian model and the multifarious nature of religion in nineteenth-century Britain, the thought of Michael Faraday would seem to provide an interesting test case. His absence from Brooke's book is apparently a consequence of the author's decision to concentrate on the relation between Christianity and the geological and biological sciences in this period.[13] With the thought of Johannes Kepler, whose not quite orthodox brand of Lutheran Protestantism is seamlessly continuous with his natural philosophy, as Hübner showed in his *Die Theologie Johannes Keplers zwischen Orthodoxie und Naturwissenschaft* (1975), the Darwinian model breaks down equally spectacularly. One might perhaps expect this model to be unimportant for historians of the Scientific Revolution, but that is to accept too simple a rational reconstruction. Historians of science on the whole appear to have adopted a modified Brewsterianism which allows them to see Galileo as by far the most important natural philosopher of his time. Kepler, a Protestant, whose excommunication for unorthodox beliefs concerning the relation of spirit and matter in the doctrine of the Eucharist is presumably considered irrelevant, is seen as a relatively unimportant figure, merely a *mathematicus* (which is to ignore the fact that, like Galileo, Kepler also claimed to be a natural philosopher).[14] Their distinguished contemporary William Harvey, into whose concerns matters of religion are apparently deemed to enter hardly at all, provides an even weaker example of Religion *versus* Science and gets even shorter shrift. This, at least, is the picture of the historiography of the Scientific Revolution that one obtains if one looks at the balance in the literature as a whole. It is faithfully reflected in Brooke's *Science and Religion*, and is also visible in the volume of essays edited by Lindberg and Westman, entitled *Reappraisals of the Scientific Revolution* (1990). The crude measure provided by the number of index entries in the volume is confirmed by reference to the main text. Ernan McMullin, writing on 'Conceptions of science in the Scientific Revolution',[15] devotes three and a half pages to Kepler,[16] followed by eight to Galileo, and the section on Galileo begins

> In this and the next section we come finally to the two scientists whose names most easily come to mind when the Scientific Revolution is men-

[13] For Faraday's religion, see G. N. Cantor, *Michael Faraday: Sandemanian and Scientist. A Study of Science and Religion in the Nineteenth Century* (London, 1991).

[14] Field is grateful to Owen Hannaway for pointing out, in a private conversation in 1989, that Kepler's quasi-chemical problem with the Eucharist was closely similar to his problems of relating force and matter in his physics.

[15] Ernan McMullin, 'Conceptions of science in the Scientific Revolution' in Lindberg and Westman, *op. cit.* (note 1), 27–92.

[16] These pages are about Kepler's *A Defence of Tycho against Ursus*, an unfinished work which remained unpublished in the author's lifetime, but was printed in the original Latin and in English translation in N. Jardine, *The Birth of History and Philosophy of Science: Kepler's A Defence of Tycho against Ursus, with essays on its provenance and significance* (Cambridge, 1984), to which numerous references are given. No attempt is made to use Kepler's scientific works as evidence, as is done for Galileo.

> tioned. Galileo began a mathematization of Nature that Newton carried to its triumphal conclusion.[17]

There is no mention of Harvey. He is, however, referred to in Cook's 'The new philosophy and medicine in seventeenth-century England'[18] a section of which considers the general neglect of history of medicine.[19] We are, of course, not suggesting that most historical work is actually Brewsterian, but the fact remains that Brewster and his ilk seem to have played a large part in determining historians' choice of problems.

Nor is the balance greatly righted by the simultaneous influence of a more recent historiography which has emphasized the revolutionary effect of Platonism (usually understood as involving increased use of mathematics, but sometimes as an injection of Neoplatonic magical reasoning) against the entrenched Aristotelianism of conventional natural philosophers. This style again tells against Harvey and, because it was at the time more usual than it is now to take Galileo's word for how un-Aristotelian his opinions were, again emphasizes Galileo. Kepler, whose personal letters happen to survive in large quantity, frequently emerges as having a personality problem, being a Platonic cosmologist with an Aristotelian physics, and thus, in appropriately humanist style, is sometimes described as Janus-faced. We do not wish to suggest that any serious historian working today would ignore the important work that the late Walter Pagel and others have done on Harvey's philosophical beliefs (and indeed the connections between his work and religious beliefs),[20] or would fail to take account of the valuable aid to understanding provided by recent studies that show the continuing vigour of Aristotelian work in natural philosophy.[21] Moreover, as Martins' contribution to this volume makes clear, Aristotelian ideas continued to play an important part in determining what constituted an acceptable form of scientific reasoning, as is seen in the philosophical obstacles Huygens encountered in reading Newton's *Principia*. The continuing influence of Aristotelianism, which is also illustrated in Dragoni's essay, is well recognized in the work of most historians today. None the less, analysis in terms of Platonism does seem to have played a part in reinforcing Brewsterian

[17] McMullin, *op. cit.* (note 15), 62.

[18] Harold J. Cook, 'The new philosophy and medicine in seventeenth-century England', in Lindberg and Westman, *op. cit.* (note1), 397–436.

[19] *Ibid.*, 401–5.

[20] W. Pagel, 'Religious motives in the medical history of the seventeenth century', *Bull. Inst. Hist. Med.*, 1935, **3**: 97–128, 213–31, 265–312; reprinted in W. Pagel, *Religion and Neoplatonism in Renaissance Medicine*, edited by M. Winder (London, 1985). Also Pagel's 'The reaction to Aristotle in seventeenth-century biological thought: Campanella, Van Helmont, Glanvill, Charleton, Harvey, Glisson, Descartes', in E. A. Underwood (ed.), *Science, Medicine and History: Essays on the Evolution of Scientific Thought and Medical Practice written in Honour of Charles Singer*, (2 vols., Oxford, 1953), **1**: 489–509; and Pagel, 'Harvey and Glisson on irritability with a note on Van Helmont', *Bull. Hist. Med.*, 1967, **47**: 497–514; both of these papers are reprinted in Pagel, *From Paracelsus to Van Helmont: Studies in Renaissance Medicine and Science*, edited by M. Winder (London, 1986). See also, C. Hill, 'William Harvey and the idea of monarchy', in C. Webster (ed.), *The Intellectual Revolution of the Seventeenth Century* (London, 1974), 160–81, esp. pp. 169–73 (first published in *Past Present* in 1964), but see ensuing discussion G. Whitteridge, 'William Harvey: a Royalist and no Parliamentarian', *ibid.*, 182–8, and C. Hill, 'William Harvey (no Parliamentarian, no heretic) and the idea of monarchy', *ibid.*, 189–96.

[21] See, for example, the accounts of Jesuit work on electricity and magnetism in J. Heilbron, *Electricity in the Seventeenth and Eighteenth Centuries: A Study in Early Modern Physics* (Berkeley, 1979), and Heilbron, *Elements of Early Modern Physics* (Berkeley, 1982).

predilections in historians' choice of subjects. Curiously enough, further reinforcement has recently come from the Vatican. In connection with an Einstein centenary, in 1979, Pope John-Paul II instituted a series of publications to deal with 'the Galileo case'.[22] Since Galileo's condemnation appears to have been connected with his Copernicanism,[23] this papal intervention tends to support the historiographic tendency that may also partly derive from the importance given to mathematical sciences in the writings of scholars interested in the influence of Platonism or some form of Neoplatonism, namely the tendency to see astronomy as particularly important. What has come to be called the 'Copernican Revolution'[24] is thus sometimes presented as if it provided a paradigm for (or picture in little of) the processes at work in the Scientific Revolution as a whole.[25] Historians of other subjects in this period are left to cope as best they can with what, adapting Steven Jay Gould's parody of Freud, one might call 'astrophysics envy'.[26] Such envy is not a problem for today's scientists, since astrophysics is now heavily encumbered with proliferating subatomic particles, thus putting it in the unenviable state summed up by Fermi's alleged remark 'If I could remember all those names I'd be a botanist.'[27]

That 'botany envy' appears to be unknown in our own time no doubt encourages historians – and perhaps particularly those whose background lies in science – to see natural history as 'backward' in the seventeenth century. In any case, it does not fit very well into the larger theoretical frameworks for describing the development of natural philosophy designed by Brewsterians or Platonizers. However, the striking changes in some parts, dependencies or appendages of natural history, describable as 'the life sciences', changes which may very reasonably be regarded as 'revolutionary', have until relatively recently been seen as coming within the remit of the interpretive system epitomized in the title of Dijksterhuis's *The Mechanization of the World Picture* (1958).[28] Unfortunately for the coherence of such an interpretive system, the group of seventeenth-century theories of matter that are still usually designated as 'the mechanical philosophy' have increasingly been shown to include crucial elements that are incompatible with the standard interpretation of 'mechanical' as meaning explanation purely in terms of matter and motion.[29] It seems likely soon to become an orthodox opinion to hold that the only true mechanist was probably Descartes, with Aristotle as the next-best candidate. In particular, it has been clear for some time that any mechanical philosophy encountered problems with explaining attraction –

[22] The series is called *Studi Galileiani, Research Studies Promoted by the Study Group Constituted by John-Paul II*, the publisher is Vatican Observatory Publications (Vatican City). The first volume appeared in 1984.

[23] Vigorous advocacy for an alternative view is to be found in P. Redondi, *Galileo eretico*, (Turin, 1983); English translation, *Galileo Heretic* (Princeton, 1987).

[24] From the title of T. S. Kuhn, *The Copernican Revolution* (Cambridge MA, 1957).

[25] T. S. Kuhn, *The Structure of Scientific Revolutions* (Chicago, 1962).

[26] Gould contrasts 'historical sciences' (such as geology and palaeontology) with experimental and mathematical sciences (such as physics). The former sciences are regularly accorded lower intellectual status. Their practitioners are thus diagnosed as subject to 'physics envy'. S. J. Gould, *The Mismeasurement of Man* (New York, 1981).

[27] Enrico Fermi (attrib.). Fermi had a reputation for wit.

[28] First edition, in Dutch, Amsterdam, 1950, German translation 1956, English 1961.

[29] J. Henry, 'Occult qualities and the experimental philosophy in pre-Newtonian matter theory', *Hist. Sci.*, 1986, **24**: 335–81; and Henry, 'Robert Hooke, the incongruous mechanist', in Hunter and Schaffer *op. cit.* (note 9), 149–80.

electric, magnetic or chemical – and the concomitant notion of 'force' as action at a distance. It is accordingly significant that in his contribution to this volume Gad Freudenthal is able to show that one widely accepted 'mechanical' explanation for attraction appears to be derived from a Stoic notion incorporated into the Aristotelian theory of matter in order to deal with the related problem of cohesion. Freudenthal's argument thus tends – though this is clearly not its main purpose – to lend further support to the view that much of the mechanistic explanation of the seventeenth century was in fact derived from an Aristotelian tradition. This fits in well with the modern picture of Harvey as deeply indebted to Aristotle, and thus a 'mechanist' in rather the same way, for instance in believing that form must follow function. Since Aristotle's ablest pupil, Theophrastus, is now chiefly remembered for his (thoroughly Aristotelian) work on systematizing the study of plants, one may perhaps hope that botany, and natural history as a whole, will eventually be drawn into this larger picture of the continuing influence of Aristotelian styles of thought. It may be that what is required for the revival of historians' interest in natural history is a better integration of history of medicine into the history of science proper. The contributions of both Cook and Westfall could be read as providing arguments for such an integration, since each demonstrates that it would be in accord with the way the subjects were perceived at the time concerned.[30]

Support for the same view can also be deduced from Nutton's analysis of the printing history of Greek scientific texts. Though there is a slight time lag – mathematics, in which we include the mathematical sciences, got off to a slower start than medicine – the story is essentially the same for both. Greek texts seem to have found a relatively restricted readership and were followed by Latin translations and commentaries, which proved to be more popular. This aspect of the history of mathematics is largely ignored in Rose's *The Italian Renaissance of Mathematics* (1975), which is concerned with humanists and translators rather than their readers. Moreover, it has been almost completely neglected by most historians of mathematics. Possibly for reasons of technical competence – since it is, to put it bluntly, rather difficult to get to grips with the mathematics underlying the presentational rhetoric in many Renaissance mathematical texts – the history of mathematics has tended to be written by mathematicians for their peers, with emphasis on the mathematical content. The style of such work tends to look rather old-fashioned to most historians of science. However, as can be seen in the Platonizing-mathematization model for the Scientific Revolution, the history of mathematics undoubtedly has substantial connections with developments in natural philosophy as well as in the mathematical sciences. In this volume, Westfall's contribution makes clear that there were many 'scientists' whose everyday activities engaged or depended on their mathematical skills. Field's contribution discusses how mathematical skill can be seen to be embedded in the craft of the painter, and Willmoth's contribution examines the institutionalization of mathematical skills in the Ordnance Office. It is beyond the

[30] See also Cook, *op. cit.* (note 18), 401–5.

scope of this introduction to consider the wider questions that seem to be raised in regard to the history of mathematics, but there can be little doubt that both history of science and history of mathematics would benefit from closer integration. As with the history of medicine, which also has connections with mathematics through astrology, the modern fashion for specialization not only cuts across intellectual areas once seen as continuous but also fails to recognize boundaries that were, sometimes, important for contemporary practitioners.

A real problem concerning boundaries which is not addressed by any of the contributors to this volume, except indirectly by Gabbey, Westfall and Willmoth, is that of the division between what would today be called 'science' and 'technology'. This silence is acknowledged by the absence of the word 'engineers' from the subtitle.

In any case, this volume cannot pretend to be a complete survey of the present state of historical studies of the Scientific Renaissance or the Scientific Revolution. Nor would we wish it to be seen as constructing an argument for any particular viewpoint, though, as already mentioned, some community of outlook does seem to emerge.

The contributors have been put in an order that is partly chronological and partly thematic. We begin with Nutton's dissection of the received notion that what was reborn in the Renaissance was 'classical learning' or 'Greek science' and that this rebirth impinged on science. Since Nutton deals primarily with medical texts, we then turn to Eamon's account of the opinions of a sixteenth-century medical practitioner, Leonardo Fioravanti, whose comment that Hippocrates, Galen and Avicenna did not write in Latin but in their mother tongues (see p. 33) shows the characteristic Renaissance trait of enlisting the Ancients on one's own side while at the same time, as it turns out, claiming to subvert their teaching, in this case by an appeal to direct contact with nature. The following chapter, by Cook, discusses work in natural history carried out in Northern Europe in the seventeenth century and seen at the time as belonging to the mainstream of scientific development. Cook notes that such work has been largely ignored by recent historians of science.

In the following chapter, Westfall examines the spread of interests in the 'scientific community' defined as the group of individuals whose biographies are included in the *Dictionary of Scientific Biography*. The number of medical men tends to confirm Cook's point about the relative neglect of their scientific interests by historians. Westfall's purpose is, however, to investigate the relations between practitioners of what we should now call science and practitioners of what we should now call technology or crafts. There follow three chapters on people who were practitioners of both: Field looks at Piero della Francesca as mathematician and painter; Winkler and Van Helden look at Hevelius as astronomer, draughtsman and publisher, and show how this combination of activities allowed him to make an important contribution to the development of astronomy; Willmoth looks at the English Ordnance Office's concern with both the practical business of warfare and the mathematical theory connected with some of the problems it presented.

It seemed appropriate that a chapter much concerned with gunnery should be followed by Gabbey's historiographic discussion of the subject to which gunnery has the greatest claim to have contributed, namely 'mechanics'. Gabbey having exposed further ambiguities in the standard phrase 'mechanical philosophy', the next four chapters deal with some of its most distinguished practitioners: Hunter discusses Boyle's concern with moral and religious problems; Gad Freudenthal discusses the nature and background of a favourite 'mechanical' solution to the problem of attraction; Figala and Petzold, in a chapter that might also be classified as dealing with the relations between scholars and craftsmen, discusses Newton's personal contacts with a chemical/alchemical practitioner in the first decade of the eighteenth century; Home discusses Newton's beliefs about subtle matter, which partly relate to the problem of attraction discussed by Freudenthal.

The next three chapters, coming just before the Afterword, deal with problems encountered by Newton's readers. Martins shows that Huygens's non-acceptance of Newton's gravitational theory was a matter of philosophical concern with the foundations of science. Casini shows how the quarrel over calculus provided an incentive for reopening the apparently already 'closed' debate over Newton's *Opticks* and how political and social as well as scientific and religious factors played a part in the production of Algarotti's elegant popular account of Newton's work. Dragoni discusses how an enlightened series of Papal interventions furthered the cause of the new science in the city of Bologna in the mid eighteenth century.

The volume concludes with an Afterword by Rupert Hall, entitled 'Retrospection on the Scientific Revolution'. Hall is not inclined to read any great significance into the title *The Scientific Revolution* that he gave to his book in 1954 (though he notes that Kuhn apparently took the phrase as a 'term of art'). He is chiefly concerned with the results of the different ways in which historians of the period have approached their task over the years.

It seems to us that the phrase 'the Scientific Revolution', whatever its status may have been in the 1950s, has now become a term of art. Future generations will presumably regard it as a compound term, and pay less attention than we do to considering the separate significance of the noun and the adjective – unless asked to do so in a university entrance examination. That is, 'Scientific Revolution' seems to be on its way to becoming the same kind of term as 'Renaissance', though perhaps with something of the polymorphism associated with the more standard partner to 'Renaissance', namely 'Reformation'.[31]

What then of the ringing phrases in which Herbert Butterfield spoke of this revolution? The other contributors to this volume, whether in courtesy or malice we do not care to speculate, have left it to the editors to cite Butterfield's description:

> it outshines everything since the rise of Christianity and reduces the Renaissance and the Reformation to the rank of mere episodes, mere internal displacements, within the system of medieval Christendom.[32]

Interpreting this passage presents a problem of a kind familiar to historians. The 'it' refers to a 'revolution' located in the sixteenth and seventeenth centuries, that is, to

[31] For example, G. R. Elton (ed.), *Renaissance and Reformation: 1300–1648* (London, 1963).

[32] H. Butterfield, *The Origins of Modern Science: 1300–1800* (Cambridge, 1949), vii.

what we should now call the Scientific Revolution. However, to Butterfield this revolution is identified with the subject of his book, whose title is *The Origins of Modern Science*. With this identification, Butterfield's statement seems highly defensible. Science has not come to seem less important since 1949. The Scientific Revolution of the sixteenth and seventeenth centuries has, however, been reduced to the rank of an episode, though a very important episode, in the rise of modern science. At least, that is the status it would probably be accorded if any of today's historians were minded to write a book – or (more likely) edit a collection of essays – that purported to deal with a subject so large as the history of science from the Middle Ages to the present day. It is now clear that very significant contributions were made in the eighteenth and nineteenth centuries. As Rupert Hall put it in 1983, one may speak of

> a 'second scientific revolution' in the nineteenth century through which, at last, the age-old vision of a harmony between the microscopic and the macroscopic worlds came within reach of attainment.[33]

Few would dispute that the philosophical and methodological changes that took place in the sixteenth and seventeenth centuries were crucial for what was to happen next. But that recognition does not make them exactly equivalent to Saint Denys's first step – and, as we remarked at the beginning of this Introduction, there is some argument about what is to be taken as the end point of the journey.

The comparison with a miracle, even filtered through the mind of Voltaire, is presumably blasphemous, but to a man or woman of 1300 our technological advance, much of which, since the late nineteenth century, has been based on science, would surely seem little short of miraculous. In the eighteenth and early nineteenth centuries practising men of science, such as David Brewster (whose chief interest was in optics), William Whewell (who worked mainly in geology, chemistry and physics), and Francis Baily (an astronomer now best remembered for 'Baily's beads'), spent considerable time and effort studying the history of their subjects.[34] Today's scientists attach less importance to history. On occasion, however, for anecdotal purposes, they do choose to trace their ancestry back to a 'popular' (probably heavily mythologized) account of the Scientific Revolution. For instance, when the magnificent Voyager II photographs of Jupiter's satellites were first shown to a meeting of the Royal Astronomical Society, in 1979, the speaker began by showing 'the discovery document', that is a slide of the title page of Galileo's *Sidereus Nuncius* (Venice, 1610). Galileo would have liked looking at the other slides too.

Naturally enough, historians are on the whole better at doing justice by long-dead natural philosophers than scientists are. Not all former scientific achievements can easily be sufficiently separated from their historical context to make sense as pieces of science to people without some historical training. However, it is necessary for a piece

[33] A. R. Hall, *The Revolution in Science* (London, 1983), 19.
[34] D. Brewster, *The Martyrs of Science; or, the lives of Galileo, Tycho Brahe and Kepler* (London, 1841), and *Memoirs of the Life, Writings and Discoveries of Sir Isaac Newton* (2 vols., Edinburgh, 1855). W. Whewell, *History of the Inductive Sciences from the Earliest to the Present Time* (3 vols., London, 1837). F. Baily, *An Account of the Rev. John Flamsteed* (London, 1835; reprinted 1966).

of science to be useful that it shall, for succeeding generations of specialists in that field of science, be separable from the context of discovery. Rather the same is true for a work of literature to become a 'classic', or for a sculpture or a painting to still be recognizably beautiful many years after it is made. Accumulation in science is not unique in kind. However, the consequences of scientific accumulation have achieved unique importance in our civilization. This in turn gives cultural importance to the origin of the 'modern' phase of that accumulation. As Westfall writes in his contribution to this volume, in a passage Hall cites in his Afterword:

> The concept of a Scientific Revolution rests on the radical reordering of the understanding of nature that did in fact take place in the sixteenth and seventeenth centuries. I am convinced that there is no way to understand the history of science without the recognition of this reality . . .[35]

Perhaps not all our contributors would be prepared to voice their opinion so bluntly, but the essays in this volume testify to the reality of their commitment to elucidating at least some of the questions raised by such a formulation.

[35] See below pp. 72 and 246.

1

Greek Science in the sixteenth-century Renaissance

VIVIAN NUTTON

There can be few more arid, or more acrimonious, themes for a paper than one that conjoins science and the Renaissance. Both terms are frustratingly imprecise alike in their meaning and in their chronological span. Nancy Siraisi, in her recent survey of the medical Renaissance, discusses the career and writings of Taddeo Alderotti, who died in 1295, well before most scholars would be prepared to concede a Renaissance, let alone a Scientific Revolution.[1] Add to that the different chronologies for the Renaissance as applied to art and to literature, and for Northern or Eastern Europe as contrasted with Italy, and the whole topic degenerates into little more than a series of unanswerable questions of definition. To proceed further would seem to demand either an overarching concept, like Kristeller's humanism or Frances Yates' hermeticism, or a desire for academic self-immolation.

But on closer reflection, there is a period when both historians of the Renaissance and of the Scientific Revolution appear comfortable with each other, the roughly five generations that separate Niccolò Leoniceno's attack on the Elder Pliny in the early 1490s from the publication of William Harvey's *Exercitatio anatomica de motu cordis et sanguinis in animalibus* in 1628. Within this period, there is the suspicion, to put it no higher, that, in one aspect at least, the revival of classical learning, the concerns of those interested in literature and art impinged also on science. But what these concerns were, and how they impinged, are complex questions, rarely raised either by historians of science or by scholars interested in the fate of the classics.[2] The simple affirmation that, say, the rediscovery of Galen or Dioscorides inspired the rebirth of

[1] Nancy G. Siraisi, 'Some current trends in the study of Renaissance medicine', *Ren. Quart.*, 1984, **37**: 585–90.

[2] Most studies of the rediscovery of the classics end in 1500, and concentrate on more 'literary' authors. U. von Wilamowitz in his *History of Classical Scholarship* (Eng. tr., London, 1982), devoted a page, p. 45, to zoology and medicine, but said nothing about mathematics, astronomy, or pharmacology. More recent authors are equally neglectful: Alexander Demandt, 'Was wäre Europa ohne die Antike', in P. Kneissl and V. Losemann, eds., *Alte Geschichte und Wissenschaftsgeschichte. Festschrift für Karl Christ*, (Darmstadt, 1988), 113–29, devotes only five lines of a valuable survey to medicine and science. The volume edited by A. Buck and K. Hartmann, *Die Antike-Rezeption in den Wissenschaften während der Renaissance*, (Weinheim, 1983), is filled with generalities and scarcely touches on the *Naturwissenschaften*. More to the point is G. Oestreich, 'Die antike Literatur als Vorbild der praktischen

anatomy or botany masks significant changes within the reception process, and leaves out altogether some teasing and much-neglected problems. In this chapter it is proposed to look in detail at this process in Renaissance medicine, and then to transfer the investigative model to two other areas of Renaissance science.

Medicine

It should be stated at the outset that, in medicine and science, the classical sources that mattered were Greek, not Latin, and that the problem with which this chapter is concerned is the reception of Greek texts and Greek ideas. This is stated so bluntly because, although the influence of the new Galen and Hippocrates was diffused mainly through the medium of Latin translations, these translations only became possible through the prior existence of Greek texts, whether in manuscript or in printed form.[3] As a consequence, also, the type of medical scholarship that is associated with Greek is very different from that founded on the Latin translations – a point to which I shall return later. It might also be objected that by concentrating on the Greeks, this chapter unwittingly excludes two major Latin writers, the Elder Pliny, and Cornelius Celsus, both available in many incunable editions and hence accessible in print a generation or more before any Greek author. But Pliny is a dubious ally for the doctor. His view of the medical profession was notoriously, and outrageously, damning, and, as became clear in the botanical controversies of the 1490s, his medical information was far from accurate. Indeed, it was little better than that of the Arabs, and frequently worse.[4] Above all, it was based on that of the Greeks, and whatever advantages might be gained from its stylish Latin were outweighed by its misunderstandings of the Greek. Medical education, save perhaps in Leipzig, seems to have neglected Pliny entirely, and not without reason.[5] A similar argument can be made against Celsus. One can trace his stylistic influence on Italian doctors, but his Renaissance *fortuna* is still far from clear.[6] Besides, his work is openly

Wissenschaften im 16. und 17. Jahrhundert', in R. R. Bolgar, ed., *Classical Influences on European Culture A.D. 1500–1700* (Cambridge, 1976), 315–24, but, for reasons of space, medicine, botany, and mathematics are deliberately left out. The only detailed survey, that of George Sarton, *The Appreciation of Ancient and Medieval Science During the Renaissance (1450–1600)* (Philadelphia, 1955), is an unreadable compilation of frequently dubious data.

[3] I say 'new' to distinguish it from the knowledge of Galen and Hippocrates derived, directly or indirectly, from medieval Latin translations from the Arabic (and, occasionally, from the Greek).

[4] For Pliny's prejudices, see V. Nutton, 'The perils of patriotism: Pliny and Roman medicine', in Roger French and Frank Greenaway, eds., *Science in the Early Roman Empire: Pliny the Elder, his Sources and his Influence* (London and Sydney 1986), 30–58 (repr. with same pagination in V. Nutton, *From Democedes to Harvey: Studies in the History of Medicine* (London, 1988)). For the botanical controversy, see, with bibliography, Roger French, 'Pliny and Renaissance medicine', in French and Greenaway, 252–81. For editions and commentaries, see C. G. Nauert, 'Caius Plinius Secundus', in P. O. Kristeller and F. E. Cranz, eds., *Catalogus translationum et commentariorum*, (4 vols., Washington, 1960–80), 4: 297–422.

[5] Johann Lange was lecturing on Pliny at Leipzig in 1516, see the preface by Nicholas Reusner to Lange's *Epistolae medicinales* (Hanover, 1589), but whether this was to medical students is unclear. Bruce Eastwood, 'Plinian astronomy in the Middle Ages and Renaissance', in French and Greenaway, *op. cit.* (note 4), 197–235, esp. 219–20, argues that Pliny's relatively simple synthesis was more suitable for students than the (translated) Greek texts on which he drew.

[6] There has been little work done on the Renaissance Celsus since Friedrich Marx in the preface to his editions of the *De medicina* (Leipzig, 1915).

dependent on Greek originals.[7] So, paradoxically, the presence of these two Latin authors only increased the authority of the Greeks from whom they drew most of their medical information. Hence the slogan that without Greek all medicine was a mere imposture, a slogan coined by a Netherlander at a time when very few Greek printed texts of medicine were readily available for those who wished to drink from a pure Hellenic spring.[8]

The floodtide of this Hellenicizing movement in medicine spans a single generation, from 1525, when the first collected edition of Galen in Greek appeared from the Aldine press in Venice, until 1557, the year of publication by Martinus Juvenis at Paris of a medical text by the Byzantine author Johannes Actuarius. After this date, Greek editions of the classical medical authors become very rare, with the exception of Hippocrates, whose collected works were issued in sumptuous bilingual editions by the Juntine press in 1588 and again by the Wechel press in 1595.

To set the end of this movement as early as 1557 may occasion some surprise,[9] for acquaintance with Greek continued to increase, and there were many devoted followers of the classical authorities long after this date, William Harvey included. But, firstly, these tended to read their Greek authors in Latin, not Greek. The annotations made by Harvey in his copy of Goulston's selection of Galenic texts, published in 1640, were all made on the Latin translation, not on the Greek.[10]

Secondly, the type of scholarship involved changes. By the 1570s, there are fewer discussions of the meanings of individual words, fewer attempts to correct the text by resorting to older and better manuscripts, and fewer searches for new and unpublished texts. Instead, one finds a wider discussion of the medical principles involved, in which the accuracy of the translation and of its underlying Greek text is assumed valid. The philological approach to saving the credibility of Galen, associated perhaps most clearly with John Caius (1510–1573), gradually disappears, and the errant Galen is himself replaced as figurehead by Hippocrates.[11] Thirdly, investigations along philological lines, one might argue along lines already followed and advocated by Galen himself in his own commentaries on Hippocrates, became more and more the province of the antiquarian, and less that of the doctor. Take, for example, Girolamo

[7] He is not, as some have believed, largely a Latin translator of a (lost) Greek medical encyclopaedia, for he often sets out an independent line (so, rightly, Philippe Mudry, 'La médecine romaine: mythe et réalité', *Gesnerus*, 1990, **47**: 133–48), but there can be no doubt that much of his material comes from the Greek or that the debates in which he participates are with Greek physicians of both past and present.

[8] Jason Pratensis (van de Velde), *De pariente et partu* (Antwerp, 1527), sig. A.5r–v. Van de Velde (1486–1558) was doctor to Maximilian of Burgundy.

[9] A roughly similar pattern of publication can be seen for Latin translations of Galen and Hippocrates: for Galen, the fullest years are from 1533 until 1550, after which there is a dramatic decline to a level of two or three printings a year: for Hippocrates, the best years are from 1535 to 1555, after which there is an average of five printings a year (ten in 1579) until the end of the century.

[10] V. Nutton, 'Harvey, Goulston and Galen', *Koroth*, 1985, 8: 112–22 (repr. in Nutton, *From Democedes to Harvey op cit.* (note 4) with identical pagination).

[11] V. Nutton, 'John Caius and the Eton Galen: medical philology in the Renaissance', *Medizinhistorisches J.*, 1985, **20**: 227–52 (repr. in Nutton, *From Democedes to Harvey op cit.* (note 4) with identical pagination); Nutton, 'Prisci dissectionum professores: Greek texts and Renaissance anatomists', in A. C. Dionisotti, Anthony Grafton and Jill Kraye, eds., *The Uses of Greek and Latin. Historical Essays* (London, 1988), 111–26; Nutton, 'Hippocrates in the Renaissance', in Gerhard Baader and Rolf Winau, eds., *Die hippokratischen Epidemien. Theorie – Praxis – Tradition, Sudhoffs Archiv, Beiheft* **27** (Stuttgart, 1989), 420–39.

Mercuriale (1530–1606), the most eminent Italian professor of medicine in the last quarter of the sixteenth century, and editor of the *Opera omnia* of both Galen (1576, in Latin translation only) and Hippocrates (1588, Greek and Latin). His lectures on the diseases of women and children (Venice, 1587; 1583), on dermatology (1572), and on the plague (1577), show a deep acquaintance with the classics, but they are very different in kind from his *Variae lectiones*, short essays on antiquarian problems (Books 1–4, 1570; 5, 1576; 6, 1585; repr. with further additions, 1588, 1598). In the former, the classical texts serve as a basis for wider discussion; in the latter they are the very focus. The difference can best be seen in one of the monuments of Greek-inspired medicine and Mercuriale's most famous work, the treatise *De arte gymnastica*, whose six books first appeared at Venice in 1569, when the author was thirty-nine. Its message had an avowedly medical purpose. It was an attempt to introduce into practical medicine the physical exercises and training of antiquity. In it Mercuriale carefully identified and explained the various ancient exercises, technical terms, and the like, quoting from an abundance of Greek sources, and listing the recommendations of the ancients for each exercise. He does not go much beyond this, and, in a sense, by concentrating on collecting the relevant classical texts, he was only doing for gymnastics what Manardi, half a century earlier, had done for skin diseases, and another contemporary, Matthioli (1500–1577), had been doing for botany. But with each subsequent edition – there were three more in Mercuriale's lifetime, in 1573, 1587, and 1601, and others in 1644 and 1672 – the emphasis switches gradually from medicine to antiquarianism.[12] In the posthumous editions, notably that of 1672, Mercuriale's book, now beautifully illustrated with engravings based on ancient coins and sculptured reliefs, takes its place as a compendium of information about the past, not a collection of advice for the present. In its format, it is aimed at connoisseurs, not at doctors.

Only the Greek Hippocrates escapes this process of medical marginalization, for two reasons. The first is that, in part under the influence of the whole Galenic tradition, Hippocrates had come to be recognized as the formulator of the true principles of medicine. Whatever errors or shortcomings there might be in the classical tradition of medicine could thus be attributed to the misjudgments of later physicians and interpreters, Galen and the Arabs in particular, without detracting from Hippocratic principles. Secondly, as generations of medical teachers had insisted, the authoritative sayings of Hippocrates were frequently enigmatic, and required all the resources of scholarship in their explication. In his own commentaries, Galen had increasingly emphasized the need to approach the wording of the Hippocratic text with great care, and to be alert to interpolations, misreadings, and misunderstandings. In his view, the true meaning of Hippocrates had frequently been obscured by textual corruption and even by forgery. His sixteenth-century followers

[12] For the life and career of Mercuriale, see Christine Nutton, introduction to the reprint of *De arte gymnastica* (Stuttgart, 1978); and Vivian Nutton, 'Les exercices et la santé: Hieronymus Mercurialis et la gymnastique médicale', in Jean Céard, Marie Madeleine Fontaine and Jean-Claude Margolin, eds., *Le corps à la Renaissance* (Paris, 1990), 295–308.

took him at his word, and in their commentaries and discussions often examined in great detail the Hippocratic Greek in order to authenticate and clarify the dark sayings of the sage of Cos.[13] In this way philological methods and Hippocratic Greek gained a foothold within medicine, albeit one that was not without its difficulties. In this respect, the quarrel at Paris in 1578–79 over Hippocrates' book *On wounds in the head* is instructive. The greatest classical scholar of the age, Joseph Justus Scaliger (1540–1609), ostensibly assisting a medical man, brought out his own commentary on this Hippocratic text, in which he showed beyond doubt that much that had been said about this text by the medical men in Paris was philologically inept and textually worthless. The response of the Parisians was twofold; Scaliger was firmly told that interpreting Hippocrates was none of his business, and that, whatever the erroneous linguistic basis of the medical commentators, the facts of medicine proved the truth of the Hippocratic first principles, and it was these, not emendations or deletions, that should be emphasized. In short, even if the Parisian professors agreed with his conclusions – and, on the whole, they did not – they considered them irrelevant to medicine.[14]

The picture was very different fifty years earlier. Then the philological efforts of the Aldine team in Italy, and, north of the Alps, of Cornarius, Leonhard Fuchs, and the rest were greeted with enthusiasm. Phrases such as 'new birth', 'bringing the Greeks out of darkness', and 'the opening up of pure springs' marked the arrival of the long-lost classical texts. Such medical errors as there were might be put right with new manuscripts to hand, or by emendation; even the young Vesalius consulted manuscripts when editing a Latin version of Galen's anatomical texts and he provided some philological solutions to some of their difficulties. The arguments in medical letters, commentaries, and treatises also focused on linguistic problems: the meaning of words, and their changes of meaning over the passage from Greek through the Arabic to the Latin of the Middle Ages. Giovanni Manardi (1462–1536), the pupil and later successor of Leoniceno at Ferrara, collected together all the relevant passages on skin diseases, trying to identify in his classical texts diseases found in medieval compendia and in his own day. Others, like Janus Cornarius (1500–1558), professor at Marburg and Jena, were concerned with the problems of translation, removing what seemed to be medical errors by providing what, in their eyes, was a more accurate translation. This done, others could then use the new Latin versions, confident that they reproduced correctly the meaning of Galen or Paul of Aegina. And, like Vesalius, if they then found errors, they might look for different explanations and different approaches to the truth.[15]

[13] An early example of such a commentary is that of Leonhard Fuchs on *Epidemics* VI (Hagenau, 1532; 2nd edn, Basle, 1537), but similar discussions of specific Hippocratic passages can be found in Italian lectures and letters of the early sixteenth century.

[14] The Parisian debate over the edition of Vertunien and Scaliger (Paris, 1578), can be followed in R. L. Hawkins, 'The friendship of Joseph Scaliger and François Vertunien', *Romantic Rev.*, 1917, **8**: 117–44, 307–27; Antony Grafton, *Joseph Scaliger; a Study in the History of Classical Scholarship* (Oxford, 1983), **1**, 180–84; and V. Nutton, 'The Legacy of Hippocrates: Greek medicine in the Library of the Medical Society of London', *Trans. Med. Soc. of Lond.*, 1986–7, **103**: 30–31.

[15] See Nutton, 'Prisci dissectionum . . .', *op. cit.* (note 11).

But this enthusiasm for the rebirth of Galen and Hippocrates, with the Aldine editions of 1525 and 1526, has tended to obscure other no less significant features of the process by which Greek scientific texts came to influence the Renaissance. In particular, few scholars have commented on the relatively late date at which these scientific texts became accessible in print, whether in the original Greek or in Latin translations newly made from the Greek. Before 1500, only three scientific authors had been printed in Greek: Aristotle and Theophrastus in 1497, and Dioscorides in 1498, all by the Aldine press.[16] Two works by Galen followed in 1500, and then there was a gap of a generation until the Aldine firm put out a series of medical authors, Galen in 1525, Hippocrates in 1526, Paul of Aegina in 1528, and the first half (only) of Aetius in 1534.[17] For Aretaeus and Alexander of Tralles, as well as parts of Rufus of Ephesus, Soranus, and Oribasius, one had to wait until the 1540s and beyond.[18] If this list is contrasted with that of other first editions,[19] one can see that, compared with literary and historical Greek texts, medicine lagged behind by half a generation and more, and mathematics still more so. This gap in part explains why a literary Renaissance based on Greek preceded a similarly inspired scientific Renaissance. The point is still valid if we turn to the new humanist Latin translations based on Greek manuscripts, for, before 1525, with the notable exception of Linacre's unsurpassed Latin versions of Galen's *Method of healing* and *On the preservation of health*, the new versions merely presented a more accurate translation of texts already in the medical curriculum.[20] An exception might also be made for the new versions of Dioscorides, but, as we shall see, Dioscorides is a very special case. Only after 1525 can one really talk of these translations bringing to general attention new medical material and areas of Galenic scholarship largely unknown.[21] I say largely unknown, because many of Galen's minor works had been turned into Latin in the first half of the fourteenth century by a remarkable doctor-translator, Niccolò da Reggio, and had been printed in the four or five collections of the Latin *Opera omnia* published between 1490 and 1525. But manuscripts of these minor works were few and far between, and references

[16] The Aldine Aristotle of 1497 also contained a text of Galen's *Philosophical enquiry*, but I have excluded this little work from consideration for two reasons: it is a tract of philosophical doxography, and it is not by Galen.

[17] For the 1500 Galen, published by Callierges and Blastos, and for the 1525 Aldine Galen, see V. Nutton, *John Caius and the Manuscripts of Galen* (Cambridge, 1987), 26, 34, 39–42. The papal privilege for the 1526 Aldine Hippocrates makes it clear that the firm planned to publish a complete Aetius and an Oribasius.

[18] Alexander (and a Greek version of Rhazes *On plague*), Paris, R. Estienne, 1548; Aretaeus of Cappadocia, Paris, A. Turnebus, 1554; three works of Rufus and two chapters of Soranus, Paris, A. Turnebus, 1554; Oribasius, *Medical collections*, Bks. 24–25, Paris, G. Morel, 1556.

[19] No easy task, admittedly. The most accessible list is given by the majestic Beriah Botfield, *Prefaces to the First Editions of the Greek and Roman Classics and of the Sacred Scriptures* (London, 1861), but it is far from complete, e.g. it omits the 1500 Galen, and excludes all Greek patristic writings. Nor does it enable one to judge the frequency of Greek printings of subsequent editions.

[20] For Linacre's translations, see Richard J. Durling, 'Linacre and medical humanism', in F. Maddison, M. Pelling, C. Webster, eds., *Essays on the Life and Work of Thomas Linacre* (Oxford, 1977), 84–103. For Galenic editions and translations, Richard J. Durling, 'A chronological census of Renaissance editions and translations of Galen', *J. Warburg Courtauld Inst*. 1961, **24**: 230–305, is fundamental.

[21] The distinguished Greek scholar Demetrius Chalcondylas (*c*. 1424–1511) prepared a version of the previously unknown *Anatomical procedures*, but it remained in manuscript; *c*. 1522 the Bolognese anatomist Berengario da Carpi embarked on a revision of this translation, which appeared only in 1529 (Bologna, G. B. Phaelli).

to Niccolò's versions in medical writings, of any date, are very few.[22] The arrival of the new Greek Galen marginalized these older versions still further, with the result that the few genuine texts of Galen which today survive only in medieval Latin, e.g. *On the various parts of the medical art*, and which, in part, were printed in the Latin *Opera omnia*, disappeared from general view, to be rediscovered only at the end of the nineteenth century.

These exceptions, however, do not significantly change the picture of a medical Renaissance beginning with the publication of the Greek Galen in 1525,[23] or of this accessibility of Greek medicine coming half a generation or more after that of more literary and philosophical texts. Why so late? Especially as, from the 1490s onwards, there was a recognizable group of physicians in Venice, associated with Alessandro Benedetti (d. 1512), whose fondness for Greek and Greek learning, acquired both in Venice and in Venetian settlements in the Aegean, brought upon them the scorn of more traditionalist opponents.[24] By the end of the fifteenth century such propagandists for Greek as Niccolò Leoniceno (1428–1524) and Giorgio Valla (1447–1500) had given ample proof of the advantages to be gained from Greek. As the Bolognese professor Urceo Codro (1446–1500) put it in the mid 1490s, only the scholar with Greek could reveal the truths of Hippocrates and Galen; and without the Greeks, the laws of anatomy, surgery and clinical medicine would not exist, and the achievements of Antyllus, Heliodorus and Oribasius would remain unknown.[25] The potentiality was there, as well as the propaganda, a generation before the actual *editiones principes*. Why then the delay?

I suggest three reasons, all interlocking. The first is that there were very few manuscripts of Greek medicine available. Diomedes Bonardus, who edited the 1490 edition of the Latin translations of Galen, complained in his preface that he had had to scour the whole of Italy in search of Latin manuscripts, frequently without success.[26] The number of Greek manuscripts was even fewer. Those that were by this stage in Italy seem, for the most part, to have been in private libraries, whose owners were not always keen to allow them to be used as the basis of a printer's copy. One of the largest collections of medical texts, that bequeathed by Cardinal Bessarion to Venice, was still in its boxes as the Venetians argued about their proposed new library, and how accessible the Vatican's collections were is still an open question. The Medici library in Florence seems, around1500, to have been the place to go for a manuscript copy of a rare Greek text, but relations between the Florentines and the Venetian printers were not entirely cordial. The lack of manuscripts available to printers is a partial explanation for this delay, and it is by no means impossible that the outburst of Greek texts in 1525 was the direct result of the coming onto the market in late 1524 of the

[22] For Niccolò's versions and the Latin *Opera omnia*, see Nutton, *op. cit.* (note 17), 20–1.

[23] The listings in Durling, 'census', *op. cit.* (note 20), might even suggest that it was not until the mid 1530s that the impact of the new Galen began to be felt.

[24] Roger French, 'Berengario da Carpi and the use of commentary in anatomical teaching', in A. Wear, R. K. French, I. M. Lonie, eds., *The Medical Renaissance of the Sixteenth Century* (Cambridge, 1985), 42–74, esp. 46–7.

[25] Urceus Codrus, *Sermones* ix–x (Paris, 1515), cited by Nutton, *op. cit.* (note 17), 22, 31, with a discussion of his sources.

[26] For the truth of Bonardus' claim, see Nutton, *op. cit.* (note 17), 21.

greatest private library of Greek medical texts, that owned by Niccolò Leoniceno, the Nestor of medical humanism, who died that year at the age of 95 or 96. Just as it was Leoniceno's possession in the 1490s of rare, and perhaps unique, medical manuscripts in Greek that occasioned his attack on Pliny, so, after his death, the release of his manuscripts from the clutches of the old 'book-burier', as his pupil, Giovanni Manardi, called him, may have brought about their appearance in print.[27]

But the availability of manuscripts is only one facet; the availability of readers is equally important. The study of Greek as an academic subject does not become at all widespread, even within Northern Italy, until the 1520s, and it is perhaps only with the (re)foundation of grammar schools in England and in Germany in the middle years of the sixteenth century that Greek there became a subject at all widely taught in schools.[28] The smaller the number of Hellenists, the more likely that their requirements could be met by manuscript copying, even in an age of print. Besides, the initial impact of Greek was on literature and philosophy, whose *appassionati* had the leisure and the education to learn Greek, or on theology, where the sacred texts themselves were originally written in Greek.[29] How many doctors in 1520 knew any Greek is unknown, but it may be easily surmised that the number was smaller than those of the literary men attracted to Homer and the tragedians, and that the need to know Greek was less pressing for them than for the theologians.[30] Despite indications to the contrary, the word of Galen is not the word of God.

Publishers and printers have never been shining examples of altruism: their productions are designed to make money, as well as to instruct or entertain. Hence a peculiar difficulty with scientific texts in Greek, with an appeal, at best, to a small market.[31] In medicine, the capital investment involved in publishing the whole, or even most, of Galen or Hippocrates, was huge, in both paper and labour.[32] It was, in all likelihood, the sheer costs of setting up two texts by Galen, *On the method of healing*, and *On the method of healing, for Glaucon*, that effectively bankrupted the

[27] Compare the lists of Greek MSS. in Italy before 1500 in R. Bolgar, *The Classical Heritage* (Cambridge, 1973), 455–505: he gives the following figures for scientific texts Apollonius of Perge 4; Archimedes 3; Dioscorides 9; Galen 9; Hippocrates 8; Nicomachus 5; Hero 3; Ptolemy 15. I have discussed the evidence for manuscripts of Galen in Nutton, *op. cit.* (note 17), 22–8. For the importance, and remarkable range, of Leoniceno's library, see Daniele Mugnai Carrara, *La Biblioteca di Nicolo Leoniceno*, Florence, 1991.

[28] Useful material can be found in the essays by Clough and Schmitt in Maddison *et al.*, eds., *op. cit.* (note 20), 1–23, 36–75.

[29] It should be remembered that the main function of Greek in the new 'humanist' schools (e.g. St Paul's, Merchant Taylors', Oundle, and Oakham in England; the Melanchthon-inspired *gymnasia* in Germany; or the Louvain *Collegium trilingue*) was propaedeutic to theology. This purpose is expressed at both Oakham and Oundle in inscriptions on the school-house written in Latin, Greek, and Hebrew.

[30] Non-medical men, e.g. Politian, Ermolao Barbaro, Bessarion, Lianori, Bonamico, figure as often in the lists of owners of Greek medical manuscripts as do medical men; but this argument should not be pressed too far, for in part it merely reflects the greater survival (and publication) of material relating to literary humanists.

[31] It is no coincidence that the first texts translated from the new Greek Galen, the *Exhortation to medicine*, *The best type of instruction*, and *That the best doctor is also a philosopher*, by Erasmus and published at Basle by Froben in May 1526, were those most likely to appeal to a wider, non-medical audience: see B. Ebels-Hoving and E. J. Ebels, 'Erasmus and Galen', in J. Sperna Weiland and W. Frijhoff, eds., *Erasmus of Rotterdam* (Leiden, 1988), 132–42.

[32] The only comparable enterprise would be the printing of the *Corpus Aristotelicum* (including Theophrastus and ps.-Galen); which, in terms of words, was some 30% larger than the Hippocratic Corpus, but some 30% less than the Galenic. And, from the point of view of a market, Aristotle was far more saleable.

printing patnership of Callierges and Blastos in 1500, and put an end to their plan to produce a complete edition of Galen. Their choice was a good one, their typography excellent, but they misjudged the market.[33] Even the Aldine Galen may not have been a runaway success. At 25–30 gold sous in France; 30 florins or gulden in Germany; and 14 scudi in Rome, it was beyond the pocket of all save the rich.[34] Furthermore, once Galen and Hippocrates had appeared, the incentive to publish more Greek medicine was much less. The later encyclopaedists, like Oribasius, made up in bulk for what they lacked in originality, which made them even less attractive to a printer, let alone to those who might judge a Greek text on its literary merit. The Aldine Press broke off its series of medical folio imprints halfway through Aetius, almost certainly through lack of demand, and no other printer took up the challenge;[35] the Basle printers stayed safe with Galen (1537–8), Hippocrates (1538) and Paul of Aegina (1538), the most useful of the encyclopaedists;[36] the Parisians preferred single texts and selections to large and expensive tomes of *Opera omnia*. For the rest, printers put out small-format editions of, at most, a handful of texts, which involved only a small initial outlay, or like Gryphius in publishing the selection of Hippocratic texts made by Rabelais in 1532, they appended the Greek to a more saleable volume of Latin translations.[37] Later editions of *Opera omnia*, of Hippocrates in 1588 and 1599, and of Galen in the next century, come provided with a facing Latin translation.

As far as medicine is concerned, the following model may be suggested for the reception process of classical learning.

(1) A pre-printing stage, when a handful of Greek manuscripts become available to a few cognoscenti or collectors; their importance is guessed at, but access to these manuscripts is not always assured. Few scholars are able to read or understand their technical language. New humanist Latin translations are, on the whole, concerned with providing a more accurate version of standard texts.[38] Questions are mainly of

[33] See Nutton, *op. cit.* (note 17), 29, 36–7, where some other abortive plans to publish a complete Galen are noted; it is unclear whether it was lack of a market or a failure of interest that prevented Georg Agricola from publishing his substantial list of corrections to the Aldine Galen that he had amassed by 1528, *ibid.*, p. 42.

[34] *Ibid.*, p. 48; the figure of *c.* 25 gold sous for Paris comes from the *Commentaires de la Faculté de Médecine de l'Université de Paris*, in 1526/7, ed. M. L. Concasty, 113, 128. K. Jurina, *Vom Quacksalber zum Doctor medicinae* (Cologne, 1985), 7, reports that in 1526 a Nuremberg doctor paid 30 gulden, a third of his annual stipend as civic doctor, for the Aldine Galen.

[35] The papal privilege for the 1526 Aldine Hippocrates refers to forthcoming volumes of Paul of Aegina, Aetius, and Oribasius, and one of the Aldine team, Georg Agricola, certainly collated an Oribasius and at least two manuscripts of Aetius, Nutton, *op. cit.* (note 17), 39, 46.

[36] The Basle printer Johann Walder put out the *editio princeps* of the Greek *Hippiatrica* in 1537, but this was his only Greek medical text.

[37] *Hippocratis et Galeni libri aliquot* (Lyons, S. Gryphius, 1532); to a reprint of new humanist translations of the standard set texts of the medical curriculum, Hippocrates, *Aphorisms, Prognostic, The nature of man, Diet in acute diseases*, and Galen, *Art of medicine*, was added Rabelais's text of the *Aphorisms*. According to M. A. Screech and S. Rawles, *A new Rabelais bibliography* (Geneva, 1987), n. 105, pp. 521–8, the volume proved a best-seller. Its second printing, by Gryphius in 1543, again sold well, *ibid.*, n. 106, pp. 529–34. Interestingly, the third printing, by Gryphius in 1545, *ibid.*, n. 107, pp. 535–8, dropped the Greek, which, in some copies, was later bound in from an earlier edition. The copies I have seen, however, show little sign that anyone read the Greek.

[38] It is perhaps worth insisting on this point, for there is a tendency to associate the spread of the new learning with the introduction of printing. In fact, as Marie Hall has well stated, *The Scientific Renaissance, 1450–1630*, (London, 1962), 23–30, the first half-century of printing largely perpetuated the standard authorities and books of the Middle Ages.

identification and translation of words and phrases, rather than of wholesale interpretation, and are often resolvable by philological means.
(2) A relatively brief stage of printing, in which new material is published in Greek, and then given greater accessibility by being provided with a Latin translation. This requires the pre-existence of a market sufficient in both numbers and learning, and hence comes only at a late date in the publication of classical texts.
(3) The assimilation of new texts, leading to changes, e.g. in anatomy and physiology, either taking their inspiration from classical models or reacting against them. The deficiencies of a philological explanation for apparent errors of the ancients become clearer; debate shifts to a discussion of the evidence and argument, rather than to the meaning of individual words. The Greek texts themselves become marginalized.[39]

Botany and mathematics

With this tripartite scheme in mind, one may look briefly at two other areas of Renaissance science, botany and mathematics, and compare them with medical printing. Firstly botany. Here, the printing of the two main texts, Theophrastus' *History of Plants* and Dioscorides' *Materia medica*, came very early, both before 1500; Theophrastus, on the coat tails of Aristotle, in 1497, Dioscorides in 1499, both by the Aldine press in Venice.[40] The first of my three stages is thus short, and precedes any other type of scientific printing. It was followed by a large gap. The second edition of Dioscorides, put out by the Aldine press in 1518, incorporated many changes suggested by readings in manuscripts owned by Leoniceno. Dioscorides continued to be printed in Greek throughout the sixteenth century; the first edition with an accompanying Latin translation was that published by J. Soter at Cologne in 1529. Janus Cornarius brought out his edition at Basle in the same year with Johann Bebel, and there were further bilingual editions at Paris (in 1549, by P. Haultin and A. Birkmann) and at Frankfurt (in 1598, by Wechel). This steady, if lacunose, printing history may indicate that in botany, at least, the third stage, the marginalization of the Greek, may not even have come to be. As Karen Reeds has argued, much of sixteenth-century botany can be seen, both literally and metaphorically, as a series of commentaries on Dioscorides, in which questions of Greek continued to matter.[41]

Why these differences? Dioscorides is printed early in Greek, for three main reasons. Firstly, a market was available. The earliest manifesto of the new humanist

[39] The claim of Sebastian Singkeler in the preface to his bilingual *Schola medicorum*, incorporating the (pseudo-Galenic) *Introduction* and *Medical definitions* (Basle, Platter and Lasius, 1537), that without Greek one would be unable to break the shell and get to the very kernel of medicine, was already being rejected in favour of the Latin.
[40] See Karen Meier Reeds, 'Renaissance humanism and botany', *Ann. Sci.*, 1976, **33**: 519–42.
[41] *Ibid.* Note also John Riddle, 'Dioscorides' in Kristeller and Cranz, *op. cit.* (note 4), **4**: 1–143.

Greek-based science, Leoniceno's assault on the errors of Pliny the botanist, had stressed the immense superiority of Dioscorides, as well as the need to correct Pliny's fallible understanding with the aid of the Greek.[42] Secondly, botany was not an academic subject, like medicine; it attracted a wider audience, of garden-lovers as well as apothecaries.[43] And thirdly, it does not require a substantial acquaintance with Greek to gain a reputation as an expert on Dioscorides. Save for his preface, Dioscorides' Greek is restricted in its vocabulary, and even more so in its grammar and syntax; it is descriptive, not argumentative. Compared with Galen, let alone with Aretaeus, Hippocrates, and more literary authors, Dioscorides offers few hazards of grammar and syntax. The problems of interpretation that his text raises are concerned with specific words, specific substances, and with questions of translation and identification that can be solved only philologically or by the addition of practical knowledge of the plants themselves. In this sense, even the late-sixteenth-century academic herbals acquire a philological base. Matthioli's commentaries are filled with disquisitions on Greek words, and even vernacular botanical authors were liable to display their erudition with a carefully dropped Greek word. By contrast, the much more theoretical discussion of plants in Theophrastus is rarely taken up, and his information on plants is largely subsumed under that of Dioscorides. By contrast with Dioscorides, his text is neglected, and reprinted only along with Aristotle.[44] When scholars, such as Antonius Musa Brasavola in the 1540s, finally began to study the way in which drugs and plants worked, they were influenced by Galen rather than Dioscorides. In addition, the theoretical approach they adopted did not demand close attention to the Galenic text, and could thus rest on the new humanist Latin versions.[45]

If Renaissance botany seems never to have reached my third stage, Renaissance mathematics omitted the second stage almost entirely. Apart from the handful of short astronomical texts included in the Aldine edition of the *Astronomi veteres* (Venice, 1499), printed Greek mathematical texts are confined to a small number of German printings, Euclid in 1533, Theon in 1538, Archimedes in 1544, all relatively late for an *editio princeps*, and, what is equally significant, all small in size: they did not demand such a great expenditure of capital as, say, an edition of Galen or Aetius. Many of the mathematical texts that later on in the century gained importance, the *Conics* of Apollonius of Perge, Diophantus, Pappus, and the *Pneumatica* of Hero of Alexandria, were never printed in Greek in the sixteenth century.[46] They were known in manuscript to only a few mathematicians with access to the relevant libraries or private collections, and their impact in printed form depended entirely on a

[42] See French, *op. cit.* (note 4), for details of the controversy in the early 1490s.

[43] For the use of Dioscorides in the development of botany as an academic subject, see Reeds, *op. cit.* (note 40), 533–9.

[44] Charles Schmitt, 'Theophrastus', Kristeller and Cranz, *op. cit.* (note 4), **2**: 239–322.

[45] Compare Reeds, *op. cit.* (note 40), 541.

[46] Sarton, *op. cit.* (note 2), omits the *Astronomi veteres* (and many of the mathematicians), largely because he considers them as shading off into astrology, on which see his comments on the German printings of Ptolemy (1535, 1538, 1553), p. 147. On the *Pneumatica*, see Marie Boas, 'Hero's *Pneumatica*. A study of its transmission and influence', *Isis*, 1949, **10**: 38–48.

circulation in Latin translation.[47] Besides, as Paul Rose has shown, for all that mathematics was taught in many universities, notably Bologna, the number of mathematicians with a good knowledge of Greek and, even more important, who knew where Greek mathematical manuscripts were to be found, was very, very small. Hence, the significance of Maurolico (1494–1575) and, still more, Commandino (1509–1575) as moderators of this learning to others.[48] In mathematics, then, it is the third stage that takes overwhelming precedence, the assimilation of classical discoveries through the medium of Latin, and the marginalization of Greek.

Discussion and argument also rarely focus on philological methods. True, both Maurolico and Commandino went to great lengths to secure what they thought was a good base text from which to proceed, but they criticized their predecessors as translators of Euclid for mistakes in mathematics, not in Greek. Indeed, Maurolico was prepared to admit that Campanus was such a faithful translator that he did not bother to correct obvious mistakes in his Greek text of Euclid but let the Greek error shine through the Latin. Where he went wrong was in knowledge of practical mathematics, which should have saved him from mistakes in understanding.[49] Contrast this with John Caius, or still more, with Cornarius, who were both convinced that Vesalius's attack on Galen was the result of Vesalius's poor knowledge of Greek, not of anatomy.[50] One has only to glance at Cornarius's criticisms of Vesalius's Latin translations of Galenic anatomical texts, criticisms he inserted into the margin of his copy of the 1542 Froben edition that he was preparing for the printer of the next (1549) Basle edition and which can now be seen in the British Library, classmark 774. n. 13. Here Cornarius offers a damning indictment of Vesalius's Greek learning, and suggests that it is this deficiency that has inspired his attack on the great Galen. By contrast, the Latin translators of mathematical texts direct their attention to the mathematics, not to the words themselves.

Conclusions

In this short chapter, it has been argued that merely to talk in broad terms of the influence of Greek science in the Renaissance hides major differences in the way in which the Greek science was known, made accessible, and interpreted. These three stages are often telescoped in modern accounts, with resulting confusion, and insufficient attention is paid to the way in which each discipline used the classical texts. To put it crudely, in botany and, to some extent, in medicine, practical

[47] Compare the absence of Greek mathematical texts in such an important humanist mathematical university as Vienna, H. Grössing, *Zur Geschichte der Wiener mathematischen Schulen des 15. und 16. Jahrhunderts* (Baden Baden, 1983).

[48] P. L. Rose, *The Italian Renaissance of Mathematics* (Geneva, 1975). His discussion of Commandino's manuscripts, mainly available to him in Rome and Urbino, is exhaustive; that of Maurolico's sources still leaves many puzzling gaps.

[49] *Ibid.*, 165.

[50] Compare Nutton, 'Prisci dissectionum . . .', *op. cit.* (note 11).

problems could easily be categorized as textual problems; in mathematics, mathematical problems remained mathematical.

Secondly, there is the paradox that, as the sixteenth century wore on, and the claims for the need for the physician or mathematician to know Greek became ever more strident, so the actual usefulness of that knowledge diminished. Mathematicians, as we have seen, did not, for the most part, use Greek texts of mathematical authors. Discussions of Galenic medical theory rarely involved discussions of textual minutiae, or could be controverted by the discovery of a new reading; they were conducted on the basis of the Latin versions, not the Greek. This is not to say that some physicians did not also practise medical philology – North Italians, like Mercuriale and Donzellini, the Basle physicians around Zwinger, and, not least, the Paris Hippocratics, all hunted for new manuscripts, collated, and compared.[51] But the results of their labours were increasingly irrelevant to medical practice; the various philological classifications of the genuine works of Galen or Hippocrates did not prevent reliance on non-Galenic or less authentic Hippocratic writings as evidence for sound therapy. This increasing irrelevance of Greek comes at a time when changes in school curricula, not least in Northern Europe, had made knowledge of Greek less rare than it once was. But acquaintance with Demosthenes' *Orations* or with Homer is no guarantee of being able to understand Galen, let alone Hippocrates, and one may well wonder how much of the Greek learning found in medical commentaries of the 1580s is just window-dressing, taken from Latin translations or, worse still, from the compendia of others. If William Harvey, no mean Galenist, could in the 1640s annotate only the Latin side of a bilingual edition of selected works of Galen, we may doubt how often others less gifted or less involved with Galen consulted the works of the master in Greek.[52] The truth is hard to come by; and a combination of rhetorical prefaces, university and college statutes, and the injunctions of Grecian writers can easily offer the overwhelming impression of a medical science based on Greek ideas discussed and understood in the original. Indeed, despite the ever-increasing claims for Paracelsianism and for a medicine based on empiricism rather than on theory, the overwhelming majority of medical writers, professors, and educated practitioners continued recognizably in a Galenic or Hippocratic tradition well into the seventeenth century.[53]

But there are occasional hints that all was not well on Olympus. In a funeral oration of *c*. 1627, Balthasar Venator praised the Dutch mother of one of the foremost scholars of the day for her knowledge of French, Italian, Latin, English, and Greek, which enabled her to read Galen in his own tongue, something that scarcely one in a

[51] Compare Nutton, 'John Caius . . .', *op. cit.* (note 11), esp. 243–52. For Paris, see Iain M. Lonie, 'The "Paris Hippocratics": teaching and research in Paris in the second half of the sixteenth century', in Wear *et al.*, *op. cit.* (note 24), 155–74.

[52] See Nutton, *op. cit.* (note 10).

[53] See, most recently, on the shift away from this classical tradition, Harold J. Cook, 'The new philosophy and medicine in seventeenth-century England', in David C. Lindberg and Robert S. Westman, eds., *Reappraisals of the Scientific Revolution* (Cambridge, 1990), 397–436. This is not to imply that the Galenic/Hippocratic tradition was in any sense uniform; rather, that from a series of principles derived from classical texts, Renaissance writers on medicine could produce Protean transformations and markedly individual discourses.

thousand physicians could do.[54] Significantly perhaps for my thesis, her son did not become a physician, but the leading antiquary of the early seventeenth century. His name was Jan de Groot, Janus Gruter, one of those Dutchmen who did so much for intellectual links between Britain and the Netherlands in the age of the Scientific Revolution. Classicists and historians of science should take note of a melancholy truth. This aside from the funeral pulpit marks the death not only of a distinguished Dutch scholar but also of a strictly Greek-language influence on European science.[55]

[54] Balthasar Venator, *Panegyricus*, [1631?], cited by L. W. Forster, *Janus Gruter's English years* (London, 1967), 88–9.

[55] The rediscovery of ancient medical texts in the nineteenth and twentieth centuries from papyri, manuscripts, and, increasingly, Arabic or medieval Latin versions has excited little or no interest among physicians (or even classicists), with two possible exceptions; the 'lost' books of Galen's *Anatomical procedures* (in Arabic), and the so-called Anonymus Londinensis papyrus, in part a doxographical record going back, it is supposed, to the school of Aristotle, and offering new information on Hippocrates. In the 1990s Hippocratic medicine has still to make its come-back among fashionable 'holistic' therapies. M. Simon, *Sieben Bücher Anatomie des Galen*, Leipzig, 1906; Eng. tr. by W. H. L. Duckworth, M. C. Lyons, B. Towers, *Galen, On anatomical Procedures. The Later Books*, Cambridge, 1962; most recent edition by I. Garofalo, *Galeno, Procedimenti anatomici*, 3 vols, Milan, 1991. H. Diels, *Anonymi Londinensis ex Aristotelis Iatricis Menoniis et aliis Medicis*, Berlin, Reimer, 1893; Eng. tr. by W. H. S. Jones. *The Medical Writings of Anonymus Londinensis*, Cambridge, 1947; new editions, with substantial revisions, are promised by D. Manetti and by J. Pigeaud.

2

'With the rules of life and an enema': Leonardo Fioravanti's medical primitivism

WILLIAM EAMON

> I confess the truth, that I believe there is no better thing than to toil at experience and to imitate those first ones, who did not know physic, nor any method at all, but just had good judgement.
>
> Leonardo Fioravanti, *Capricci medicinali* (1561)

One of the most characteristic expressions of Renaissance culture was its captivation with nature and experience. From Leonardo da Vinci to Montaigne, Renaissance intellectuals found in nature and experience a dual standard against which to measure what was to them a sterile, bookish, and bankrupt scholastic tradition. 'He who wishes to explore nature must tread her books with his feet', wrote Paracelsus. 'Writing is learned from letters. Nature however (by travelling) from land to land: One land, one page. Thus is the Codex Naturae, thus must its leaves be turned.'[1] Paracelsus's Danish disciple, Peter Severinus, exhorted naturalists to

> sell your lands, burn up your books, buy yourself stout shoes, travel to the mountains, search the valleys, the deserts, the shores of the sea, and the deepest depressions of the earth; . . . Be not ashamed to study the astronomy and terrestrial philosophy of the peasantry. Purchase coal, build furnaces, watch and operate the fire. In this way and no other you will arrive at a knowledge of things and their properties.[2]

If empiricism was a recoil against the inflexible rationalism and bookishness of scholastic thought, often it was also a reaction against the political control that academics maintained over institutions of learning and the professions. The medical empiric's familiar pronouncement that practical experience was worth more than all the books of Hippocrates and Galen taken together was, among other things, a missile aimed at the monopoly physicians held over the medical marketplace. When he urged going 'back to nature', making nature and not some abstruse philosophical system the

[1] *Defensiones un Verantwortungen wegen etlicher verunglimpfung seiner Missgönner*, qu. Walter Pagel, *Paracelsus: An Introduction to Philosophical Medicine in the Era of the Renaissance* (Basle and New York, 1958), 56–7.

[2] Petrus Severinus, *Idea medicinae philosophicae* (The Hague, 1660), qu. Allen Debus, *The English Paracelsians* (New York, 1966), 20.

measure of proper therapeutics, the empiric was proposing a standard of authority that was accessible, so he claimed, to all who had eyes to see.[3]

In arguing that nature and experience were the sole sources of truth, the early modern empiricists fabricated various new fictions of their own. They also revived some old fictions, one of the most compelling of which was the idea of a time when natural philosophy existed in a 'pure' state, uncorrupted by scholastic logic-chopping, a pristine age when 'science' was nothing more than direct, 'hands-on' contact with nature. A number of different versions of the doctrine of scientific primitivism emerged during the period of the Scientific Revolution. Francis Bacon, for example, deployed primitivism to contrast the fruits of empiricism to the sterility of logic. 'In the invention of arts', he wrote,

> it would seem that hitherto men are rather beholden to a wild goat for surgery, to a nightingale for music, to the ibis for clysters, to the pot lid that flew open for artillery, and in a word to chance, or anything else, rather than Logic.[4]

One of Bacon's most fascinating primitivist images was his metaphor of Pan's hunt, in which he described experimentalism as 'a sagacity and a kind of hunting by scent rather than science'. Bacon interpreted the myth of Pan to mean that

> the discovery of things useful to life . . . is not to be looked for from the abstract philosophies, as it were the greater gods, no not though they devote their whole powers to that special end – but only from Pan; that is from sagacious experience and the universal knowledge of nature, which will often by a kind of accident, and as it were while engaged in hunting, stumble upon such discoveries.[5]

We are also reminded of the English Baconians' pledge to study the 'natural philosophy' of unlettered craftsmen and common people.[6]

It should come as no surprise that some of the most enthusiastic sixteenth- and seventeenth-century proponents of scientific primitivism were not philosophers but medical empirics, surgeons, and 'charlatans'. As outsiders *vis-à-vis* the medical establishment, they had the least to lose and the most to gain by challenging the official medical canon. In this chapter, I want to take a closer look at the doctrine of primitivism in medicine and to explore some of its cultural and political ramifications. The focus of my paper is the sixteenth-century Italian surgeon and popular healer, Leonardo Fioravanti (1518–1588), who was Italy's most ardent champion of the doctrine of medical primitivism. A charismatic healer and a vociferous critic of the physicians, his unorthodox methods made him the focus of a cult-like following. Although he was scorned and persecuted by the medical establishment, he had many

[3] On Renaissance 'empiricism' in general, see Hiram Haydn, *The Counter-Renaissance* (New York, 1950), 176–92.

[4] Francis Bacon, *De augmentis scientiarum* (London, 1623), in *The Works of Francis Bacon*, ed. J. Spedding, R. L. Ellis, and D. D. Heath (London, 1870; repr. New York, 1968), 4: 409.

[5] Bacon, *De sapientia Veterum* (London, 1609) in *ibid.*, 6: 713. See also *ibid.*, 4: 421. For discussion, see Paolo Rossi, *Francis Bacon: From Magic to Science*, trans. S. Rabinovitch (Chicago, 1968), 98, 155; William Eamon, 'Science as a *Venatio*', in *Festschrift for Willy Braekman*, ed. C. De Backer (Brussels, forthcoming).

[6] Charles Webster, *The Great Instauration: Science, Medicine, and Reform, 1626–1660* (Harmondsworth, 1976), 15–19, 324–483.

disciples who propagated his doctrine throughout Italy and Europe. Fioravanti was also a highly successful writer whose voice was heard by countless sixteenth-century readers. Although he championed a 'new way of healing' that was radically different from the 'canonical way', he insisted that it was in reality a return to the most ancient medicine of all, that of 'the first physicians'.

Charlatan, prophet, or natural philosopher?

Although Fioravanti is fairly well known to historians of early modern medicine, posterity has been neither sympathetic nor understanding. His poor reputation among modern historians is in large measure an inheritance from the negative judgements of him rendered by sixteenth-century academic commentators. In modern scholarship he is usually represented either as a charlatan or as Italy's foremost Paracelsian.[7] Both characterizations, I believe, widely miss the mark. As for the first, Fioravanti was a much more serious threat to orthodox medicine than the *ciarlatani* who peddled their nostrums in the *piazze* of Italy's major cities. Not only did he compete with the physicians, he attacked them in print. Moreover, his books were exceedingly popular and influential. With disciples throughout Italy and Europe, his ideas circulated widely, and became the basis of what was essentially an alternative medical movement. Nor is it satisfactory to label him a Paracelsian. Although he advocated chemical treatments and praised Paracelsus (along with a host of other alchemists and empirics), his writings contain none of the mystical flavour one usually associates with Paracelsianism. He did not give much importance to such characteristically Paracelsian ideas as the theory of correspondences between the microcosm and macrocosm or the doctrine of signatures. While he shared Paracelsus's contempt for the medical establishment and his commitment to empiricism, beyond that I have found few important instances of the influence of Paracelsus in his writings.

A more fruitful approach to understanding Fioravanti's ideas is to examine what he said in his own writings. Before doing so, however, it will be helpful to look briefly at his career. Most of what we know about Fioravanti's life comes from the autobiography he published in his *Tesoro della vita humana* (1570), a record of his travels, discoveries, and travails, mostly with the physicians.[8] In many ways the work resembles the *Apology and Treatise* of Fioravanti's more famous French contemporary, the surgeon Ambroise Paré. Like Paré's work, the *Tesoro* gives detailed accounts of some of the more remarkable cases Fioravanti treated. Added to these adventures are the testimonials of those whom Fioravanti treated or who followed Fioravanti's 'new way of healing'.

Fioravanti's autobiography begins in the year 1548, when, at the age of thirty, he

[7] The best biography of Fioravanti, though still inadequate, is Davide Giordano, *Leonardo Fioravanti Bolognese* (Bologna, 1920). In addition, see Domenico Furfaro, *La vita e l'opera di Leonardo Fioravanti* (Bologna, 1923); Nicola Latronico, 'Leonardo Fioravanti bolognese era un ciarlatano?', *Castalia*, 1965, **31**: 162–7; and Giancarlo Zanier, 'La Medicina paracelsiana in Italia: Aspetti di un'accoglienza particolare', *Riv. Stor. Fil.*, 1985, **4**: 627–53.

[8] Leonardo Fioravanti, *Tesoro della vita humana* (Venice, 1570).

left his native Bologna and embarked on an extended period of travel throughout Italy and the Mediterranean. It is significant that Fioravanti tells us nothing whatsoever about his early life, nor about his medical training. His silence suggests that he regarded those first thirty years as irrelevant and unimportant as far as his professional development was concerned. It is as if his real training began when he left Bologna, 'solely with the intention of going out into the world in order to gain knowledge of natural philosophy.'[9] Despite his repeated claims that he had a medical degree, no record of his ever having matriculated at the Bologna medical school has turned up. It was not until 1568 that Fioravanti, hounded by the Venetian physicians, returned to Bologna and obtained – some said he bought – degrees in medicine and philosophy.[10] Fioravanti reported rather vaguely that he 'practised' for fifteen years in Bologna, that is, since the age of sixteen, which would seem to indicate an apprenticeship to a surgeon. Whatever his training may have been before he left home, Fioravanti plainly thought his education really began only when he 'went out into the world'. 'To tell the truth,' he wrote, 'there is no better way to learn than to go out into the world, because every day there are new things to be seen, and various and diverse important secrets to be learned.'[11]

Fioravanti wrote that he was following in the footsteps of the ancient empiricists. More likely, he was appropriating the familiar literary stereotype of the wandering empiric, which we encounter in authors ranging from Paracelsus to Cervantes. In Italy, the topos was popularized in such works as the famous *Secreti del reverendo donno Alessio Piemontese* (1555), one of the most celebrated sixteenth-century 'books of secrets'. Alessio (*alias* Girolamo Ruscelli) wrote that out of a 'singular devotion to philosophy and to the secrets of nature', he had spent his entire life travelling in search of secrets. He gathered secrets 'not only from men of great knowledge and profound learning, and noblemen, but also from poor women, artisans, peasants, and all sorts of men'.[12] Fioravanti's narrative bears an unmistakable resemblance to Alessio's preface. 'Wherever I travelled', he recalled,

> I never grew weary of study or of searching for the choicest experiments (*esperimenti*), whether from the most learned doctors of medicine, or even from empirics and every other sort of folk, whether peasants, shepherds, soldiers, religious people, country women, and every other quality of person.[13]

Fioravanti's travels took him to Sicily, where an old alchemist taught him the art of distillation. In Calabria he observed the techniques of plastic surgery invented by the famous Vianeo brothers. He learned the 'rules of life' observed by the peasants of the

[9] 'mi parti della mia dolce patria Bologna, solamente con intentione di andare caminando il mondo per haver cognitione della natural filosofia', *ibid.*, 17v. See also the Proemio to the same work.

[10] For details concerning Fioravanti's Bologna degree, see Giuseppe A. Gentili, 'Leonardo Fioravanti Bolognese alla luce di ignorati documenti', *Riv. Stor. Sci.*, 1951, **42**: 16–41.

[11] Fioravanti, *op. cit.* (note 8), 26v–27r.

[12] Alessio Piemontese, *Secreti del reverendo donno Alessio Piemontese* (Venice, 1555), 'I Lettori'. On this work, see William Eamon, 'The *Secreti* of Alexis of Piedmont, 1555', *Res Pub. Litt.*, 1979, **2**: 43–55. On the books of secrets, see Eamon, 'Science and popular culture in sixteenth-century Italy: the "Professors of Secrets" and their books', *Sixteenth Cent. J.*, 1985, **16**: 471–85.

[13] Fioravanti, *op. cit.* (note 8), 2.

south of Italy. For two years he served as a military surgeon in the Spanish navy, accompanying the fleet in its north African campaign and witnessing the bloody battle of Monastir. Returning to Italy, he practised in Naples, Rome, and Milan, everywhere astonishing people with his miraculous cures. In Palermo, he wrote, 'the people gathered around me as to an oracle'.[14] Despite his fame, or more likely because of it, he was continually embroiled in disputes with the medical establishment. The Roman physicians, jealous of his growing reputation at the papal court, barred him from practising in the city. The Venetian physicians conspired against him to prevent him from selling his medications, and tried every possible means to discredit him, including challenging the legitimacy of his Bologna medical degree.[15] In Milan, the physicians had him thrown into prison because he did not 'medicate in the canonical way'. From prison, the pugnacious Fioravanti issued a public challenge: 'that there be consigned to me alone twenty or twenty-five sick people with diverse ailments, and an equal number with the same infirmities to all the physicians of Milan, and if I do not cure mine faster and better than they do theirs, I am willing to be banished forever from this city.'[16] Unfortunately, the record does not tell us whether (as is unlikely) the physicians took up Fioravanti's challenge. However, despite the physicians, he continued to practise and to propagate his medical doctrine.

Fioravanti's fame reached all the way to Madrid, and to the court of Philip II, who in 1576 welcomed the famous old empiric into his court. Fioravanti, the king remembered, had rendered distinguished service as a surgeon in the Spanish army in Italy. But even though Fioravanti earned the king's respect, he was attacked by the royal physicians. Only a few months after his arrival in Madrid, the king's prosecutor brought charges against him of malpractice and of using unorthodox and unapproved medicines.[17] According to the charges, Fioravanti had obtained a licence to practise medicine at the court on the basis of falsified documents. Furthermore, he was ignorant of Latin, gave his patients drugs solely of his own invention, and prescribed poisons, killing several of his patients. In his defence, Fioravanti explained that his 'new way' of healing was drawn from experience and from ancient Hippocratic doctrine, which was completely misunderstood by the modern physicians. Although he admitted that one of his patients had died, it was not on account of his extreme remedies. For death is natural; when God ordains that a man's hour has come, no physician can help him. As for being ignorant of Latin, had not Hippocrates, Galen, and Avicenna written in their own mother tongue and not in Latin? Evidently Fioravanti's defence succeeded. He remained at Philip's court for about a year, experimenting on tobacco and other New World plants, studying Ramon Lull's alchemical theories, and propagating his medical doctrine.

[14] *Ibid.*, 27r. [15] Gentili, *op. cit.* (note 10), 29–33.

[16] Eugenio Dall'Osso, 'Due lettere inedite di Leonardo Fioravanti', *Riv. Stor. Sci.*, 1956, **47**: 283–91, p. 288. See also Nicola Latronico, 'Una disavventura milanese di Leonardo Fioravanti', *L'Ospedale magg.*, 1941, **29**: 481–2.

[17] This is revealed in Fioravanti's defence against legal proceedings taken against him in Madrid by the king's prosecuting attorney Martin Ramón: British Library Add MS. 28,353, ff. 57–61. I am grateful to David Harley, who at the Oxford meeting gave me a photocopy of this manuscript. For a brief discussion, see David Goodman, *Power and Penury: Government, Technology and Science in Philip II's Spain* (Cambridge, 1988), 248.

The scourge of the physicians

Fioravanti's years of wandering and of observing the techniques of empirics convinced him of one thing: official, academic medicine was intellectually bankrupt, a tissue of lies and obfuscations. When he surveyed the present state of medicine, he saw nothing but confusion and disorder. The physicians, he asserted, all had conflicting systems they continually argued over:

> One says he understands Galen's system very well, and will practise along that road. Another wants to comment on the subtleties of Avicenna, and another to read from a professorial chair. And so each and every one believes he knows more than the others, and no one wants to yield to his companion. And in this way the world always gets so muddled that it's impossible to know the truth.[18]

Fioravanti believed that man originally learned medicine directly from nature. In ancient times knowledge of medicine was held in common by everyone. Over the ages, however, the physicians 'usurped' medicine and arrogated to themselves the authority to practise. Creating a monopoly for themselves, they deprived the world of the benefits of unlettered empirics, who followed nature and the rules of life. Fioravanti blamed this unhappy state of affairs on bad laws regulating medical practice,

> by means of which legislators have divided medicine and surgery into two parts, theory and practice, such that those who use theory call themselves 'rational physicians', while those who use practice are called 'empirics', and for them it's not legal to practise medicine because they can't chatter like the physicians.[19]

Having pulled off this 'swindle' (*gabbaria*), the physicians fabricated complex and esoteric theoretical systems which they used to further separate themselves off from the rest of society. As they competed among themselves, sects arose, like the scattering of tongues, eclipsing the true medicine of the 'first physicians'. The injustice of this was manifest, Fioravanti thought, because medicine was 'a gift of God given to all the creatures of the world; it is neither right nor honest that anyone should be prohibited by law from seeking to use it for his happiness'.[20]

For Fioravanti, the reform of medicine was a matter of going back to 'natural' methods of healing, which are learned not by studying theories, but by keen observation. 'Good judgement' and common sense, he believed, were worth more than scientific or theoretical knowledge. He ridiculed the physicians' insistence on knowing the causes of diseases before treating them. How foolish, he thought, because you never treat the causes, you only treat the disease: 'Why should we bother knowing this damned "cause" when it's never cured? To treat a wound you have to know what's important to the wound, not why the man was wounded.'[21]

Fioravanti believed that medicine had gained more from unlettered experience than

[18] Leonardo Fioravanti, *Della fisica* (Venice, 1582), 'Ragionamente', b1v.
[19] Leonardo Fioravanti, *Cappricci medicinali* (Venice, 1561, 1582), 32r. [20] *Ibid.*, 32v.
[21] Leonardo Fioravanti, *Del compendio de i secreti medicinali* (Venice, 1564), 26.

from book learning. The physicians, guided only by theory, were like the navigator who gazes at maps in his study, and then when he goes out to sea gets lost or shipwrecked. He had seen it happen many times:

> The physicians will study a very pretty theory; they will think they find the causes of infirmities and remedies to cure them, and then when they come upon some difficult case, they won't know how to bring about a cure . . . Then truly, some old experienced hag will come along, who with the rules of life and an enema will make the fever cease, or with some unction will make the pain go away, or with some fomentation will make the patient sleep. And in so proceeding, the old hag will know more than the physicians.[22]

Rejecting what he regarded as the abstruseness and the needless complexity that had beset medicine with the establishment of competing sects and schools, Fioravanti urged a return to the pristine and simple system of the 'first physicians' – 'who otherwise knew no medical system, nor any method at all, but just had good judgement'.[23] The first physicians, he thought, learned medicine from the brute animals, who knew by instinct how to cure themselves:

> It's quite true that nature gave all the animals a very great gift, which was that each animal, all by itself, without aid or counsel from anyone, knew how to cure its infirmity . . . The dog, when it feels itself sick, goes to the forest, and finds there a certain sort of herb, which it recognizes by natural instinct, and eats it, and that herb immediately makes it vomit and evacuate from behind; and it is cured at once. The ox, horse, and mule, when they feel themselves aggravated by some infirmity, bite the end of their tongue until blood flows out, and are healed. Hens, when they are sick, take out a certain membrane under the tongue and the blood flows from it, and immediately they are healed. And many other animals do similar things to cure various infirmities. . . . The animals therefore really know how to doctor themselves, and haven't previously studied medicine; they don't have it by science (*per scientia*), but by experience and the gift of nature . . . And so each time men saw these things they observed it, and in this way came to know that evacuation and bloodletting were very useful.[24]

From these observations Fioravanti drew the following conclusion, which became the basis for his entire medical system: all diseases result from two principal causes, the 'bad qualities and indisposition' of the stomach, and the 'alteration and corruption' of the blood. Accordingly, he reduced therapeutics to two means: drugs to purge the body of corruption, and bloodletting to release bad humours from the blood. Fioravanti rarely recommended bloodletting, because blood is our soul (*anima nostra*); it is better to save the blood and purify it of bad qualities. However, for certain ailments he found it efficacious to draw a little from under the tongue, following the example of the animals. In general he preferred robust purges to cleanse the stomach:

> The first cause of all infirmities is the indisposed and corrupt stomach, from which follows the corruption of the entire body, and by reason of this cause

[22] Fioravanti, *op. cit.* (note 18), b2v. [23] Fioravanti, *op. cit.* (note 19), 31r. [24] *Ibid.*, 29–30.

> the blood along with all the interior parts suffers, and for this reason it follows that to be able to liberate the body from all kinds of infirmities it is necessary to evacuate it of these corrupt humours, whether by vomiting or by purgation. And the truth of this is verified every day by experience, which shows that those medicines which provoke vomiting, evacuating a great deal, cause much better effects than any other for the health of the sick body.[25]

Rejecting the physicians' reliance on diet and weak medications, he advocated a therapeutics of 'direct action' in the form of violent emetics and purgatives to 'remove the sickness and return the body to its pristine health'.[26]

The Golden Age and the rule of nature

It will be apparent that in many respects Fioravanti's medical primitivism was an evocation of the ancient legend of the Golden Age, which originated in Hesiod. The Roman poet Ovid popularized the legend, and Tacitus transformed a version of it into history when he contrasted the noble barbarian kings to the corrupt Roman emperors. Central to this legend was the theme of progressive decadence from an age of innocence and perfect happiness, when man lived close to nature and the Earth provided all his needs, down to the present age of savagery, greed, tyranny, and moral corruption.[27] Roughly speaking, the Golden Age is all that the present is not. The myth was widely appropriated by sixteenth-century Italian writers, who longed to escape from the intolerable conditions of contemporary society. The contradiction between reality and the world as it might be made the lure of Utopia irresistible.[28] The Utopians discovered in pristine nature a sufficient reason that could serve as a guide and standard of human action. 'Virtue', wrote Thomas More, means 'living according to nature. To follow nature is to conform to the dictates of reason.'[29] Anton Francesco Doni, who in 1548 published the first Italian translation of More's *Utopia*, imagined a 'New World' in which 'everyone lived by their own sweat and did not drink the blood of the poor', where everything was held in common, and where people followed a religious instinct that was natural to them.[30] Similarly Nicolò Franco contrasted primitive man, living peacefully and honestly according to precepts discovered by natural reason, to modern man, corrupted by the arts and sciences and completely given over to greed and luxury.[31]

[25] Pietro and Ludovico Rostino, *Compendio di tutta la cirurgia*, ed. and amplified by Fioravanti (Venice, 1588), 169–70.

[26] At the Oxford meeting, Nancy Siraisi reminded me that the idea that medicine was discovered by the animals had become conventional in academic medical treatises by Fioravanti's time. However, the uses Fioravanti made of the doctrine of medical primitivism were radically different from those found in conventional academic treatises. As far as I am aware, no academic medical writer ever used primitivism to undermine academic medicine, as Fioravanti did.

[27] Harry Levin, *The Myth of the Golden Age in the Renaissance* (New York, 1972).

[28] Paul F. Grendler, *Critics of the Italian World, 1530–1560* (Madison, 1969), 162–77. In addition, see Lauro Martines, *Power and Imagination: City-States in Renaissance Italy* (New York, 1979), 322–31; Luigi Firpo, *Lo Stato ideale della Controriforma* (Bari, 1957).

[29] Thomas More, *Utopia* (London, 1516), trans. Edward Surtz (New Haven, 1964), 92–3 (with minor revisions).

[30] Grendler, *op. cit.* (note 28), 170–77.

[31] *Ibid.*, 92–96.

The juxtaposition of a society arranged according to nature's laws against the real conditions of the day presented a stark contrast. Looking to nature for norms of the perfect society – whether in the form of refurbished pagan myths of the Golden Age, popular tales of a land of plenty (*paese di coccagna*), or the Stoic doctrine of happy innocence in a state of nature – proved to be one of the few avenues of social criticism open in Counter-Reformation Italy. But like any 'safe' form of social criticism, it could be pushed too far. The picture of man *in puris naturalibus* – as he is in nature – was not especially shocking. Indeed, it easily blended with the Biblical account of the creation and with scholastic portrayals of the condition of man in the Garden of Eden. What was shocking was the thesis (implicit in such a picture) of the self-sufficiency of the *lex naturae*, against which all deviations could be measured.

Medical primitivism had much in common with these other forms of cultural primitivism. A manifestation of widespread intellectual demoralization, it was not merely a critique of the doctors, but an expression of a deeply-felt longing for social, political, and moral reform. Projecting the metaphor of disease as a corruption of the human body onto the body politic, Fioravanti wanted to purge society of wickedness and return it to a pristine state of moral and political justice.

The politics of purgation

Fioravanti claimed that his therapeutic system was a return to the 'natural way' of healing, a methodology discovered by the earliest physicians but lost because of the corruption of medical practice by theory. In combing the cities and countryside of Sicily, he discovered that these natural methods still survived among the common people as unwritten 'rules of life'. Indeed, they were the secret of longevity. In Messina, he met a hundred-and-four-year-old man who reported that the only medicine he ever took was some soldanella (*Convolvulus* or bindweed, a mild cathartic) in the springtime, 'and every time I take it,' the old man said, ' it makes me vomit thoroughly and leaves my stomach so clean that for a year I cannot fall ill.' An eighty-year-old man from Naples told Fioravanti that he occasionally took white hellebore with a cooked apple, 'which he then vomited several times and thus was excellently purged'.[32]

Indeed, the most prominent feature of Fioravanti's therapeutics is the frequency with which he recommended violent emetics and purgatives. He was, it seems, obsessively concerned with purifying the body of putrid and corrupting substances. He had an imposing armoury of emetics and purgatives whose active agents included hellebore, veratrum, antimony, and mercury. He gave them catchy trade-names like 'angelic electuary' (*elettuario angelico*), 'magistral syrup' (*siroppo maestrale*), 'blessed oil' (*olio benedetto*), and his powerful and trusty standby, *dia aromatica*, the 'fragrant

[32] Fioravanti, *op. cit.* (note 19), 53–54.

goddess' he prescribed as the first course of action against almost every ailment he encountered.

It is not just that Fioravanti believed passionately in the efficacy of vomiting and purging. Equally striking was his obsessive fascination with the strange and revolting matter that spewed forth from the bodies of his patients. He described these effluents with an interest bordering on wonderment. In 1558, he attended a woman who suffered terrible attacks of indigestion. After giving her two ounces of his *dia aromatica*,

> she began to vomit, . . . and spent the whole night vomiting the rubbish that lay in her belly. And amongst the other things that she brought up, there was an object like a uterine tumour, but round and hairy in form, and alive, which filled me with wonder, because I had never seen the like before. I washed it and placed it in a box in cotton wadding so that I might show people this portentous object; but [later] upon inspection, I found that it had dissolved and that so small a quantity remained that it had no shape; nonetheless, when the woman vomited, it was large and wondrous.[33]

Another woman vomited 'a great quantity of putrid matter including a great tumor as large as a hand, alive, and it lived in tepid water for another two days.'[34]

Nor were emetics and purges always sufficient to cleanse the body of its pollution. Sometimes Fioravanti combined them with diuretics, expectorants, and sudorifics. In the summer of 1569, while he was living in Messina, he was asked to treat the inhabitants of the neighbouring villages, who were dying one by one from some sort of 'putrid fever'.

> The first thing I did was to give them a bolus that made them vomit greatly. After this I gave them every morning for three to four days a syrup solution which moved their bowels strongly and followed this with a cupping-glass, whilst I annointed their bodies with oil of hypericum [to make them sweat] . . . Of the three thousand I treated, only three died, and they died of old age.[35]

One of Fioravanti's patients had to endure vomiting, bloodletting from underneath the tongue, fomentations, and eight consecutive days of being purged. Not surprisingly, the falling sickness called for equally extreme measures. Fioravanti purged one unfortunate epileptic for ten days, and then made him follow 'the rules of life, but no diet'.[36]

Fioravanti's obsessive concern with purgation was by no means unique. Many

[33] Fioravanti, *op. cit.* (note 8), 80v–81r. See also the chapter, 'Hypercatharsis', in Piero Camporesi, *The Incorruptible Flesh: Bodily Mutilation and Mortification in Religion and Folklore*, trans. Tania Croft-Murray (Cambridge, 1988), 106–30.

[34] Fioravanti, *op. cit.* (note 8), 32.

[35] Fioravanti, *op. cit.* (note 19), 40r–v.

[36] Fioravanti, *op. cit.* (note 8), 91r–v. The breadth of symptoms to which Fioravanti applied his purgationist therapeutics is suggested by his belief that ailments of head originate from the 'bad quality of the stomach', which transmits great quantities of humid vapours to the head, filling and corrupting all its occult parts. From this it followed that when confronted with an earache, for example 'it is necessary not only to treat the ear, but also the stomach and the head, because it's a general rule that if you want to cure any infirmity, you have to remove the cause, . . . and since the cause [of this ailment] comes from the stomach, you have to purge the stomach with two drams of my *dia aromatica*': L. Fioravanti, *La cirurgia* (Venice, 1570), 24r–v.

popular healers of the day advocated a similar therapeutic approach. Tommaso Zefiriele Bovio (1521–1609), a Verona lawyer-turned-empiric advocated purging with a fervour that at least equalled Fioravanti's. Bovio attacked the orthodox physicians in books with titles like *Scourge of the Rational Physicians* and *Thunderbolt Against the Supposed Rational Doctors*. He claimed that the physicians killed their patients with their strict diets and weak medicines. Like Fioravanti, he preferred strong vomits and purges, especially his *Hercules*, which he reported made one of his patients 'vomit a catarrh as big as a goose's liver, and emit from above and below loathsome excrement'. Bovio urged the same course of therapy for every internal ailment: 'chase away the evil, then maintain nature'; that is, use 'robust' purges followed by healthy foods.[37] According to the conception of disease that Fioravanti, Bovio, and I think many other empirics shared, illness was not some benign imbalance of humours that could be rectified by diet and regimen; it was an invasion of the body by 'corruptions' that had to be forcefully expelled with potent medicines. Only then could the body be restored to its 'pristine health.' In the struggle between sickness and health that these popular healers engaged in, therapeutic intervention necessarily took on heroic dimensions.

Several factors, I believe, explain the prevalence of this conception of disease among vernacular healers. Foremost among them was the changing medical marketplace.[38] Money lubricated exchanges in the medical marketplace no less than in the economy at large. By reducing the cost of exchange, it increased the volume of exchanges, and tended to accelerate the demand for certain types of medical services. In particular, it furthered the demand for active remedies for specific ailments as opposed to passive regimens for living. Patients grew impatient with regimen; they wanted a quick dose for what ailed them. Popular healers delivered with an imposing arsenal of drugs that produced dramatic results. And they touted a theory of disease to match: that of disease as a corruption of the body that had to be forcefully expelled.

Another factor working in favour of this 'ontological' conception of disease was the religious climate of post-Tridentine Italy. For we cannot help noticing the resemblance of the therapeutics of purgation to the Church's extremest medicine, exorcism. Piero Camporesi has observed that the therapeutics of purgation, the use of vomitory and expulsive techniques, reached its peak in popularity during the second half of the sixteenth century:

> The decades which knew the highest rate of diabolism, the golden age of witch-hunts, the Tridentine 'reconquest', Catholic supremacy, the hegemony of theocracy, happened to coincide with the age of . . . superpurges, of ostentatious and vehement cleansings of the sullied flesh, of obsession with individual catharsis and collective purification.[39]

[37] Tommaso Zefiriele Bovio, *Flagello contro dei Medici communi detti Rationali*, in *Opere di Zefiriele Tomaso Bovio* (Venice, 1626), 4, 52: 'scaccio il male, sostento la natura'. On Bovio, see Alfonso Ingegno, 'Il Medico de' disperati e abbandonati: Tommaso Zeffiriele Bovio (1521–1609) tra Paracelso e l'alchimia del seicento', in *Cultura populare e cultura dotta nel seciento* (Milan, 1985), 164–74; and Antonio Dal Fiume, 'Un Medico astrologo a Verona nel '500: Tommaso Zefiriele Bovio', *Crit. Stor.*, 1983, **20**: 32–59.

[38] See in particular Harold J. Cook, *The Decline of the Old Medical Regime in Stuart London* (Ithaca, 1986), 28–69.

[39] Camporesi, *op. cit.* (note 33), 123.

The conception of disease as a contamination of the body by worms, foul-smelling creatures, and other corrupting agents resonated with the tense political and religious climate of post-Tridentine Italy.

Camporesi's idea of a 'medico-ecclesiastical ideology of supercatharsis' provides a valuable framework for looking at Fioravanti as a medical practitioner. First of all, it should be noted that exorcisms were more widely practised during the Counter-Reformation than ever before.[40] Exorcisms, which were often performed publicly, demonstrated in a dramatic fashion the Church's jurisdiction over supernatural forces. This made them effective instruments of anti-Protestant propaganda. Moreover the exorcist, who protected society from a pollution within, held an office of considerable spiritual power. On the other hand, there was a great deal of ambiguity surrounding the office of the exorcist, because not only priests but also laymen could perform the rite. Girolamo Menghi, who wrote one of the standard texts on the subject, allowed that 'certain devout persons with or without [the office of] exorcism can undo maleficial infirmities and chase away demons from tormented bodies'.[41]

Since the line between demonic possession and diseases of natural origin was often difficult to draw, exorcists and physicians competed with one another in contesting jurisdictions. This ambiguity of social roles and professional identities opened up possibilities for popular healers to appropriate the role of exorcist, and for exorcists to play the role of the doctor. Adding to the confusion, there was a tendency to physicalize demonic possession. Hellebore, highly praised and much used by empirics like Fioravanti and Bovio, was a staple anti-maleficial agent. It was recommended by authorities such as Menghi, who reported having seen all manner of maleficent objects 'emitted by the body in vomit or from beneath', during exorcism. Menghi also noted that there were exorcists who abused the art by indulging in carnivalesque, self-aggrandizing displays of power over demons. He compared them to the quacks and charlatans who promised to cure anyone. 'From being honest, straightforward exorcists, many become doctors and charlatans, seeking this or that herb against the demons, offering medicines, powders, and similar things to the possessed, temerariously usurping to themselves the office of physicians.'[42] On the other hand Scipione Mercurio, a physician, angrily rebuked exorcists who 'prescribe purging medicines and very strong solvents in order to purge the spiritual body of wicked humours, not being familiar with the quality of these medicines nor, much less, the patient's temperament.'[43] In that world, where it was often impossible to tell whether diseases were of natural or demonic origin, the figure of the exorcist and the charlatan merged.

I do not think it is going too far to suggest that in cleansing the stomach and driving out its polluting sickness, Fioravanti was performing a kind of physiological exorcism. His treatments, acts of purification, mimicked the exorcist's rite of 'chasing away

[40] Mary O'Neil, *Discerning Superstitions: Popular Errors and Orthodox Response in Late Sixteenth Century Italy* (Diss, Stanford University, 1981).

[41] Girolamo Menghi, *Compendio dell'arte essorcistica* (Venice, 1576), qu. O'Neil, *op. cit.* (note 40), p. 336.

[42] *Ibid.*, 351. [43] Scipione Mercurio, *De gli errori populari d'Italia* (Venice, 1603), 139v.

demons from tormented bodies'. Rarely did Fioravanti merely heal his patients; he nearly always 'healed and saved' them.[44] Nor, perhaps, is it surprising that he should have projected the metaphor of catharsis onto the body politic, urging the rulers of the day to expel the polluting 'flatterers' from their courts, to restore the moral order of ancient times, and to return society to its pristine condition of equity and justice. According to the anthropologist Mary Douglas, the physical body is a microcosm of society: 'The physical experience of the body, always modified by the social categories through which it is known, sustains a particular view of society. There is a continual exchange of meanings between the two kinds of bodily experience so that each reinforces the categories of the other'.[45] Fioravanti believed that the cause of Italy's moral and political decline was an internal pollution that began in the courts and spread throughout the commonwealth. As the 'bad quality of the stomach' spreads its contagion to all the body's organs, so corrupt rulers and their fawning courtiers ruined the whole body politic. He chastised princes for giving themselves over to flatterers who made them 'see black for white'. He reminded them that the duties of princes were similar to those of physicians, to care for their subjects with compassion and love. By abandoning the people, were they not behaving like the physicians? How different were modern princes, he lamented, from those ancient rulers, like Octavian, who loved his subjects and treated them with justice, liberality, and clemency.[46]

Nor did Fioravanti shrink from giving princes some words of advice directly. In 1570, emboldened by the rightness of his world view, he wrote to Alfonso II, the Duke of Ferrara, to whom he had dedicated his *Capricci medicinali*, to express the 'great sadness and sorrow' he felt over 'the ruin of your most noble city of Ferrara, and over your personal troubles'. Fioravanti was moved to write the letter by news of a series of violent earthquakes that struck Ferrara in November–December 1570, causing massive destruction to the city and killing hundreds of people.[47] Such terrible events did not happen by chance, he was convinced; they were God's punishment to a wicked world.[48] And who in late sixteenth-century Italy would have doubted the misery and misfortune of that unhappy land? Alfonso's court had been a place of dazzling splendour. The duke lavished his patronage upon some of Italy's leading poets, musicians, artists, and humanists. Yet to sustain this magnificent show of wealth, he ground the people down with oppressive taxes, monopolies on grain and salt, and callous game laws.[49] It was plain to Fioravanti that the earthquakes were the visible sign of God's wrath. He could not contain his moral outrage over the rotten state of Ferrara; but even there he saw an opportunity:

[44] E.g., Fioravanti, *op. cit.* (note 8), 26v, 'fù sana & salva'.
[45] Mary Douglas, *Natural Symbols: Explorations in Cosmology* (New York, 1973), 93–112; p. 93. See also her *Purity and Danger* (London, 1966), esp. ch. 7, pp. 114–28.
[46] Fioravanti *op. cit.* (note 19), 237v–238; also his *Dello specchio di scientia universale* (Venice, 1572), 140–5.
[47] On the earthquake, see Angelo Solerti, *Ferrara e la corte estense nella seconda metà del secolo decimosesto* (Città di Castello, 1900), CLXI–CLXXIII.
[48] Leonardo Fioravanti, *Del regimento della peste* (Venice, 1565), cap. I.
[49] J. A. Symonds, *Renaissance in Italy* (7 vols, London, 1875–86), 2: 30–2. See also Solerti, *op. cit.* (note 47).

> To speak plainly, they say in these parts that Your Most Illustrious Lordship has let himself be bribed by the evil men with whom he has involved himself, and in doing so every year has allowed many taxes to be levied on the poor people in his magnificent city; besides this, that Ferrara has become a Sodom and Gomorrah of public sodomy, and that in joking about this, everybody scorns God with blasphemy and filthy words, sins so great that God's Majesty cannot suffer them. For all this they blame Your Most Illustrious Lordship, as prince and patron, saying you should remedy the whole situation with the arm of your justice. And so My Illustrious Lord, I have taken it upon myself to make all these things known to you, because I am certain that as a Christian and Catholic prince you will turn to God for help and counsel, and will see to all these things, so that our Lord God will calm his anger and return your city to its pristine state, as he did to the city of Nineveh. And if it is Your Illustrious Lordship's wish, after all this has come to pass, I will show him the true way to quickly restore and augment the city.[50]

It is fascinating to see Fioravanti posturing as a modern-day Jonah, urging the duke to drive out the 'flatterers' from his court and to return the city to its 'pristine state'. Just as God sent Jonah to warn the corrupt city of Nineveh, he was now sending Fioravanti to help restore the ruined Ferrara.

Obviously, purging the body politic did not entail making fundamental changes in the political order. Fioravanti was no revolutionary. He believed that absolute rule by a prince was the ideal form of government, 'confirmed by the rules of nature'.[51] Indeed, the concept of pollution tends to support moral and political values, not to subvert them.[52] Pollutions have a very simple remedy for undoing their effects. They are rites of purification, whether by purgation, exorcism, or ritual cleansing, which erase offences against the body or group and restore it to its original pure state. Fioravanti had no doubt something was terribly wrong with society and government in Italy. 'I feel the world crying out for a change, and all the people rising up,' he warned, 'they cannot endure any longer being so miserably afflicted and oppressed'.[53] But for him the solution was not to tear down the system; it was to rid it of moral pollution. Purge the body of corruptions using powerful emetics, then fortify it with good, wholesome food was the course of action Fioravanti recommended in his therapeutics. He advocated a similar 'therapeutics' for moral and political reform.

Yet while Fioravanti's political doctrines were safely within the margins of the acceptable, it was not long before his teachings were put to more radical purposes. Among those who followed his doctrine was a distiller and one-time court jester from Bologna named Costantino Saccardino. In 1622, Saccardino was convicted along with his son and two others of having soiled with excrement the sacred statues of the city and affixed to them placards full of blasphemies and vague threats against the political and religious authorities. Saccardino and his co-conspirators were publicly hanged in

[50] Dall'Osso, *op. cit.* (note 16), 289–90. [51] Fioravanti, *op. cit.* (note 46), 144–5.
[52] Douglas, *Purity and Danger*, *op. cit.* (note 45) 135–7. [53] Fioravanti, *op. cit.* (note 46), 141.

the market square in Bologna, an example to all nonconformists.[54] An enthusiastic proponent of 'spagyrical' medicine, Saccardino had also written a book of medical recipes. Besides peddling his own medical secrets, the book promised 'practical arguments revealing the many deceptions which for reasons of self-interest frequently occur in both medicine and medicinal materials'.[55] Borrowing from Fioravanti, Saccardino asserted that the animals, from whom man originally learned medicine, still have the ability to heal themselves by nature, as do women, peasants, and empirics. How radically different were the 'idle modern physicians' from the wise physicians of old, who

> used to visit the languishing infirm and bring them medical relief prepared with their own hands, without ambition or pride, humbly visiting poor and rich alike. They refrained from arguing or engaging in words, which is what some physicians do now, so that often the poor patient succumbs, in their very presence, before such bickering and dies.[56]

Saccardino's aggressively original reading of his sources led him to even more radical conclusions where religion was concerned. 'Only baboons believe hell exists,' he would say, 'Princes want us to believe it so they can have things their own way; but now, at last, the whole dovecote has opened its eyes'.[57] Here Saccardino lifted another page from Fioravanti, who had declared in his *Specchio di scientia universale* that the times were ripe for a great medical reformation because the invention of printing made it possible for ordinary people to see through the lies perpetrated by the physicians. In ancient times, he wrote,

> they got people to believe anything they wanted, because in those days there was a great shortage of books, and whenever anyone could discourse even a little about *bus* or *bas* he was revered as a prophet, and whatever he said was believed. But ever since the blessed printing press came into being, books have multiplied so that anyone can study, especially because the majority of them are published in our mother tongue. And thus the kittens have opened their eyes.[58]

For all his megalomania, Fioravanti was a kind of prophet – not, of course, of a medical reformation or a social revolution that actually came about, but of a style of empiricism that would play an important role in the Scientific Revolution. The idea of a restauration of science to be accomplished by following the path of pure empiricism, returning to nature unsoiled by scholastic precepts and theories, was a favourite theme of the virtuosi. Whenever they opposed the uncorrupted knowledge of unlettered

[54] On the Saccardino case, see Carlo Ginzberg and Marco Ferrari, 'La colombara ha aperto gli occhi', *Quad. Stor.*, 1978, **38**: 631–9.

[55] Costantino Saccardino, *Il Libro nominato la verità di diverse cose, quale minutamente tratta di molte salutifere operationi spagiriche et chimiche* (Bologna, 1621), t.p.: 'utili ragionamenti quali scuoprono molti inganni, che per interesse spesso, quanto nella medicina, quanto nelle materie medicinali, intervengano'.

[56] *Ibid.*, 14, trans. J. Tedeschi and E. Lerner, in Carlo Ginzburg, 'The dovecote has opened its eyes: popular conspiracy in seventeenth-century Italy', in *The Inquisition in Early Modern Europe: Studies on Sources and Methods*, ed. Gustav Henningsen and John Tedeschi (DeKalb, 1986), 190–8, pp. 192–3.

[57] Quoted in *ibid.*, 193. See also Carlo Ginzburg, 'High and low: the theme of forbidden knowledge in the sixteenth and seventeenth centuries', *Past Present*, 1976, **73**: 29–41, pp. 35–6.

[58] Fioravanti, *op. cit.* (note 46), 41r–v. This allusion was identified by Piero Camporesi, 'Cultura popolare e cultura d'élite fra medioevo ed età moderna', in *Stor. Ital.*, 1981, **4**: 79–157, pp. 87–8.

artisans and peasants to the pretentious studies of the learned, they indulged in a kind of scientific primitivism. In Bacon's polemic against the indifference of men of letters to the work of illiterate mechanics, in Robert Boyle's praise of the usefulness of the mechanical arts to philosophy, and in the claim, increasingly heard in the seventeenth century, that scientific knowing means knowing how to do, we hear the echo of primitivism.[59] In his *Micrographia*, Robert Hooke evoked the image of mankind's pure empirical knowledge as it existed in the Garden of Eden before the Fall. He proposed the microscope, which allowed investigators to gaze upon things unknown to the scholastics, as a means of recapturing that pristine empiricism innocent of theory, and to release mankind from the original sin of having wilfully abandoned the 'prescripts and rules of nature'.[60] Ironically, in the polemic waged on behalf of the new science against the old, one of the most effective strategies was to urge going back to the oldest science of all.

[59] Antonio Pérez-Ramos, *Francis Bacon's Idea of Science and the Maker's Knowledge Tradition* (Oxford, 1988).

[60] Robert Hooke, *Micrographia* (London, 1665; repr. New York, 1961), Preface, 9. In addition, see Catherine Wilson, 'Visual surface and visual symbol: the microscope and the occult in Early Modern Europe', *J. Hist. Ideas*, 1988, **49**: 85–108; Michael Aaron Dennis, 'Graphic understanding: instruments and interpretation in Robert Hooke's *Micrographia*', *Sci. Context*, 1989, **3**: 309–64.

3

The cutting edge of a revolution? Medicine and natural history near the shores of the North Sea

HAROLD J. COOK

When taking stock of work on the Scientific Revolution, it is sometimes helpful to note what has been lost from discussion over the years as well as what has been added. While no one would wish to advocate a simple return to the tradition of writing now often termed positivistic, it is notable that historical work on early modern science done with a simple progressivism in mind gave a much more central place to medicine and natural history than work done by more recent generations, who have discussed what they have labelled the Scientific Revolution. Until roughly the end of the 1940s, a great deal, perhaps even the vast bulk, of work on sixteenth- and seventeenth-century science was about 'discovery'. The cataloguing of the accumulation of positive knowledge illustrated the superiority of the unfettered rational thought of the post-Renaissance period *vis-à-vis* the restricted and superstitious philosophizing of the medieval past.[1] There are a great many obvious problems with this approach to history. But its main proponents – scholars such as Charles Singer and George Sarton – did encourage excellent work on medicine and natural history by, for instance, Agnes Arber, F. J. Cole, Charles S. Sherrington, Charles Raven, and M. F. Ashley Montague.[2]

From the later 1940s onwards, however, the contribution of natural history to the

I would like to thank the Fulbright Commission and the Netherlands America Foundation for Educational Exchange, the National Endowment for the Humanities, the National Library of Medicine (NIH Grant LM 05066), and the Graduate School of the University of Wisconsin for jointly supporting research in the Netherlands during 1989–90, when this paper was written.

[1] See, for example, Charles Singer, *A Short History of Anatomy and Physiology from the Greeks to Harvey* (New York, 1957) (First publ. as *The Evolution of Anatomy* (London and New York, 1925)); George Sarton, *The History of Science and the New Humanism* (Cambridge, MA, 1937); Sarton, 'The quest for truth: scientific progress during the Renaissance', in *Sarton on the History of Science. Essays by George Sarton*, selected and edited by D. Stimson (Cambridge, MA, 1962), 102–20.

[2] Singer or Sarton are mentioned in the following major studies as offering much encouragement: Agnes Arber, *Herbals, Their Origins and Evolution: A Chapter in the History of Botany, 1470–1670* (Cambridge, 1912; 2nd edn enlarged, Cambridge, 1938); F. J. Cole, *A History of Comparative Anatomy: From Aristotle to the Eighteenth Century* (London, 1944); Charles S. Sherrington, *The Endeavour of Jean Fernel* (Cambridge, 1946); Charles E. Raven, *English Naturalists from Neckham to Ray: A Study of the Making of the Modern World* (Cambridge, 1947); M. F. Ashley Montague, *Edward Tyson, M.D., F.R.S., 1650–1708, and the Rise of Human and Comparative Anatomy in England: A Study in the History of Science*, Memoirs of the American Philosophical Society no. 20 (Philadelphia, 1943).

natural philosophy of the early modern period has been neglected by most historians of science. Those who created and used the term 'Scientific Revolution' had in mind something other than mere positive discovery or the rise of rationalism when they described the sixteenth and seventeenth centuries: they had in mind a revolution in the conceptual outlook of Europeans that was rooted in a mathematical and mechanical world view. Clearly, the enormous power and philosophical interest associated with modern physics changed much in the writing of the history of science.[3] The new generation saw the mathematization of nature and the key role of thought experiments in modern physics as originating in the Scientific Revolution of the early modern period. A. Rupert Hall has noted how much the work of Alexandre Koyré strove to provide an historical background to what seemed like the greatest revolution of all time: relativity theory and quantum mechanics.[4] Hall might have added that this was also the purpose of Alfred North Whitehead, E. A. Burtt and E. J. Dijksterhuis as well.[5] But he is right to point to the enormous influence of Alexandre Koyré during the post-world-war generation, when the subject was becoming institutionalized as a discipline in Britain and the United States. An avowed Platonist, Koyré steadfastly resisted any deviation from a history of pure scientific ideas. For instance:

> I do not see what the *scientia activa* has ever had to do with the development of the calculus, nor the rise of the bourgeoisie with that of the Copernican, or Keplerian, astronomy. And as for experience and experiment – two things which we must not only distinguish but even oppose to each other – I am convinced that the rise and growth of experimental science is not the source but, on the contrary, the result of the new *theoretical*, that is, the new *metaphysical* approach to nature.

In short, he went on, science emerged from 'the mathematization (geometrization) of nature' and from no other source but this shift in pure thought.[6]

In the idea that true science stems from the mathematization of nature, various historians have found a moral and ideological framework for the defence of liberal democratic society, issues well articulated by A. Rupert Hall.[7] However, a man with an equal interest in moral concerns, the Anglican canon Charles E. Raven, noted this direction of the history of science with dismay:

> To represent the history of science, as is done in almost all the text-books as a papal succession, Copernicus, Kepler, Galileo, Newton, with Boyle and

[3] See, for instance, the changing outlook on Isaac Newton's achievement after the development of quantum theory, in *The Royal Society Tercentenary Celebrations 15–19 July 1946* (Cambridge, 1947).

[4] A. Rupert Hall, *The Revolution in Science 1500–1750* (London, 1983), p. 147.

[5] Alfred North Whitehead, *Science and the Modern World* (Cambridge, 1926; New York, 1967); Edwin Arthur Burtt, *The Metaphysical Foundations of Modern Physical Science* (2nd edn, Garden City, New York, 1954); E. J. Dijksterhuis, *The Mechanization of the World Picture*, C. Dikshoorn, trans. (Amsterdam, 1950; Oxford, 1961). See Lorraine Daston, 'History of science in an elegiac mode: E. A. Burtt's 'Metaphysical Foundations of Modern Science' revisited', *ISIS*, 1991, **82**: 522–31; 'Inleiding' to E. J. Dijksterhuis, *Clio's Stiefkind*, ed. K. van Berkel (Amsterdam, 1990), 11–21.

[6] Alexandre Koyré, *Newtonian Studies* (London, 1965; Chicago, 1968), p. 6: emphasis and parenthesis in the original. For a discussion of arguments for a scientific 'metaphysics', see Gary Hatfield, 'Metaphysics and the new science', in *Reappraisals of the Scientific Revolution*, David C. Lindberg and Robert S. Westman (Cambridge, 1990), 93–166.

[7] A. Rupert Hall, 'On Whiggism', *Hist. Sci.*, 983, **21**: 45–59; also see Michael Polanyi, *Science, Faith and Society* (London, 1946; Chicago, 1964); J. Bronowski, *Science and Human Values* (New York, 1965).

> Hooke and a few others wedged into the series, is only possible on the assumption that the important contributions are those which led up to the dominant mechanism and determinism of the late nineteenth century, and that the astonishing achievements of zoologists and botanists in the sixteenth and seventeenth centuries can be ignored.[8]

More recently, there have been a number of well-aimed attacks on the idea that a mathematical and mechanical world view can encompass the philosophical ideas of the seventeenth-century virtuosi.[9] But with the notable exception of a few scholars such as Walter Pagel, P. M. Rattansi, Allen Debus, and Charles Webster, few have been worried that the concept of the Scientific Revolution has virtually excluded attention being given to much of the most lively work done in the period: that related to medicine and natural history.[10] It is remarkable that books on the subject of sixteenth- and seventeenth-century natural history written by scientists or by historians of science cease almost entirely after about 1950 – about the time that Raven complained of the new orthodoxy.[11] In the several attempts to write organized accounts of early modern science, some authors, such as Richard S. Westfall, have had difficulty incorporating into their accounts the tremendous amount of excellent work that went into subjects like natural history.[12] Marie Boas and A. Rupert Hall have been the most successful in writing accounts that give some attention to the subject, but even Rupert Hall tried to distinguish between natural history and biology in ways that undermine the significance of the former.[13] It has been left rather to people such as Keith Thomas, Jack Plumb, Joseph Levine, and other historians of the early modern period whose concern has been rather with cultural history than with the history of science as such, to write books and essays on changing ideas of the world of

[8] Charles E. Raven, *Natural Religion and Christian Theology*, The Gifford Lectures 1951, First Series: Science and Religion (Cambridge, 1953), 7.

[9] Among the stronger trumpets, see Keith Hutchison, 'What happened to occult qualities in the Scientific Revolution?', *ISIS* 1982, **73**: 233–53; John Henry, 'A Cambridge Platonist's materialism: Henry More and the concept of soul', *J. Warburg Courtauld Inst.*, 1986, **49**: 172–95; Henry, 'Occult qualities and the experimental philosophy: Active principles in pre-Newtonian matter theory', *Hist. Sci.*, 1986, **24**: 355–81; Henry, 'Robert Hooke, the incongruous mechanist', in *Robert Hooke: New Studies*, ed. Michael Hunter and Simon Schaffer (Woodbridge, Suffolk, 1989), 149–80; Peter Dear, *Mersenne and the Learning of the Schools* (Ithaca, 1988); Steven Shapin, 'The house of experiment in seventeenth-century England', *ISIS*, 1988, **79**: 373–404; Steven Shapin and Simon Schaffer, *Leviathan and the Air Pump: Hobbes, Boyle, and the Experimental Life* (Princeton, 1985); and Simon Schaffer, 'Godly men and mechanical philosophers: souls and spirits in Restoration natural philosophy', *Sci. Context*, 1987, **1**: 55–85.

[10] For example, Walter Pagel, *From Paracelsus to Van Helmont: Studies in Renaissance Medicine and Science*, ed. Marianne Winder (London, 1986); P. M. Rattansi, 'Recovering the Paracelsian milieu', in *Revolutions in Science: Their Meaning and Relevance*, ed. William R. Shea (Canton, MA, 1988), 1–26; Allen G. Debus, *Man and Nature in the Renaissance* (Cambridge, 1978); and Charles Webster, *The Great Instauration: Science, Medicine and Reform 1626–1660* (New York, 1975).

[11] One of the great exceptions is Howard B. Adelmann, *Marcello Malpighi and the Evolution of Embryology*, 5 vols. (Ithaca, 1966). I also except the continuing tradition of important work on eighteenth-century natural history, pursued by those such as Jacques Roget, Shirley Roe, and Phillip Sloan.

[12] For an attempt to merge the 'mechanical world view' with the natural historical inquiries of the period, see 82–104 of Richard S. Westfall, *The Construction of Modern Science: Mechanisms and Mechanics* (New York 1971; Cambridge, 1977).

[13] Marie Boas Hall, *The Scientific Renaissance, 1450–1630* (London, 1962; New York, 1966), 50–67; A. Rupert Hall, *The Scientific Revolution, 1500–1800: The Formation of the Modern Scientific Attitude* (London, 1954), 275–302; revised edition as *The Revolution in Science 1500–1750* (London, 1983) 331–46; A. Rupert Hall, *From Galileo to Newton* (London, 1963; New York, 1981), 175–215.

plants, animals, and even man.[14] In the current scholarly climate, matters are changing rapidly[15] – but this might be a good moment to re-evaluate the place of natural history in what has come to be called the Scientific Revolution.

Early modern natural history

Natural history in the sixteenth and seventeenth centuries can be defined as investigation into *res naturae*: the things of nature. Since the time of Pliny and Dioscorides, natural history included a description of all heaven and earth (thus covering cosmology, astronomy, geography, chorography, mineralogy, meteorology, and other such subject-matter), and the living things of this earth and their purposes (thus including fishes, birds, insects, and other animals, herbs, shrubs and trees, their economic and medical uses to man, and investigation of the tools, costumes, customs, and beliefs of people). As Phillip Sloan has recently pointed out, this ancient tradition of natural history is virtually identical to the encyclopedic tradition; but he also notes that in the modern period there were new stimuli coming from what he calls the hermetic tradition.[16] Certainly, if one adds chemistry to the list of topics included under natural history, one finds the subject-matter to be virtually identical to what Robert Boyle called 'physiology', or what David Lux has found the French virtuosi calling 'physique'.[17] By the mid seventeenth century, chemical methods had been taken up as perhaps the most important tool of analysis of the composition of minerals, plants and animals. I would add that by the sixteenth century, natural historical investigations were also receiving an added impetus from the revived Hippocratic tradition, then thought to be a method of close observation which even many chemists consciously followed.[18]

It might be useful for heuristic purposes to say with Sloan that natural history had description as its end, while natural philosophy had causal analysis as its goal, so that

[14] Keith Thomas, *Man and the Natural World: Changing Attitudes in England 1500–1800* (London, 1983); J. H. Plumb, 'The acceptance of modernity', in *The Birth of a Consumer Society: The Commercialization of Eighteenth-century England*, ed. Neil McKendrick, John Brewer, J. H. Plumb (Bloomington, 1982), 316–34; Joseph M. Levine, *Doctor Woodward's Shield: History, Science and Satire in Augustan England* (Berkeley, 1977); Levine, 'Natural history and the history of the scientific revolution', *Clio*, 1983, **13**: 57–73.

[15] For example, see Katherine Park and Lorraine J. Daston, 'Unnatural conceptions: The study of monsters in 16th and 17th-century France and England', *Past Present*, 1981, **92**: 20–54; Paula Findlen, 'Jokes of nature and jokes of knowledge: the playfulness of scientific discourse in early modern Europe', *Ren. Quart.*, 1990, **43**: 292–331; William B. Ashworth, Jr., 'Natural history and the emblematic world view', in Lindberg and Westman, *op. cit.* (note 6), pp. 303–32; the Society for the Bibliography of Natural History was renamed to the Society for the History of Natural History in 1983; and the new *Journal of the History of Collections* was first published in 1989.

[16] Phillip R. Sloan, 'Natural history, 1670–1802', in *Companion to the History of Modern Science*, ed. R. C. Olby, G. N. Cantor, J. R. R. Christie and M. J. S. Hodge (London, 1990), 295–313.

[17] Boyle's works consistently use the word 'physiology' as a description of what the natural philosopher investigates; David S. Lux, *Patronage and Royal Science in Seventeenth-Century France: The Académie de Physique in Caen* (Ithaca, 1989).

[18] Wesley D. Smith, *The Hippocratic Tradition* (Ithaca, 1979); Iain M. Lonie, 'The "Paris Hippocrates": teaching and research in Paris in the second half of the sixteenth century', in *The Medical Renaissance of the Sixteenth Century*, ed. Andrew Wear, Roger K. French, and Iain M. Lonie (Cambridge, 1985), 155–74, 318–26; Vivian Nutton, 'Hippocrates in the Renaissance', in *Sudhoffs Archiv*, 1991, **27**: 420–39.

the two can be separated. Such a distinction would allow us to say that since contemporaries understood natural philosophy to be a search for the causes of things, natural history was an aid to knowledge rather than knowledge itself. But whether a clear and firm boundary between natural history and natural philosophy can be drawn along Sloan's lines is somewhat doubtful. Tendencies toward the one or the other were there, to be sure, just as Pliny tends to one and Aristotle to the other. But Pliny does write of causes as well as things, and while Aristotle emphasized analysis more than discovery, he naturally also considered the latter to be important.[19] Many natural philosophers in early modern Europe were arguing that one really must get the facts straight before theorizing about causes. In one seventeenth-century French dialogue on acids and alkalis, for example, it is the philosopher who introduces 'facts', and the empiricist who raises intellectual doubts.[20] And as historians are all too aware from their own practice, theory and description influence one another profoundly, a fact of which people in the seventeenth century were also well aware.

Francis Bacon, of course, made the collection of natural history experiences and experiments the foundation of his philosophy of science. For many years now, so-called 'Baconian' empiricism has come to seem rather more naive than truly scientific.[21] Yet in later years, Sir Hans Sloane, soon to be the President of the Royal Society, wrote that 'matters or fact' were the very essentials of knowledge, expressing his excitement at being able to uncover such matters.

> It may be ask'd me to what Purposes serve such Accounts, I answer, that the Knowledge of *Natural-History*, being Observations of Matters of Fact, is more certain than most Others, and in my slender Opinion, less subject to Mistake than *Reasonings*, Hypotheses, and *Deductions* are . . . These are things we are sure of, so far as our Senses are not fallible; and which, in probability, have been ever since the Creation, and will remain to the End of the World, in the same Condition we now find them.[22]

To large numbers of people in the seventeenth century, then, the discovery of true facts and the sorting out of details seemed wonderfully fresh and new (for instance, Sir Thomas Browne's language in trying to sort out true things from false is beautiful), but it also presumed certain attitudes towards knowledge. The philosophical movement toward what has been described in medical history as 'ontology' may seem much too simplistic or 'positivistic' to us today, but it was greeted with much interest

[19] On Aristotle and empiricism, see G. E. R. Lloyd, *Magic, Reason and Experience* (Cambridge, 1979), 200–25; the best overview of Pliny remains Lynn Thorndike, *A History of Magic and Experimental Science* (New York, 1923), 1: 41–99.

[20] François André, *Entretiens sur l'acid et l'alcali, où sont examinées les objections de M. Boyle contre ces principes* (Paris, 1672); transl. by 'J. W.' as *Chymical Disceptations: Or, Discourses upon Acid and Alkali. Wherein are Examined the Objections Of Mr. Boyle against these Principles* (London, 1689).

[21] For more appreciative recent appraisals of Bacon's philosophy, see Peter Urbach, *Francis Bacon's Philosophy of Science: An Account and Reappraisal* (La Salle, Illinois, 1987); A. Pérez-Ramos, *Francis Bacon's Idea of Science and the Maker's Knowledge Tradition* (Oxford, 1988); John C. Briggs, *Francis Bacon and the Rhetoric of Nature* (Cambridge, 1989).

[22] Hans Sloane, *A Voyage To the Islands Madera, Barbados, Nieves, S. Christophers and Jamaica, with the Natural History of the Herbs and Trees, Four-footed Beasts, Fishes, Birds, Insects, Reptiles, &c.* (London, 1707), sig. Bv. Also see G. R. De Beer, *Sir Hans Sloane and the British Museum* (Oxford, 1953).

throughout Europe: even in places where Sir Francis Bacon's works were not well known.[23]

So perhaps it is well to look at work in natural history again. Opening Thomas Birch's *History of the Royal Society* almost at random, the report of the meeting of 6 December 1682 shows well how central natural history was for the Royal Society. The meeting began with a discussion of the injection of tincture of indigo into the intestines of a live dog. The colour was found to pass into the lacteals after three hours. This led to a request for Drs Ablionby, Tyson, and Slare to try the experiment themselves and report about it. Dr Smith reported that perfume given into the lacteals was retained in the chyle; and a paper of Dr Lister's on the chyle, with speculations about why it was white, was then read. Grew objected to Lister's speculations, which led to a discussion of the 'milk' in plants, and further to a discussion on the intestines of ruminant animals. The next subject was reading a letter to the Society from Willem Ten Rhijne, a Dutch physician in Batavia in the East Indies, and the discussion of the letter led to someone declaring that the camphor tree, growing in the Chelsea physic garden, had been sent by Ten Rhijne, and furthermore that Van Munting (probably the father Abraham, but possibly the son, Albertus) had been able to grow tropical plants, even cinnamon and nutmeg, in 'Friesland' (in fact, in Groningen, also in the north of the Netherlands). This caused someone to remark that the Prince Borghese had promised the Royal Society a 'plumbus marinus', said to be able to cause rain, thunder, and subterranean heat. A further discussion of minerals ensued, with reports of black lead from parts of England and New England, and a request for Dr Lister to send down a piece of petrified ash that had magnetic properties. The discussion continued on pyrites and medicinal springs, bringing matters back to another report of the injection of tincture of indigo into the duodenum, in two dogs. Finally, as usual, the meeting concluded with an experiment by Hooke: an unspecified body was placed so as to be just covered in water. Then the vessel it was placed in was struck, and the body sank further down into the water.[24]

This kind of general discussion of curiosities of nature, physiology and anatomy, botany, and a closing demonstration, was quite typical of the work of the Royal Society – 'The Royal Society of London for the Promoting of Natural Knowledge' as it was called in its 1663 charter.[25] The English members were clearly very well aware of the excellent work being done in the Netherlands and the Dutch colonies (as well as in Italy, France, and elsewhere) along similar lines. The discussion of the chyle in the gut was stimulated by Dutch work that had led to the discovery of the lymph vessels. These were the same years in which Leeuwenhoek's letters to the Royal Society created such lively interest. Moreover, Ten Rhijne's communication from Batavia led

[23] On medical ontology, for example, Peter H. Niebyl, 'Sennert, Van Helmont, and medical ontology', *Bull. Hist. Med.* 1971, **45**: 115–37.

[24] Thomas Birch, *The History of the Royal Society of London* (4 vols., London, 1756–1757), **4**: 119.

[25] Robert G. Frank, Jr., 'Institutional structure and scientific activity in the early Royal Society', *Proc. XIVth Int. Cong. Hist. Sci.* (Tokyo, 1975), **4**: 82–101; R. P. Stearns, 'The relations between science and society in the later seventeenth century', in *The Restoration of the Stuarts: Blessing or Disaster?* (Washington, D.C., 1960), 67–75; Michael Hunter, *Establishing the New Science: The Experience of the Early Royal Society* (Woodbridge, 1989).

the Royal Society to publish his book, which included the first European account of acupuncture.[26] Ten Rhijne's book also included an account of moxibustion, which the Royal Society had already heard about from Constantijn Huygens (1596–1687), who had heard about it in turn from the son of the East Indies minister Hermann Busschof. Huygens had recommended the treatment of moxibustion to Sir William Temple, the English ambassador in The Hague, to relieve his gout, and Temple found the treatment so successful that Thomas Sydenham – Temple's neighbour in Pall Mall – later took up the subject in his book on the gout.[27] The Royal Society's interest in the treatment caused them not only to publish Ten Rhijne's book, but to commission an English translation of Busschof's Dutch book on moxibustion.[28]

Swammerdam and his milieu

Another example of the range and importance of natural historical investigations being done in the seventeenth century can be obtained by very briefly surveying Jan Swammerdam's highly skilled and wide-ranging work.[29] His father owned an apothecary's shop bordering the Amsterdam dockyards, where ships from around the world tied up, and like many others in the period the elder Swammerdam began to collect things: Chinese porcelain and other objects, including exotica and naturalia which he often bought from sailors returning from the East Indies. The cabinet of the elder Swammerdam became quite well known. The son began to create his own cabinet when still young, concentrating on collecting insects. Within a few years, he had assembled no less than 1200 of them, mounted and arranged.

But the opportunities available in Amsterdam allowed Swammerdam to expand his horizons further. He received an excellent education in the local Athenaeum – a university in all but the awarding of degrees – which had recently been established by

[26] Willem Ten Rhijne, *Dissertatio de arthritide* (London, 1683).

[27] See George Rosen, 'Sir William Temple and the therapeutic use of moxa for gout in England', *Bull. Hist. Med.*, 1970, **44**: 31–39. Rosen was unaware that 'Zulichem' was Huygens: see J. A. Wrop, ed., *De Briefwisseling van Constantijn Huygens (1608–1687)*, (6 vols., 's-Gravenhage, 1911–17), vol. 6, Letter No. 6995, 'Aan H. Oldenburg', A la Haye, ce 16/26 Nov. 1675', pp. 368–9; Thomas Sydenham, *Tractatus de podagra et hydrope* (London, 1683), transl. into English in 1684 by James Drake, and several times thereafter: the passage occurs at the third paragraph from the end.

[28] Herman Busschof and Hermann Roonhuis, *Two treatises, the One Medical, of the Gout, . . . the Other Partly Chirurgical, Partly Medical* (London, 1676).

[29] The summary of Swammerdam is based on Cole, *op. cit.* (note 2); G. A. Lindeboom, ed. and comp., *Ontmoeting met Jan Swammerdam* (Kampen, 1980); Lindeboom, 'Jan Swammerdam als microscopist', *Tijd. Gesch Geneesk. Natuurw. Wisk. Tech.*, 1981, **4**: 87–110; Lindeboom, ed. and comp., *Het Cabinet van Jan Swammerdam (1637–1680)* (Amsterdam, 1980); Lindeboom, 'Jan Swammerdam (1637–1680) and his *Biblia Naturae*', *Clio Med.*, 1982, **17**: 113–31; Marian Fournier, 'Jan Swammerdam en de 17e eeuwse microscopie', *Tijd. Gesch. Geneesk. Natuurw. Wisk. Tech.*, 1981, **4**: 74–86; Fournier, *The Fabric of Life: The Rise and Decline of Seventeenth-century Microscopy* (Ph.D. dissertation, Twente University, 1991), 82–97; R. P. W. Visser, 'Theorie en Praktijk van Swammerdams wetenschappelijke methode in zijn entomologie', *Tijd. Gesch. Geneesk. Natuurw. Wisk. Tech.*, 1981, **4**: 63–73; Daniel C. Fouke, 'Mechanical and "organical" models in seventeenth-century explanations of biological reproduction', *Sci. Context*, 1989, **3**: 365–81; Frans P. M. Francissen, 'Vroege nederlandse bijdragen tot de kennis van Ephemeroptera of Eendagsvliegen', *Tijd. Gesch. Geneesk. Natuurw. Wisk. Tech.*, 1984, **7**: 113–28. Swammerdam also had a fictionalized biography devoted to him: Olga Pöhlmann, *Jan Swammerdam: Natuuronderzoeker en Medicus*, trans. H. W. J. Schaap (Amsterdam, 1944).

the learned and tolerant regents of the city.[30] Swammerdam also learned from one of the four burgomasters of the city, Johannes Hudde, how to make lenses for microscopes from tiny drops of glass – Leeuwenhoek also learned this technique from Hudde before finding his own methods to improve upon it. The tiniest of these lenses made with Hudde's methods have been found to provide a magnification of 2200 times, although Swammerdam probably ordinarily worked with ones magnifying 800 to 900 times.[31] Aided by such an instrument, Swammerdam developed techniques to do some very remarkable anatomical work on insects, so that when the young Cosimo de' Medici, Grand Duke of Tuscany, visited Swammerdam's cabinet in 1668, Jan dissected a caterpillar to show how the wings of the future butterfly were already contained in the body of the caterpillar. The Grand Duke was so impressed with the skill and the intellectual power of Swammerdam's work that he offered him 12,000 gilders for his collection of insects if he would bring it to Florence and enter his service, but Swammerdam turned him down. When he soon thereafter published (in Dutch) his magnificent *Historia insectorum generalis* (1669), he dedicated the book to the burgomasters of Amsterdam, who rewarded him with an honorarium of 200 gilders.[32]

By this time, Swammerdam had also become an excellent anatomist. Working with several experimentally inclined physicians in Amsterdam, he acquired quite excellent dissecting skills and developed new techniques to preserve and investigate larger specimens. Even before he went to the University of Leiden, where the professors were pursuing anatomical, botanical, chemical, and clinical work with vigour, Swammerdam had developed skills that very much impressed Professor Van Horne (who had discovered the thoracic duct, among other organs). At Leiden, with the encouragement of Van Horne, Swammerdam did work of a pioneering nature on muscles and respiration. He anatomized live dogs and frogs at his rooms or occasionally at Van Horne's house. His medical thesis showed him to be a physiological mechanist – like his teachers in the 1660s and early 1670s being strongly influenced by Cartesianism.

When he returned to his father's house in Amsterdam, he continued his intensive comparative-anatomical work as part of a small group of physicians led by Gerard Blasius – a friend of the City regents, a chemist and a botanist, and one of the most notable comparative anatomists of the day. This group of physicians, the Collegium

[30] H. Brugmans, ed., *Gedenkboek van het Athenaeum en de Universiteit van Amsterdam 1632–1932* (Amsterdam, 1932); Chris L. Heesakkers, 'Foundation and early development of the Athenaeum Illustre at Amsterdam', *Lias*, 1982, **9**: 3–18.

[31] Lindeboom, 'Jan Swammerdam als microscopist', *op. cit.* (note 29), 96–101; Fournier, 'Jan Swammerdam', *op. cit.* (note 29), 75–6. The statement about the power of the lenses is from Lindeboom, 98–101, based in large part on P. Harting, *Het microscoop, deszelfs gebruik, geschiedenis en tegenwoordige toestand*, (3 vols., Utrecht, 1848–1850).

[32] For about 200 gilders at Amsterdam in 1670 he could have bought a last of wheat (2 tuns – large barrels – or about 2 metric dead-weight tonnes), or for 100 gilders a last of rye, or for 15 gilders a tun of herring. In Leiden in 1620, a 12-pound loaf of rye bread cost 6.4 stuivers (at 20 stuivers per gilder). A Captain in the Dutch navy earned 30 gilders per month, a common sailor about 10. From N. W. Posthumus, *Inquiry into the History of Prices in Holland* (Leiden 1946–1964), **2**: 770, 783; A. T. van Deursen, *Plain Lives in a Golden Age: Popular Culture, Religion and Society in Seventeenth-century Holland*, trans. Maarten Ultee (Cambridge 1991), p. 6; Charles R. Boxer, *The Dutch Seaborne Empire 1600–1800* (London, 1965), pp. 337–342.

Medicum Privatum, amounted to a research group that met over a period of twenty years, sometimes (as from October 1665 to February 1666) more than once a week.[33] Swammerdam continued to work with Van Horne in Leiden, Van Horne paying for the anatomical material, while Van Horne travelled up to Amsterdam to participate in the Collegium Medicum Privatum. Swammerdam obtained the permission of the city physician and member of the Collegium, Mathias Sladus, to do anatomical work in the city hospital, the Pietersgasthuis. After the retirement of Sladus in 1669, and the death of Van Horne early in 1670, one of the burgomasters, Coenraad Van Beuningen, obtained permission for Swammerdam to dissect those who died in the hospital. Van Beuningen had first met Swammerdam in Paris when he was the Dutch ambassador and Swammerdam had come to see Melchisédec Thévenot, who remained a close friend and steadfast supporter for the rest of Swammerdam's life. Van Beuningen, too, remained a fast friend of Thévenot's.

Swammerdam developed a new technique of preserving body parts with what he called his balsam. He even managed to 'balsam' a whole lamb and a child of one month. He also much improved the new technique of injecting wax into the vessels of the body; using coloured wax and injecting it into different vessels would make them stand out visually. He managed to make such a preparation of the chest: white wax filling the trachea to the tiniest parts, the pulmonary artery similarly filled with red wax, the pulmonary vein filled with rose wax, and the small orifices of the bronchial artery filled with a fire-red substance; a liver was similarly prepared with balsam and wax. Such techniques were soon developed further by the Leiden professor, Antonius Nuck, who used preparations of mercury to reveal the lymph system in great detail.[34] Later in his life, Swammerdam came to doubt important parts of the mechanical philosophy, and for a time he became an avid follower of a religious mystic, Antoinette Bourignon; but still his natural historical work continued at a very high level.

Although this very short summary of the life of Swammerdam cannot do justice to the range of his interests and learned activities, it touches on several important issues: the international network of scholars and patrons who considered work like his to be decidedly avant-garde (he was well known to the Fellows of the Royal Society of London and to members of the Académie des Sciences of Paris); the tremendous demands of time, money, technique, and learning necessary to carry on such work; and the mixing together of chemical, anatomical, physiological, and microscopical research, collection and preparation of specimens, and religious-philosophical interests. Should we value his work any less than that of his contemporaries who fit somewhat better our notions of the Scientific Revolution as something to do with mathematics and mechanics?

[33] G. A. Lindeboom, 'Het Collegium Privatum Amstelodamense (1664–1673)', *Ned. Tijd. Geneesk.*, 1975, **119**: 1248–54. Harm Beukers, in a paper read in Leiden in September 1991, found that the Collegium existed well into the later seventeenth century.

[34] G. A. Lindeboom, 'Dog and frog: physiological experiments', in *Leiden University in the Seventeenth Century: An Exchange of Learning*, ed. T. H. Lunsingh Scheurleer and G. H. M. Posthumus Meyjes (Leiden, 1975), 279–93; Antonie M. Luyendijk-Elshout, 'Le système lymphatique au dix-septième siècle: réalités et fantaisies', *Janus*, 1965, **52**: 283–8.

While Swammerdam's generation was the first to try to adapt Cartesianism to natural history (albeit with only modest success), the generation of his older colleague, Gerard Blasius, shows clearly the high level of work done before Cartesianism claimed the minds of so many. Blasius was the son of the architect to the king of Denmark; he studied in Copenhagen – probably with Caspar Bartholin – before coming to Leiden for his university education in 1645. He became a city physician and Extraordinary Professor in medicine in Amsterdam in 1660; shortly thereafter he became physician to the hospital, in 1666 an Ordinary Professor, in 1670 the city librarian, and in 1681 he became a member of the Russian Academy for Nature and Art.[35] In addition to working with people like Swammerdam, Blasius encouraged the young Nicholas Steno, whose first known anatomical discovery (the duct of the parotid gland in the head of sheep, the principal source of saliva for the oral cavity) was made three weeks after Steno's arrival in Amsterdam to study with Blasius.[36] Blasius wrote and edited a great number of books, almost all dedicated to the burgomasters of Amsterdam. In one of his prefaces, to a short treatise on anatomy addressed to the general public, he explains that anatomy is important for all, not only for its usefulness in medicine but for the general understanding of man and nature that it brings.[37] In fact, Blasius published not just on anatomy, but on chemistry as well. Furthermore, as professor in the Athenaeum, a leading member of the Collegium Medicum and physician to the hospital, he taught apprentice surgeons and apothecaries botany in the Amsterdam botanical garden; this was located in a large inner courtyard of the hospital.

Collecting and edification

The botanical gardens in Europe were the largest research laboratories of the day.[38] They were everywhere, absorbing huge financial resources. When the Englishman William Sherard, who later became a lawyer, made his tour of the Continent, he went first to the Jardin des Plantes in Paris, where he studied chemistry and botany with Tournefort. There he met Paulus Hermannus, the Leiden medical professor and botanist, who was paying one of his visits to Tournefort. Hermann invited Sherard to Leiden. In the book on Hermann's work edited by Sherard in 1698, Sherard portrayed the Leiden garden as a modern Garden of Eden. He also mentioned not only the extraordinary botanical gardens of Leiden and Amsterdam, but also the many private gardens he had been able to see in the Netherlands: those of the Prince of Orange, of Hans Willem Bentinck at the 'Sorgfliet' near The Hague, of Hieronymus

[35] On Blasius, see Brugmans, *op. cit.* (note 30), 163–4, 180–81, 394, 415; Cole, *op. cit.* (note 2), 330–2; Lindeboom, *op. cit.* (note 33), 1250; *BWN*; *NNBW*. On his appointment to what seems to have been a predecessor to St Petersburg Academy of Sciences, see the last two references.

[36] Gustav Scherz, 'Steno', *DSB*.

[37] Gerard Blasius, *Ontleeding des Menschelyken Lichaems* (Amsterdam, 1675).

[38] J. Kuijlen, C. S. Oldenburger-Ebbers and D. O. Wijnands, *Paradisus batavus: Bibliografie van plantencatalogi van onderwijstuinen, particuliere tuinen en kwekerscollecties in de Nordelijke en Zuidelijke Nederlanden (1550–1839)* (Wageningen, 1983).

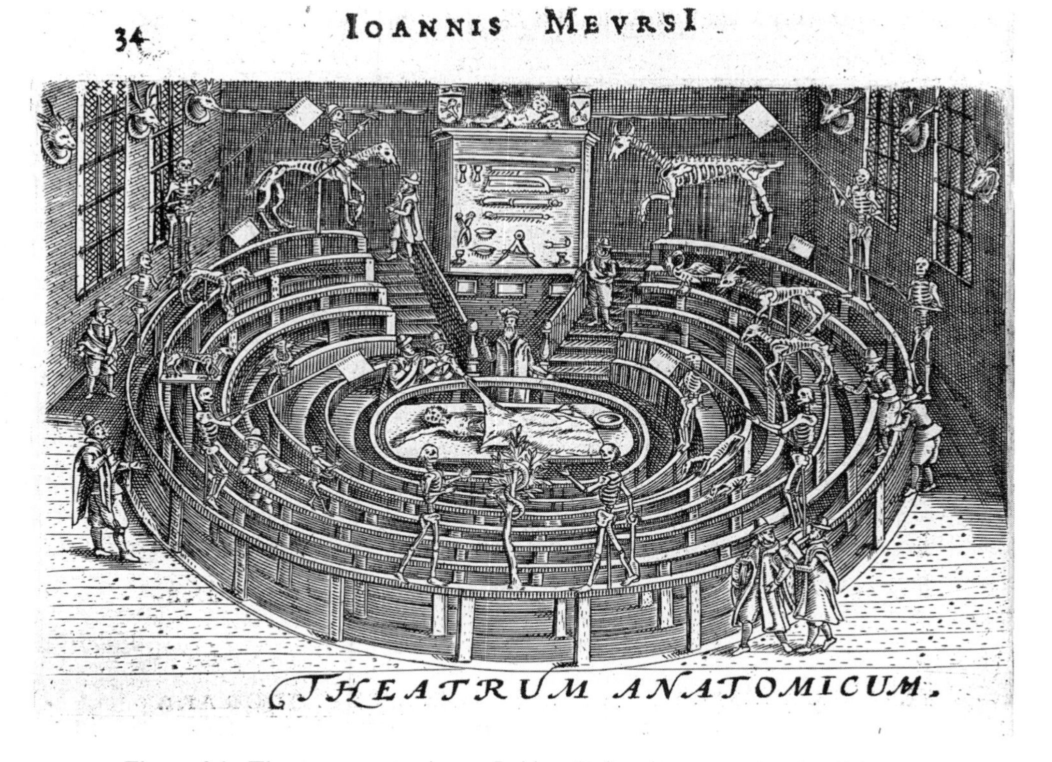

Figure 3.1. Theatrum anatomicum, Leiden University, engraving by Johannes Meursius after Crispijn de Passe, 1612. Depicted as seen during the spring through autumn, when anatomies were not in progress (but note that the artist could not refrain from placing an opened woman on the dissecting table). (Reproduced by courtesy of The Trustee of The Wellcome Trust, London.)

Beverning at 'Oud-Teylinghe' near Warmond, of Caspar Fagel at 'Leeuwenhorst' near Noord-wijkerhout, of Simon van Beaumont in The Hague, of Daniel Desmaret, of Philips de Fline at 'Sparen-Hout' near Haarlem, of Johannes van Riedt, of Agnes Block at 'Vijverhof' near Loenen, and of Frau Pullas.[39]

Like the gardens, anatomy theatres were more than a place for inculcating the latest knowledge into students. Anatomical images were rife with moral meanings about the body, the hidden structures of life, the impending death of each of us, and the vanities of this world.[40] For those who might be a bit slow to see this for themselves, during the spring to autumn months when anatomy lessons were not given in Leiden, the university theatre was filled with human skeletons (Fig. 3.1). Some held banners with

[39] Paulus Hermannus, *Paradisus Batavus* (Leiden, 1698); Johann Heniger, 'Der wissenschaftliche Nachlass von Paul Hermann', *Wiss. Z. Univ. Halle*, 1969, **18**: 527–60; John Prest, *The Garden of Eden: the Botanic Garden and the Re-creation of Paradise* (New Haven, 1981).
[40] W. S. Heckscher, *Rembrandt's Anatomy of Dr. Nicolaas Tulp: An Iconological Study* (New York, 1958); W. Schupbach, *The Paradox of Rembrandt's Anatomy of Dr. Tulp* (London, 1982).

Latin mottoes about the shortness of life, one wore a feathered cap while riding a cow to portray vanity, two held symbols identifying them as Adam and Eve while standing next to a tree, and so forth. The walls beneath were hung with works of art on various moral themes about this world and the next.[41] Perhaps the artistic traditions of each city were related to the different ways of pursuing anatomy in each, as well.[42] In short, the study of nature demanded great industry and attention to detail, but it also edified.

Perhaps in one respect we are not so far from the amusing, implausible and miraculous fourteenth-century stories in Sir John Mandeville's *Travels*: beneath the welter of appearances lay God's providence. The sentiments of Swammerdam, Leeuwenhoek, and most other Dutch natural historians and philosophers were leading to what has been called in English 'natural theology', or in Dutch, 'physico-theology'.[43] At the most general level of declaring the purpose of all these investigations, Mandeville and Swammerdam would have agreed: all is for edification. There is, however, a very important difference. For the readers of Mandeville, the difference between poetry and natural philosophy was never quite clear; for the readers of Swammerdam, the wisdom of God could be seen in the figure of the louse, but the poetry of the louse demanded getting the details unassailably correct. As he wrote to Melchisédec Thévenot, 'I present to you here with the Almighty Finger of God in the anatomy of a louse, in which you will find wonder piled upon wonder and God's Wisdom clearly exposed in one minute particle'.[44] Physical details had become essential.

This shift in values can be seen clearly in the work of yet another Amsterdam professor of anatomy, who was also a master of the Amsterdam Surgeons' Guild and Supervisor of Midwives, Frederik Ruysch. At the end of the seventeenth century, he collected at his home a large and famous cabinet. The cabinet contained a great many natural curiosities and rarities from shells and minerals to human artefacts, dried and preserved plants, mounted animals from butterflies to mammals, and a great many anatomical specimens dried, waxed, or preserved in alcohol. The five rooms of his house occupied by these exhibits had many visitors, not least among them the Tsar Peter, who upon Ruysch's death purchased the lot from his heirs and set it up in St Petersburg.

The centrepiece of Ruysch's cabinet was a large group of skeletons of human babies and foetuses, arranged in allegories portraying the transience of life. Many little skeletal hands pointed to Latin mottoes such as 'vita humana lusus' (Man's life is but a game). Nearby, embalmed babies dressed in lace lay in tiny coffins, some with glass

[41] T. H. Lunsingh Scheurleer, 'Un amphithéâtre d'anatomie moralisée', in Scheurleer and Meyjes, *op. cit.* (note 34), 217–77.

[42] Jan C. C. Rupp, 'Matters of life and death: the social and cultural conditions of the rise of anatomical theatres, with special reference to seventeenth-century Holland', *Hist. Sci.*, 1990, **28**: 263–87.

[43] See esp. Rienk H. Vermij, *Secularisering en natuurwetenschap in de zeventiende en achttiende eeuw: Bernard Nieuwentijt* (Amsterdam, 1991).

[44] G. A. Lindeboom, ed. and transl., *The Letters of Jan Swammerdam to Melchisédec Thévenot* (Amsterdam, 1975), 104–5.

Figure 3.2. Frederik Ruysch, third *Thesaurus*, from vol. 4 of his *Opera Omnia* (1721–1727), copper engraving by Huyberts. (Reproduced by courtesy of The Trustee of The Wellcome Trust, London.)

eyes staring out at the visitors. Finally, the *pièce de résistance*, were the ten treasures (*thesauri*): elaborate dioramas on the theme of mortality, constructed from kidney, gall, and bladder stones, trees made from dried veins and arteries, topped with foetal skeletons in various poses. In the third *thesaurus*, illustrated in Figure 3.2, a central figure looks heavenward, singing a lament ('Ah fate, bitter fate!') while accompanying itself on the violin; a small figure to the right of it beats time for the music with a baton set with minute kidney stones; on the far right is a skeleton girded with sheep intestines injected with wax, a spear made from a hardened male vas deferens conveying a message about man's first hour also being his last; to the left is a figure with a feather, a symbol of *vanitas*; and in front is a tiny skeleton holding in its hand a mayfly – an insect on which Swammerdam had written a famous book, centring his moral argument on the supposed fact that the creature lived in its adult form for only one day.[45]

Clearly, in Ruysch's cabinet we find exacting attention to the details of nature; but also far more. Remnants of living beings, in the form of animals, plants, and anatomical preparations, as well as depictions of living beings in illustrations, were shown in meticulous detail. But obviously, too, they also conveyed other more profound meanings. We are to learn more than simply the outward shape of the remnants of living beings. Materialistic as it was, and pursued to an ever larger extent outside the library, the investigation of nature was not yet divorced from its humanistic, literary purpose: edification.

Now, if one does not dismiss such interests as 'mere' empiricism, or as 'soft' science, or as being in some other way deficient, then one begins to notice how much of the time and energy of the early modern virtuosi was taken up with just such investigations into natural history. As a matter of fact, if we look for the 'big science' of the period, we find natural history (together with chemistry) being at the heart of the endeavours of the period. Large networks made up of thousands of people made sure that observations would be made, curiosities would be reported, and specimens would be collected from all parts of the world: from local people, jugglers and quacks, merchants and sailors, travellers and soldiers, gentlemen and aristocrats, professors and priests. Deciding what constituted 'fact' necessitated cooperative research, technical expertise, and access to collections of books and bottles, shells and stones, anatomy theatres and botanical gardens, glassware, chemicals, tools, instruments, and machines, living and dead plants and animals, paper, pictures, and postbags. To carry out much of this work required the patronage of the rich and the powerful, legal protections and permits, and money: lots and lots of money. Operating the Amsterdam botanical garden and collecting new specimens for it in the 1680s ate up about 1200 gilders per year, paid for by dues, fines, and fees levied on the doctors and apothecaries of the city.[46] And of course this big science of the period was covered in detail in the periodical press – including the *Philosophical Transactions* – and in published books (often with lavish plates), while attracting rich and ordinary visitors to the cabinets of curiosities, anatomy theatres and botanical gardens in every country in Europe.

[45] Antonie M. Luyendijk-Elshout, 'Death enlightened: a study of Frederik Ruysch', *J. Am. Med. Ass.*, 1970, **212** (1): 121–6.

[46] 'Register bevattende jaarlijkse rekening en verantwoording aan de commissarissen van de Hortus Medicus, 1683–1793', Gemeentelijke Archiefdienst Amsterdam, P.A. 27/29.

Conclusions

If we look beyond the Royal Society, we find many other informal and formal academies of science being organized around the discussion of *res naturae*. The groups in Paris surrounding aristocratic patrons like Thévenot, the royal Jardin des Plantes, and even the Académie des Sciences had such interests:[47] perhaps it is useful to remember that the first publication of the Académie was an anonymous work on animals investigated by the members, later associated with the name of Claude Perrault.[48] David Lux has recently shown how Louis XIV's minister Colbert tried to get the short-lived royal academy in Caen to spend much of its time anatomizing fish.[49] The Accademia del Cimento and other early scientific societies in Italy also published on medical and natural historical subjects, and the fame of the anatomist Malpighi soon rivalled that of Galileo.

Further north, one finds the predominantly German Academia Naturae Curiosorum established in 1652 by Dr Johann Lorenz Bausch, and the numerous natural historical publications of its members; the predominantly Danish group centred around Thomas Bartholin in Copenhagen and its publication, the *Acta Medica Hafniensia*; and the Collegium Medicorum Privatum of Amsterdam. Every university with any reputation had a botanical garden and anatomy theatre and collections of natural history specimens. Not only universities and kings, but private people also devoted large amounts of money to such work, and while they could hardly carry out human dissections on their own, public dissections reserved the front seats for the magistrates and admitted the general public as well. Perhaps by the eighteenth century one can begin to speak of a tradition of teaching natural philosophy that looks like today's methods of teaching physics, but in the seventeenth century it was natural history that dominated the curricula as well as the postbags of the learned.

Of course, it might be argued – and many historians of science have done just this – that this natural historical work was peripheral to the 'real' science of the day. But who is to say that John Ray's division of dicotyledons from monocotyledons, or Sylvius's and Boyle's division of acid from alkali, was not equally as significant as the discussions surrounding the motion of the pendulum? If we want to find an example of a bold new theory, might not the idea of the human body as being mainly liquids contained in vessels rather than solid flesh, or the idea that all living things come from

[47] Rio Howard, 'Guy de La Brosse: Botanique et chimie au début de la révolution scientifique', *Rev. Hist. Sci.*, 1978, **31**: 301–326; Howard 'Medical politics and the founding of the Jardin des Plantes in Paris', *J. Soc. Bibliogr. Nat. Hist.*, 1980, **9**: 395–402; Howard, 'Guy de la Brosse and the Jardin des Plantes in Paris', in *The Analytic Spirit*, ed. Harry Woolf (Ithaca, 1981), 195–224; Howard, *La bibliothèque et le laboratoire de Guy de la Brosse au Jardin des Plantes à Paris*, Ecole Pratique des Hautes Etudes, Histoire et civilisation du livre, n. 13 (Geneva, 1983); Alice Stroup, *A Company of Scientists: Botany, Patronage, and Community at the Seventeenth-century Parisian Royal Academy of Sciences* (Berkeley, 1990).

[48] *Mémoires pour servir à l'histoire naturelle des animaux* (Paris, 1671); also, Claude Perrault, *Description anatomique d'un cameleon, d'un castor, d'un dromadaire, d'un ours, et d'une gazelle* (Paris, 1669).

[49] Lux, *op. cit.* (note 17).

seeds or eggs, equal the notion that the Earth moves around the Sun?[50] Even Descartes spent some of his time in the Netherlands occupied with anatomical investigations.[51]

Let me therefore conclude with one final example, drawn not from the Continent but from Oxford: that of Edward Tyson. He matriculated at Magdalen Hall in 1667, where Robert Plot was then a tutor and organizing the teaching of natural history at the university. Tyson probably studied botany under the distinguished Robert Morison, and he certainly studied anatomy with Plot, who wrote of Tyson's anatomical and botanical work with great respect. After moving to London in 1677, he called on that great patron of natural history, Robert Hooke, who took to Tyson and proposed him as a Fellow of the Royal Society. Tyson soon became one of the leading figures among the virtuosi in London. It need hardly be said that he was aware of the work on comparative anatomy being done in Paris and Amsterdam, as well, which was clearly evident in his *Orang-outang*; and together with Hooke he helped to bring out an English translation of Swammerdam's *Ephemeri vita* – a book on the life of the mayfly.[52] In the words of Tyson's biographer, Ashley Montague, Tyson's anatomy of the porpoise 'sketches a method not only of General Biology but also of General Medicine, with the emphasis placed always upon a systematic and experimental methodology. Medicine is brought into the field of natural history as an experimental biology of man . . .'.[53]

In short, more than ever, the natural historians argued that the facts counted – perhaps they were all important. Their pioneering techniques of investigation took great energy, skill, time, and resources to master. But getting the facts straight was not the only thing they cared about. There were, of course, utilitarian benefits. But the deeper purpose of their work was moral edification. That edification could come through the mastery of detail rather than by expounding universal principles seemed revolutionary, even subversive, to some. To others, like Hans Sloane, finding 'matters of fact' seemed not boring but an exciting new adventure. The assemblage of things could create gardens resembling those of paradise as well as theatres reminding people of mortality. The concept of the Scientific Revolution surely must be made to include this big science, this cutting edge research of the period, or the concept is pointless.

Let us return, then, to George Sarton. Fifty years ago, he wrote a preface to Ashley Montague's book on Tyson. Sarton argued:

> The value of Tyson's anatomical analyses can scarcely be exaggerated, and I agree with the author that Tyson's *The orang-outang* is one of the

[50] Edward G. Ruestow, 'The rise of the doctrine of vascular secretion in the Netherlands', *J. Hist. Med.*, 1980, **35**: 265–87, 272.

[51] G. A. Lindeboom, *Descartes and Medicine* (Amsterdam, 1979).

[52] Edward Tyson, *Orang-outang, sive homo silvestris: or, the anatomy of a pygmie compared with that of a monkey, an ape, and a man To which is added a philological essay concerning the pygmies, the cynocephali, the satyrs, and sphinges of the ancients. Wherein it will appear that they were either apes or monkeys, and not men, as formerly pretended* (London, 1699); Swammerdam, *Ephemeri vita; or, the natural history and anatomy of the Ephemeron* (London, 1681).

[53] Ashley Montague, *op. cit.* (note 2), 100.

> outstanding landmarks in the history of science. I am not sure that I would place it quite on the same level as the *Fabrica*, the *Revolutiones*, and the *Principia*, but this is a matter which I have not the heart to discuss. It is not the historian's business to distribute prizes .... It will suffice to state that the *Orang-outang* was an epochal contribution to science and philosophy: yet it is necessary to state it loudly, because Tyson has not yet obtained the same popular fame as Vesalius, Copernicus, or Newton.[54]

Amen.

[54] *Ibid.*, p. xvi.

4

Science and technology during the Scientific Revolution: an empirical approach

RICHARD S. WESTFALL

The question of science and technology, that is, the extent to which science entered into technology at that time, has long been one of the central issues in the interpretation of the Scientific Revolution. It is hardly difficult to understand why. There is on the one hand the importance of a scientifically animated technology in the modern world; in my opinion, at least, it is the central defining characteristic of developed societies. There is on the other hand the broad issue vigorously discussed within our discipline, as analogous problems are discussed in every branch of history, expressed in the concepts of internal and external history of science; the question of science and technology stands at the very focus of that discussion. As part of a social history of the scientific community of the sixteenth and seventeenth centuries I have been collecting information about the technological involvements of members of the community; this information, an empirical approach to the question that seeks to examine, not programmatic statements, but the actual participation of scientists in technology at the time, provides the substance of my paper.

A brief description of the broader project will show what is meant by the numbers that I shall produce. As a significant part of the social history, I have been assembling what I call a catalogue of the scientific community of that period. I have based the catalogue on the *Dictionary of Scientific Biography* and include in it every Western scientist in the *DSB*, born between the decade of Copernicus' birth (the 1470s) and 1680.[1] The *DSB* offers the great advantage of supplying an extensive list of scientists, 630 in all, compiled by those most qualified to judge, a list that I did not

[1] I did purge twenty-one names, of men who did not seem to me to have been truly part of the scientific community, from the total of 651 in the *DSB* who fill the chronological criterion. Those purged include a number who appear to me to have been religious thinkers but not scientists, such as William Ames, Roberto Bellarmino, Johann Bisterfeld, Jacob Boehme, Sebastian Franck, Michael Nostradamus, Juan Villalpando, and Valentin Weigel; several who are in the *DSB* because of plagiaries and false claims, such as Thomas Geminus, Zacharias Jansen, Jean Leurechon, Georg von Löhneyss, and Marcantonio della Torre; and some who appear to me as essentially literary figures, such as François Rabelais and Luis Vives. Although the other six do not fall into ready categories, they do not have stronger claims to be considered members of the scientific community.

myself compile and cannot therefore have slanted to any purpose.[2] It is clear that my catalogue does not contain the entire scientific community of that age, and since it does contain either all or at least most of the outstanding scientists, it can hardly be representative of the whole community. Right now I have no clear idea of how typical my group is in the matter of technological involvement. Suffice it to say that for my present purposes I shall treat those 630 as the scientific community of the sixteenth and seventeenth centuries. I am currently attempting to read the best secondary literature on the members of this group, compiling reports or sketches of each one, organized under a set number of headings relevant to the social history of the scientific community. One of the headings is 'Technology', that is, involvement in projects of practical utility, and under the general heading I have fifteen different categories.

I am concerned with real technological undertakings, not with talk about the application of science to use.[3] Thus Francis Bacon appears under the category 'None'. On the other hand, my figures do not attempt to enter into the impossible, and inevitably subjective, task of measuring the extent or the intensity of an individual's involvement. For example, in 1630 the Florentine government solicited Galileo's advice about a flood control project on the plain west of Florence. In contrast, Galileo's disciple Castelli devoted his whole career primarily to similar questions. Both appear equally under the category of 'Hydraulic engineering'. Galileo developed the astronomical telescope, a microscope, a geometric and military compass (that is, a calculating device), and a thermometer, and he contributed significantly to the precision clock. Castelli, in contrast, suggested to Galileo the method of projecting the image of the Sun through a telescope onto a screen in order better to observe sunspots. Both appear equally under 'Instrumentation'. From these three examples it will appear also that only one of the fifteen categories of technology is exclusive – the category 'None'. Simon Stevin appears under seven different categories, and percentages accordingly cannot add up only to one hundred.[4]

Of the 630, one hundred and forty-eight (23.5%) appear in the exclusive category

[2] In fact I was one of the advisory editors who helped to compile the list. I cannot now remember what suggestions I sent in, and I no longer have the file of correspondence from which I could learn what they were. Obviously I do not think that that participation, more than two decades ago and long before I had even thought of my present project, compromises my claim that I cannot have slanted the list, consciously or unconsciously, to some purpose connected to the project.

[3] I need to qualify that statement. While I have excluded Baconian talk about the utility of science in general, I have not excluded discussions about individual categories of technology. Thus I have treated a book on pharmacology (or *materia medica*) as involvement in an enterprise, pharmacology, inherently directed toward human use and not solely toward increased understanding. I have treated books about improvements in agriculture and books on cartography and fortification analogously. With only a small number of exceptions, those who wrote on these subjects practised their arts at one time or another, and I have considered the books as contributions to those arts and thus as a form of technological involvement. I have also treated books on navigation in this way; almost all of the advances in navigation came from books by mathematicians who were not themselves navigators.

[4] I need to inform the reader further that the catalogue is not yet in its final form. Perhaps it never will be, because I shall be willing to add further information whenever I find it. At the moment, however, I am consciously working at improving some of the provisionally completed sketches. The final figures will therefore undoubtedly differ by a small amount from those I now have.

'None'.[5] That is, about three-quarters of the scientific community did participate in some technological enterprise. To be frank, I must say that this is a considerably higher proportion than I would have predicted when I began. However, the number needs further analysis. More than a third of my group practised medicine – 238 or 37.8% – which I take, as surely I must, to have been an activity directed toward use rather than understanding. (I refer to the practice of medicine here,and not to the pursuit of the discipline.) Add to them another twenty-nine (4.6%) who engaged in pharmacology,[6] and the total involved in medical technology amounts to a bit more than two-fifths of the whole group. There are various ways in which the catalogue can measure the size of the medical community – under the heading 'Education', those who earned medical degrees; under the heading 'Discipline', those who pursued medically related studies such as anatomy, physiology, and surgery; under the heading 'Support', those for whom a medical practice was a means of livelihood; and under the heading 'Technology', those who applied medical knowledge to maintenance and restoration of health.[7] The numbers are rather close in all measures, about three-eighths of the total. With not many exceptions, the medical community was distinct from the rest of the scientific community. Nevertheless, I would vigorously resist any suggestion that they were not part of the scientific community, and even if one were to argue that initially medicine was an enterprise separate from natural philosophy, by the second half of the seventeenth century the concepts prevalent in the new natural philosophy had invaded medicine to the extent that it could no longer be considered distinct. Be all that as it may, if we were for the moment to consider a scientific community composed solely of the rest, the percentage who participated in technological projects would drop considerably, but even in the artifically truncated community the proportion who participated in some technological project would remain

[5] I shall use percentages calculated to one decimal place. This is a legitimate calculation for the group of 630. Obviously I seek to cast light on the whole scientific community of the age, and equally obviously percentages to one decimal place are not accurate for them. I have commented above on the problem of extrapolating from the leaders of the community to the community as a whole. At most I would want to say in this case 'about one quarter', and I would insist on the qualifier 'about'.

[6] The number under 'Pharmacology' is, of course, much larger, ninety-two (14.6%). Sixty-three of that number also practised medicine, however, so that the additional number, which I want here, is only twenty-nine.

[7] A total of 224 (35.6%) earned a medical degree. The categories under 'Discipline' are not exclusive, so that frequently one man contributed to anatomy, physiology, medicine, and perhaps also pharmacology and surgery. In all, 247 (39.2%) contributed to at least one of the medically related disciplines (Medicine, Surgery, Pharmacology, Anatomy, Physiology, Embryology), while 224 (35.6% – not wholly identical to the 224 who earned medical degrees) gained part of their support from a medical practice. Anyone who, under 'Support', has 'Medical practice', has an entry for 'Medicine' under 'Technology'. The relation was not reciprocal, however; there were a number of men with an entry for 'Medicine' under 'Technology' who did not maintain a medical practice. It is well known that in the sixteenth and seventeenth centuries, a number of men who were not primarily physicians practised or expounded medicine occasionally and/or collected cures. I think of men such as René Descartes and Reiner Gemma Frisius on the one hand, and Robert Boyle and Giambattista della Porta on the other. When I reduce the 267 in the combined Medical practice and Pharmacology group under 'Technology' by 30 who were associated with these activities only secondarily, the remainder, 237 (37.6%), falls very close to the other indices of the size of the medical community. Under another heading, 'Scientific societies', I have a general category for 'Medical College', any College of Physicians. I suspect that my sources are frequently silent on this subject; in any case, since there were also numerous physicians who did not belong to a College, the number under this category, 62 (9.8%), is much smaller than the others.

about three-fifths (in the catalogue 245 of 393, or 62.3%[8]). The number is much larger than I had been prone to expect.

Again, however, the number demands further analysis. Several categories, though not nearly as large as 'Medicine' and 'Pharmacology' in the number of names they contain, are larger than the rest. Fifty of the 630 (7.9%) engaged in 'Military engineering', fifty-four (8.6%) in 'Hydraulic engineering', forty-seven (7.5%) in 'Navigation', and ninety (14.3%) in 'Cartography'. These four categories had at least two things in common: they were heavily mathematical,[9] and like medicine they did not represent enterprises that were new with the Scientific Revolution.

Military engineering

Let me pause to examine them individually. Military engineering, as it was known through the sixteenth and seventeenth centuries, came into being toward the end of the fifteenth century as the improvement and increasing use of artillery rendered the medieval keep obsolete. Military engineering had little to do with things such as the mathematical description of trajectories. Mathematicians could make their reputations with elegant solutions to the problem of the trajectory, but the assumptions on which the solutions rested had so little to do with the realities of the battlefield as to leave them irrelevant to military practice.[10] The essence of military engineering was the design of a new style of fortification in which the cannon, the erstwhile weapon of assault, became the anchor of an impregnable defence. The angle bastion gradually evolved from experience on the field of battle in the last years of the fifteenth century. The new principles of fortress design had reached the level necessary for mature statement in the Fortezza da Basso in Florence, begun in 1534 and completed in 1535, about a decade before the publication of Copernicus's *De revolutionibus*, and once the definitive solution had been realized, the principles of fortification underwent very little change during the following three centuries. In the middle of the eighteenth century, for example, as Europeans began to explore the interior of North America and to contend for its control, the forts that they built at the confluence of the Ohio and at Ticonderoga would have looked familiar to the engineers who constructed the Fortezza da Basso.[11]

[8] I reduced the 267 who practised medicine or contributed to pharmacology by 30 who are not primarily associated with those fields. When that number, 237, is subtracted from the total, 393 remain, of whom 148 had no technological involvement.

[9] Three other smaller categories shared the mathematical character – 'Mathematical applications', the majority of which were devices of one sort or another to facilitate calculations, with thirty (4.8%), 'Civil engineering', exclusive of hydraulics,with twenty (3.2%), and 'Architecture' with fifteen (2.4%).

[10] A. R. Hall, *Ballistics in the Seventeenth Century* (Cambridge, 1952), remains the definitive treatment of this issue.

[11] See John R. Hale, 'The early development of the bastion: an Italian chronology *c*. 1450–*c*. 1534', in J. R. Hale, J. R. L. Highfield and B. Smalley, eds. *Europe in the Late Middle Ages* (Evanston, 1965), 466–94; Hale, 'The end of Florentine liberty: the Fortezza da Basso', in Nicolai Rubinstein, ed. *Florentine Studies: Politics and Society in Renaissance Florence* (Evanston, 1968), 501–32; Hale, 'Terra ferma fortifications in the Cinquecento', in *Florence and Venice: Comparisons and Relations*, Acts of Two Conferences at Villa I Tatti in 1976–1977, organized by Sergio Bertelli, Nicolai Rubinstein and Craig Hugh Smyth, 2 vols. (Florence, 1979–80), 2: 169–87; Simon Pepper and

The essence of the angle bastion lay in the fact that every inch of its surface was defended by the flanking fire of cannon located in and protected by the neighbouring bastions. That is, the design of angle bastions was an exercise in applied geometry, and anyone who showed proficiency in mathematics was likely to find himself thrust into the role of military engineer. As Descartes would remark to Constantijn Huygens in recommending a protégé as a military engineer, Gillot had some experience, 'and you know that for those with a bit of intelligence [this was Descartes talking – read 'mathematical competence'] not much experience is needed to become engineers'.[12] The evidence does appear to bear his judgment out. Early in the sixteenth century Oronce Fine, already known as a mathematician, was in prison, apparently as a result of his opposition to the Concordat between the French crown and the Pope, when Francis I became aware of him. The king summarily released Fine in order that he might set him to work on the French fortifications of Milan and Pavia. A generation later, in 1552, Girolamo Cardano cured the governor of the province of Lyon of some malady. The governor promptly tried to enlist Cardano into the service of the viceroy of Milan at the considerable stipend of 1000 scudi per year. It was not as a physician that he wanted to enlist Cardano, however. He knew of Cardano's mathematical talents, and he wanted to appoint him as a military engineer. Later in the century Ludolph van Ceulen started out as a fencing instructor and a teacher of arithmetic. It was the second rather than the first that led to his appointment as an instructor on fortification in the engineering school in Leiden. Jean Charles de la Faille, a Jesuit, taught mathematics in Jesuit colleges in the Spanish Netherlands until he was appointed to the Imperial College in Madrid in 1629. There Philip IV commissioned him to instruct members of the court in fortification. He became the preceptor to the king's bastard son, Don Juan, whom he accompanied as a consultant on military campaigns, and he advised the Duke of Alva on fortifications along the Portuguese border. In the early eighteenth century, during the War of the Spanish Succession, the patrons of Edmond Halley in the Admiralty sent him to examine the fortifications of ports along the Adriatic in the territory of England's ally, Austria. At much the same time Leonty Magnitsky, the first important Russian mathematician and the only Russian who appears in my catalogue, found himself pressed into duty fortifying Tver on the occasion of the Swedish invasion in the Great Northern War.

Hydraulic engineering

Hydraulic engineering extended back some two centuries earlier than military engineering. Everywhere there were marshes and swamps that needed reclamation. In

Nicholas Adams, *Firearms and Fortifications. Military Architecture and Siege Warfare in Sixteenth-Century Sienna* (Chicago, 1986); Christopher Duffy, *Siege Warfare. The Fortress in the Early Modern World* (London, 1979). Experts cite Enrico Rocchi, *Le Fonti storiche dell'architettura militare* (Roma: Officine poligrafica, 1908) as the best authority on these matters; I have not yet been able to lay my hands on the book.

[12] Descartes to Huygens, April 1641; Descartes, *Correspondence* (8 vols., Paris, 1936–63), 4:340.

economically advanced areas such as Italy there was a demand for canals for transportation. By the early years of the fifteenth century the word 'engineer' was being applied in Italy to the men who could handle these works,[13] and, by the sixteenth, states such as Venice and Tuscany had bureaucracies with rather extensive personnel to deal with recurring hydraulic problems. Most of their operations appear to me to have been empirical, but questions of hydraulics began to attract the attention of scientists at the beginning of the seventeenth century, and from Stevin and Castelli, through Michelini, Montanari, Guglielmini and others a tradition of scientific hydraulics appeared.

Inevitably mathematics figured prominently in hydraulic engineering, so that once more anyone known to be competent in mathematics might find himself drafted into such work. Federico Grisogono was a prominent mathematician of the early sixteenth century who helped in the construction of harbour works at Rimini. Rafael Bombelli showed that the relationship could work in the opposite direction. A man without formal education who was trained by a practising engineer to drain marshes, and who spent much of his life doing just that, Bombelli came to mathematics, in which his enduring reputation resides, through hydraulic engineering. In the early seventeenth century Michael Florent van Langren, the grandson, son, and brother of engineers, designed harbours and canals in the Spanish Netherlands and planned measures to control flooding in addition to his work as an astronomer. As a young man, Blaise Pascal engaged for a time in draining a marsh. At the first meeting of the Caen Académie de Physique after royal incorporation, the Intendant of Lower Normandy let the members know his idea of what men of their ilk should do – drain swamps and build pumps that would allow the construction of fountains in the city's squares.[14] Jean Picard not only measured the Earth; he also worked extensively at supplying water for the fountains at Versailles. Ole Roemer did a similar task for his native Copenhagen (where needs more fundamental than decorative fountains were involved) when he returned home from Paris. Guido Grandi, who introduced the calculus into Italy, became the director of hydraulics in Tuscany and was involved in the drainage of the Chiana valley and the Pontine marshes and in questions involving the whole water system of the Po valley, including the perennial issue of diverting the Reno into the Po. Like Grandi, Giovanni Ceva first made his reputation as a mathematician and through it won the patronage of the Duke of Mantua, whose superintendent of waters he became. Ceva was also involved in the plans to divert the Reno into the Po; it was primarily his opposition that thwarted the scheme. A third Italian of the late seventeenth century, Geminiano Montanari, also first established himself as a mathematician and astronomer until Venice restored the chair in astronomy at Padua to attract him there from Bologna. Once he was in Padua, the Venetian state absorbed his talents in all sorts of technical work including, inevitably, the control of rivers and the protection of the lagoon.

[13] Luca Beltrami, *Vita di Aristotile da Bologna* (Bologna, 1912), 16.

[14] David S. Lux, *Patronage and Royal Science in Seventeenth-Century France: The Académie de Physique in Caen* (Ithaca, 1989), 91.

Navigation and cartography

The intimately connected technologies of navigation and cartography reached back to the same period when hydraulic engineering appeared in Europe, the thirteenth century, when the compass and the portulan chart began to revolutionize empirical methods of navigation. Then in the fifteenth century as Portugal began to explore down the coast of Africa, navigation on unknown and unbounded waters posed a wholly new set of problems. From the beginning, Prince Henry the Navigator called upon mathematicians and astronomers to answer the problems, and during the following three centuries it was the same group of men, never the practical tarpaulins who sailed before the mast, but always the astronomers and mathematicians, who taught the navigators how to determine latitude at any point on the globe, how to correct the distortions of the plain chart, how to lay out a course and estimate longitude, and eventually, with the descendant of Huygens's clock, how to determine longitude exactly.

And once again those known to be skilled in mathematics might expect to be called upon in navigation. Often they were the same men who designed fortifications and drained marshes. In 1544 the demands of navigation stood behind the nomination of Pedro Nuñez as Professor of Mathematics at the University of Coimbra. There was no place in any of the four faculties for a professor of mathematics. Never mind; the crown was going to have a professor of mathematics to promote navigation, and a professor of mathematics it had. A generation later Philip II, now King of Portugal and aware that Spain lagged in navigation, established an Academy of Mathematics in Madrid and took Lavanha and Onderiz from Portugal to staff it. In England the Muscovy Company called upon John Dee. In the Netherlands, Simon Stevin composed a manual on navigation for his patron, Prince Maurice. Galileo, who understood what was at stake and was well aware of the prize that had been offered, thought he could determine longitude by means of the satellites of Jupiter. Huygens struggled toward the same goal with his pendulum clock, and Hooke with his spring-driven watches. The Admiralty sponsored Halley's survey of magnetic declination and encouraged his prolonged study of the Moon, both of which had the determination of longitude in mind. The English government placed Newton at the head its Board of Longitude, as earlier the French government had instructed Etienne Pascal to judge the merits of a suggested method of determining longitude.

If we distinguish maps from coastal charts, modern cartography began to appear at the end of the fifteenth century in response apparently to the overseas explorations, the recovery of Ptolemy's *Geography*, and the administrative needs of the European states. From the beginning it was the child of the mathematicians and astronomers, so that every major name in cartography belonged to a man prominent in the history of science for other achievements as well – names such as Reiner Gemma Frisius, Willibrord Snel, Philippe de la Hire, Jean Picard, and the Cassinis I and II.

Not surprisingly, those who had made a name in mathematics frequently found themselves surveying and mapping. Ludovico Ferrari became known as a follower of

Cardano in mathematics; his patron, the governor of Milan, set him to mapping the territory. Cosimo I, Grand Duke of Tuscany, possibly learned about Egnazio Danti from Danti's brother, who was a sculptor at the court. Cosimo installed Danti as instructor in mathematics to the princes, appointed him Professor of Mathematics at Pisa, and set him to preparing mural maps of Italy and the world for a room in the Palazzo Vecchio. Kepler's patrons in Linz expected him to do a map of the province, though in this case nothing came of it. Galileo understood, correctly, that the satellites of Jupiter would allow the correction of maps even if it should prove impossible to use them to determine longitude at sea, and before the century was out they had proved him right by ushering in a new level of scientific cartography. Together with Peiresc, Gassendi utilized an eclipse of the Moon to correct the accepted length of the Mediterranean. Thomas Harriot mapped the coast of Virginia; the astronomer Louis Feuillée mapped the coast of Chile. Called to France to head the new Royal Observatory, Giovanni Domenico Cassini became involved, along with another astronomer, Jean Picard, in the various cartogarphic enterprises of the Académie. Luigi Ferdinando Marsigli, a military man in the service of the Austrian empire who had been well trained in mathematics by Montanari, inevitably found himself designing fortifications, building bridges, and even diverting water, but most of all he found himself mapping, apparently every region and city he saw, so that when the Peace of Carlowitz in 1699 required a precise determination of the border between the Austrian and Turkish empires, the Emperor Leopold I called upon Marsigli for that task. One in every seven scientists of the sixteenth and seventeenth centuries who are in my catalogue did some cartography. If again we think of a truncated scientific community without the physicians, it was well over one in five.[15]

Industrial science

In 1949 Benjamin Farrington attached the subtitle 'Philosopher of Industrial Science' to his biography of Francis Bacon, and it does appear to me that the image of Bacon, with his reiterated concept of the Kingdom of Man, together with the image of industrial science, the centerpiece of the scientific technology we know so well in the twentieth century, has informed studies of science and technology in the age of the Scientific Revolution. These studies appear to me further to build on the allied proposition implicit in Farrington's title that the Scientific Revolution inaugurated a new era in the relation of science to technology. My data, drawn from the technological involvements of the members of the scientific community, do not

[15] Four of those who did some cartography also practised some medicine; all of them were associated primarily with mathematical sciences. All of them belonged to the group of 30 who were not primarily associated with medicine though they appear, under 'Technology', in the categories of 'Medicine' and/or 'Pharmacology'. The 90 who engaged in cartography constituted 22.9% of the 393 who were not primarily members of the medical community. The proportion who engaged in cartography was even larger than these numbers indicate, because the 393 include a number of scholastic philosophers from the sixteenth century, quite a few naturalists (who were not involved in cartography), and the considerable number on whom there is virtually no information.

support these propositions. Economic some of the categories certainly were, especially hydraulic engineering and navigation. Industrial they were not. Three other categories that emerged, as all of my categories did, from the raw material itself do seem to fall within the precincts of industrial science – 'Mechanical devices', 'Chemical applications', and 'Metallurgy'. Their numbers are much smaller than those I have been discussing – twenty (3.2%) of the scientists in my catalogue developed new mechanical devices; fifteen (2.4%) applied chemistry to technology; nine (1.4%) made advances in metallurgy. It is relevant here to add that fourteen (2.2%) contributed to agricultural technology.

The two men who seem to me to conform best to the image of an industrial scientist, Cornelius Drebbel and Denis Papin, were ineffectual. Interestingly, both of them were caught up in the system of courtly patronage, and both of them wasted their substance trying to catch their patrons' attention with spectacular productions. Neither of them made a significant lasting impact on technology.[16] It seems pertinent as well that the governments of the European states, who were well aware of the talents of their scientists and anxious to employ them in the works I discussed above, did not turn to the scientists when they sought to encourage industry. To the extent of my knowledge they thought instead, almost exclusively, of importing technology that already existed elsewhere. When Colbert, who had founded the Académie des Sciences and set it to work mapping the country, wanted to establish a mirror industry in France, he resorted to methods not easily distinguished from kidnapping to bring craftsmen from Venice, and when he wanted to encourage the production of fine woollens in Abbeville, he imported a whole colony of artisans from the Netherlands.[17] In 1708 the Grand Duke Cosimo III of Tuscany, the state which had fostered the Accademia del Cimento, decided that he should promote the tinplate industry in his land. He did call upon a scientist for assistance, the natural historian Pier'Antonio Micheli, but he used Micheli as an industrial spy in German lands. Feigning weak-mindedness, Micheli hung around the workshops, was befriended by the guards and allowed to observe the procedures. When he observed a bit too closely, however, the guards caught on, and Micheli was lucky to escape with his life.[18]

More than the issue of industrial science, however, it is the fact that the most extensive technological involvements of scientists fell in traditions already long established that has caught my eye. Medical practice and pharmacology, military engineering, hydraulic engineering, navigation, and cartography – in all of them the application of science and mathematics to practical endeavours reached back at least to the fifteenth century and in most to the thirteenth. It might perhaps be suggested that our definition of the Scientific Revolution is mistaken and that we need also to extend

[16] The paper given by Dr Graham Hollister-Short at the conference in Oxford, at which this paper was originally presented, suggests that I need to modify this statement as far as Papin is concerned. Although he did not point to any documented connection, he argued for the role of Papin in the early history of the steam engine and expressed his conviction that Newcomen drew upon Papin's work.

[17] Josef Kulischer, *Allgemeine Wirtschaft Geschichte des Mittelalters und der Neuzeit* (2 vols., Berlin, 1954), 2: 175–7. Fernand Braudel, *Civilisation matérielle, économie et capitalisme* (3 vols., Paris, 1979), 2: 294–6.

[18] G. Targioni-Tozzetti, *Notizie della vita e delle opere di Pier'Antonio Micheli botanico fiorentino* (Florence, 1858), 78–80.

it back into the earlier centuries. There are reasons that I find compelling to reject the suggestion. The concept of the Scientific Revolution rests on the radical reordering of the understanding of nature that did in fact take place in the sixteenth and seventeenth centuries. I am convinced that there is no way to understand the history of science without the recognition of this reality, so that if we extend the concept of the Scientific Revolution to embrace the thirteenth century we will need to invent a new term to express the changes that took place in the sixteenth and seventeenth centuries. We would confront the same issue of science and technology, having altered nothing except the name.

Instrumentation

There was one other extensive category of technology that I have so far avoided discussing, a category that engaged a larger number of scientists than pharmacology, military engineering, hydraulic engineering, navigation, or cartography, a larger number than any category except medical practice. This was 'Instrumentation', the development of new scientific instruments. Given the manner in which I have defined the category, that last statement is not quite accurate, since new instruments used in surveying and navigation, of which there were quite a few, and in surgery are included. In any event, well over one scientist in every five (148, 22.7%) in my catalogue contributed to instrumentation. Scientific instruments properly speaking fall into virtually every domain of science – not just the well-known optical devices, the precision clock, the barometer, and the thermometer relevant to the physical sciences, but also such things as Swammerdam's fine scissors for dissecting insects and Ruysch's method of preserving anatomical specimens (which I classify with instruments). The beneficiary of this efflorescence of instruments, surely one of the defining characteristics, which assures us that the sixteenth and seventeenth centuries were indeed an age of Scientific Revolution, was not the economic system but science itself.

I am well aware that the perspective of my study is not the only perspective from which to view the important problem of science and technology. I do insist, however, that it is a valid perspective and an important one. To me it appears on the one hand to reveal a greater technological involvement of science than has frequently been admitted, and on the other hand unambiguously to call into question the proposition that the Scientific Revolution ushered in a new age in the relation of science to technology.

5

Mathematics and the craft of painting: Piero della Francesca and perspective

J. V. FIELD

To write about Piero della Francesca (*c.* 1412–1492) in the context of the history of science immediately exposes ambiguities in the use of the term 'Renaissance'. The humanist revival of the study of classical literature and the concomitant revival of interest in ancient art, both of which can be traced back to the late thirteenth century in the powerful works of the poet Dante Alighieri (1265–1321) and the painter Giotto di Bondone (1266–1337), are generally described by historians of art as constituting 'the Renaissance'.[1] Piero della Francesca's painting is seen as belonging to the fifteenth-century phase of this movement, known as the 'Early Renaissance', Historians of science, however, tend to use the term Renaissance to designate the sixteenth-century phase, the 'High Renaissance' to historians of art, which is the period in which the revival of interest in ancient mathematics and natural philosophy begins to gather momentum. Thus Piero della Francesca's mathematics and his optics would generally be described as 'late medieval'. As a historian of science, I find this characterization entirely reasonable: Piero's mathematics is clearly in the tradition of the manuscript 'abacus books', which derives from the *Liber abaci* of Leonardo of Pisa (*c.* 1170–1250); and Piero's concern with the science of vision seems to be related to John Pecham's *Perspectiva communis* (probably written between 1277 and 1279) rather than the more mathematical work of Witelo (Pecham's close contemporary) that was to gain favour in the sixteenth century.[2] On the other hand, Piero's use of linear perspective construction to obtain an effect of depth in his paintings, a technique he sees as derived from *perspectiva* proper, is undoubtedly characteristic of a style of painting

The problems considered here have been discussed from time to time with a number of art historians as well as with historians of science. In particular, I should like to thank Francis Ames-Lewis, Thomas Frangenberg and Martin Kemp. I am grateful to Rupert Hall for his helpful comments on an earlier draft of this article.

1 For example, George Holmes, *Florence, Rome and the Origins of the Renaissance* (Oxford, 1986) is largely concerned with the work of Dante and Giotto.

2 On Piero's use of Pecham see J. V. Field, 'Piero della Francesca's treatment of edge distortion', *J. Warburg Courtauld Inst.*, 1986, **49**: 66–99 and Plate 21c, especially pp. 80–81. As Lindberg has shown, there is considerable continuity to be found in optical work up to and including that of Kepler (1604), see D. C. Lindberg, *Theories of Vision from al-Kindi to Kepler* (Chicago, 1976).

that belongs specifically to the Renaissance. Thus Piero is using medieval mathematical know-how in the service of the Renaissance style of art, apparently in much the same way that, as Settle has suggested, Galileo seems to be using Renaissance engineering know-how in the service of the Scientific Revolution.[3]

It does not, however, seem advisable to attempt to make too much of this apparent similarity, for there is a very important difference between the two cases: Piero's work as a painter lies in the province of the craftsman (and the style of his mathematics, as we shall see, is that associated with the artisan) whereas Galileo is taking craft know-how into the realm of the natural philosopher. The significance of Piero's work lies rather in its showing the increasing use of mathematics in the crafts in the fifteenth century, which was associated – possibly as an effect rather than a cause – with the increased availability of elementary education in mathematics. This elementary education tended to concentrate upon arithmetic rather than geometry and it is to the elementary textbooks of the abacus schools that one must trace the rise of algebra to become a part of learned mathematics in the sixteenth century.[4] Moreover, the use of mathematics in the crafts seems to have encouraged thinking in three dimensions, as was required for many practical problems, rather than in the two-dimensional terms that characterize almost all the theorems of Euclid. Such habits eventually led to important changes within the learned tradition of geometry.[5] As I have suggested briefly elsewhere, it seems possible that these and other changes in the practice of mathematicians played some part in the extended use of mathematics in natural philosophy that characterizes the Scientific Revolution.[6] This practical mathematics may be seen as another element of Renaissance know-how that made a contribution to the Scientific Revolution; and it may be seen as a link between that Revolution and the humanist Renaissance.

Painting and mathematics

There is no immediately obvious link between Piero's paintings and his mathematics. It is by no means clear that he needed to be an expert mathematician to paint the way

[3] T. B. Settle, contribution to Oxford conference, 1990.

[4] The story of the rise of algebra, often taken as the whole history of mathematics in this period, has been told many times, usually relying heavily upon the classic H. G. Zeuthen, *Geschichte der Mathematik im XVI. und XVII. Jahrhundert*, ed. R. Meyer (Leipzig, 1903). For the earlier part of the story, including recent investigations of manuscript abacus treatises, see R. Franci and L. Toti Rigatelli, 'Towards a history of algebra from Leonardo of Pisa to Luca Pacioli', *Janus*, 1985, **72**: 17–82. See also C. Hay (ed.), *Mathematics from Manuscript to Print* [Proceedings of a conference held in 1984] (Oxford, 1988); and, for English translations of some interesting algebraic texts, J. Fauvel and J. J. Gray, *A Source Book in the History of Mathematics* (London, 1987).

[5] The history of geometry in this period has been much less studied than that of algebra. On the importance of the third dimension see J. V. Field, 'The natural philosopher as mathematician: Benedetti's mathematics and the tradition of *perspectiva*', in A. Manno (ed.), *Cultura, scienze e techniche nella Venezia del cinquecento*; *Atti del Convego internazionale di studio 'Giovan Battista Benedetti e il suo tempo'*, (Venice: Istituto Veneto di Scienze, Lettere ed Arti, 1987), 247–70; and J. V. Field and J. J. Gray, *The Geometrical Work of Girard Desargues* (New York, 1987).

[6] J. V. Field, 'Linear perspective and the projective geometry of Girard Desargues', *Nuncius*, 1987, **2** (2): 3–40, especially pp. 36–40.

Figure 5.1. Piero della Francesca, *The Baptism of Christ*, tempera on panel, 166 cm × 115 cm, probably 1440–45, London, National Gallery. The picture is the central panel of an altarpiece. (Reproduced by courtesy of the Trustees, The National Gallery, London.)

he did. What is chiefly visible is his mastery of the painter's craft. Small photographs inevitably fail to show the subtlety of Piero's use of colour, his delicate handling of the fall of light and his precise calligraphic treatment of details (see Fig. 5.1). However, they do, brutally, isolate some important elements that may perhaps be related to his skill as a mathematician, such as the clarity of the drawing of contours and, above all, the 'stillness' of composition for which his painting is well known. Piero tends to show any action at an apparently natural still point, and no line in the picture seems to hint at the possibility of its being, a moment later, in some other position. The smaller and uglier the reproduction, the more easily one is able to ascribe this effect to Piero's capacity for introducing quantities of subtle symmetries that provide a strong mathematical framework for the composition in the plane. For example a large number of regular geometrical figures have been suggested as underlying the composition of the *Baptism of Christ* (Fig. 5.1).[7]

Composition in the plane is not, however, all of the story, for Piero's paintings are characteristic of the Renaissance style in displaying apparently real objects and figures in space that is as apparently real as they are. This kind of naturalism compels the viewer to read the picture as a view of a real scene, thus introducing consideration of it as a composition in space as well as in the plane. The relative balance between these two kinds of composition naturally varies widely between painters (and indeed between individual works by particular painters). Piero's work is notable for his success in achieving a satisfactory relationship between the two kinds of composition. The unavoidable use of subjective words such as 'satisfactory' is an indication that this success is ultimately a measure of Piero's stature as an artist rather than as a mathematician, but one may discern a mathematical element in it to the extent that Piero's skill in constructing compositions that are easily readable in spatial terms presumably derives from his grasp of the relevant mathematics. To put it simply: he seems to have been a good enough mathematician to handle perspective with the same assurance with which he handled more traditional elements of painting.

Unfortunately, this recognition does not take us very far, for an examination of the use of perspective by artists in the fifteenth and sixteenth centuries shows that a totally convincing effect can be produced by works which are far from being mathematically correct. Moreover, it is only for relatively few works that a historian, working from the finished picture, or some reproduction of it, is able to check the correctness of the perspective. Such a check has to be carried out by looking at the shapes and dimensions given to objects whose 'true' shape and dimensions in the imagined space are known. In effect this means the would-be reconstructor of the perspective scheme generally relies upon elements of classicizing architecture, such as square coffers in a ceiling, or square tiles on a floor.

Piero has left one example of the reconstructible so well-adapted to the purpose that it has been suspected of having been designed as an exercise in correct perspective. The picture in question is his *Flagellation of Christ* (Fig. 5.2). The scene to which the

[7] B. A. R. Carter, 'A mathematical interpretation of Piero della Francesca's *Baptism of Christ*', appendix in M. A. Lavin, *Piero della Francesca's 'Baptism of Christ'* (New Haven and London, 1981), 149–63.

Figure 5.2. Piero della Francesca, *The Flagellation of Christ*, tempera on panel, 59 cm × 81.5 cm, date not known, Urbino, Galleria delle Marche. (Photograph courtesy of the Gallery).

modern title of the picture refers is apparently presented as a subsidiary element in the picture as a whole, being set far back from the picture plane, with human figures a little less than half as high as the three shown in the foreground. The original title of the picture is not known, and there is considerable dispute among art historians concerning the identity of the three foreground figures, which is very probably crucial to an identification of the subject of the picture as Piero conceived it.[8]

As with the *Baptism* (Fig. 5.1), our monochrome reproduction of the *Flagellation* serves to draw attention to the mathematical elements in the work. Even by the standards of the time (which liked its art to be pretty), the original picture, most of whose surface is well preserved, is notable for the delicacy of its finish and the brilliance as well as the subtlety of its colour. However, sharp contrasts are used in colouring the architecture, so the viewer is certainly intended to see it as important and to note how it marks out the spatial organization of the scene that is shown. It is immediately clear that in the *Flagellation* Piero is making a claim to mathematical exactness that goes somewhat beyond the naturalistic treatment of spatial relationships that we see in the *Baptism* (and many other pictures of the time). Since all Piero's pictures are well organized, it seems likely that this particular highly visible claim to exactness relates in some way to the true subject of the *Flagellation*. In any case, as Wittkower and Carter showed in 1953, Piero has supplied sufficient information to make it possible to reconstruct the ground plan of most of the scene (the buildings on the right elude exact reconstruction because we cannot see their lower edges).[9] Wittkower and Carter's reconstruction is entirely convincing and proves beyond reasonable doubt that Piero's grasp of the mathematics of perspective really was as good as it looks. There is, however, a small departure from mathematical correctness, which in the circumstances must be taken as deliberate: the broad strip of white marble that runs into the picture under the bases of the row of columns (and, with the nearest column, serves to divide the part of the picture plane containing the background scene from that containing the foreground one) has been made slightly narrower than mathematics would dictate.[10]

On this occasion, Piero presumably resisted dictation because he thought the picture looked better with a narrower band of white, and no doubt guessed that no ordinary viewer would detect this tiny departure from correctness. On at least one other occasion he resisted the demands of naturalistic perspective in a much more radical manner. The picture in question in his *Resurrection of Christ*, a fresco painted on the wall of the council chamber in the town hall of his native city of Borgo San

[8] For possible interpretations of the picture and references to the huge literature on it, see M. A. Lavin, *Piero della Francesca: 'The Flagellation'* (New York, 1972, second edition Chicago, 1990, with additional bibliography to 1986). For a reinterpretation of the subject of the picture as 'The Dream of St Jerome' see J. Pope-Hennessy, 'Whose Flagellation?', *Apollo*, 1986, **124**: 162–5 and, much more briefly, in the same author's *The Piero della Francesca Trail* (London, 1991). It should, however, be pointed out that in this short book Pope-Hennessy, who shows himself notably unimpressed by exotic historical interpretations of Piero's pictures, apparently accepts without demur the erroneous mathematics of Elkins (note 32 below) and the elaborate mathematical theorizing of Carter (*op. cit.* (note 7)).

[9] R. Wittkower and B. A. R. Carter, 'The perspective of Piero della Francesca's "Flagellation"', *J. Warburg Courtauld Inst.*, 1953, **16**: 292–302. The ground-plan found by Wittkower and Carter is reproduced in Lavin *op. cit.* (note 8), 32 and in M. H. Pirenne, *Optics, Painting and Photography* (Cambridge, 1970), 75.

[10] Wittkower and Carter, *op. cit.* (note 9).

Figure 5.3. Piero della Francesca, *The Resurrection of Christ*, fresco, 225 cm × 200 cm, probably painted in the 1460s, Sansepolcro, Pinacoteca (former Town Hall). (Photograph courtesy of the Committee for commemorating the 500th anniversary of the death of Piero della Francesca.)

Sepolcro (now called Sansepolcro). This painting, shown in Fig. 5.3, is large – the figures are about life size – and the loss of both size and colour in the reproduction rob it of the qualities one notices first in the original: solemnity and grandeur. Few pictures of any time have achieved an effect as imposing as Piero achieves in the *Resurrection*. The photograph does, however, put one in a good position for examining

Piero's use of perspective. Here we have no simple geometrical shapes to allow us to calculate the ideal viewing position and establish a ground plan, but the sense that we are looking up at the sprawling figures of the soldiers fits well with the picture's actual position on the wall, which is some way above eye level. Nevertheless, a glance at Christ's foot introduces doubt. We see the upper surface of the foot, so our eye must be higher, perhaps on a level with those of the soldier on the right whose head cuts across the edge of the drapery falling from Christ's arm. But there is little indication that we are seeing Christ himself from below. Indeed, we seem to be seeing Him straight on. Though the figure is convincingly three-dimensional, the non-naturalistic perspective in which it is seen, in contrast to the soldiers below, gives it an unearthly quality which is in accord with the religious meaning of the scene.

The use of more than one viewpoint in this picture has been noticed many times before, and can be paralleled in other fifteenth-century pictures.[11] It is significant for our present purposes as indicating very clearly that, despite his concern with mathematics, as a painter Piero is no more a candidate for a diploma in technical drawing than the next man. Like other painters, he generally relies on simple geometrical shapes as a means of establishing spatial relationships, and occasional displays of virtuosity in the use of perspective should not be mistaken for commitment to its rules as having overwhelming importance in the making of pictures.

Piero's treatises

One can speak with this degree of certainty about Piero's use of perspective because his mathematical competence is well attested not only in his practice as a painter but also by his reputation in his own time. In his Life of Piero, Giorgio Vasari (1511–1574), who was a native of Arezzo and thus may have had access to local information, tells us that from his earliest years Piero was interested in mathematics, and might indeed have made it his living, had he not decided to become a painter.[12] Vasari then goes on to say that Piero wrote 'many' mathematical treatises. Of these, three have now been rediscovered: *Trattato d'abaco*, *Libellus de quinque corporibus regularibus*, *De prospectiva pingendi*. All are available in modern printed editions.[13]

The *Trattato d'abaco* ('Abacus treatise'), as its name implies, is modelled on the texts used in abacus schools, which taught the kind of mathematics that was expected

[11] For instance Andrea Mantegna (*c.* 1430–1506) *Virgin and Child with Sts John and Mary Magdalene* in the National Gallery, London, where the framing saints are seen from below while the central group is seen straight on. The picture probably dates from the last quarter of the fifteenth century.

[12] Giorgio Vasari, *Le Vite dei piu eccellenti architetti, pittori e scultori italiani* . . . (Florence, 1550, 1568), new edn, P. Barocchi and R. Bettarini (Florence, 1971).

[13] Piero della Francesca, *Trattato d'abaco*, ed. G. Arrighi (Pisa, 1970); Piero della Francesca, 'L'Opera "De corporibus regularibus" di Pietro Franceschi detto della Francesca, usurpata da Fra' Luca Pacioli', ed. G. Mancini, *Mem. R. Accad. Lincei*, 1916, series 5, **14**, fasc. 8B: 441–580; Piero della Francesca, *De prospectiva pingendi*, ed. G. Nicco Fasola (Florence, 1942), 2nd edition with additional material and a bibliography up to 1984, ed. E. Battisti *et al.* (Florence, 1984). New editions, based on additional manuscript sources, will be published under the editorial supervision of Marisa Dalai Emiliani, Cecil Grayson and Carlo Maccagni.

to prove useful in commercial activities, that is mainly what was then known as arithmetic but would now partly be described as elementary algebra. Although conventional in its arrangement and style, Piero's *Trattato* seems to have been written for a friend, rather than for use in a school. Most of its arithmetical and algebraic problems can be found in earlier treatises, but the geometrical problems, of which there is an unusually large number, show considerable originality, notably in dealing with three-dimensional figures. In particular, Piero gives descriptions of two of the polyhedra whose discovery Pappus (fifth century AD) ascribes to Archimedes (*c.* 287–212 BC). Pappus' descriptions merely list types and numbers of faces. For instance, one solid is said to have eight triangular faces and six square ones. Piero mentions no source for the bodies he discusses, and his diagrams, which effectively show an elevation of the polyhedron with its vertices inscribed in a sphere, add considerably to what Pappus says. They show how the polygonal faces fit together to form the solid, each vertex being surrounded by faces of the same kind arranged in the same way. For instance, the solid with eight triangular faces and six square ones has vertices at each of which four polygonal faces meet, two square faces being separated by two triangular ones. Since all the vertices of the solid are alike, symmetry dictates that they must lie on a sphere. Piero clearly recognized this, since he showed each of the two solids with its surrounding sphere. There seems to be no way of knowing how Piero came to his rediscovery of these two Archimedean solids, but since it clearly involved the exercise of visual imagination it seems highly appropriate that the rediscovery should be associated with an artist.[14]

The *Libellus de quinque corporibus regularibus* ('Short book on the five regular solids') was probably written in the last decade of Piero's life. The one known manuscript is in Latin, but may be a translation from the vernacular. The work is original in being concerned only with geometry, but many of its problems are borrowed from the *Trattato*, the later version sometimes being slightly more elaborate. The first part of the *Libellus* deals with plane figures; the second considers the five regular polyhedra individually, proposing problems that are essentially numerical versions of the purely geometrical problems in Book 13 of Euclid's *Elements*; the third part deals with inscribing one regular polyhedron in another, the vertices of the inner figure usually coinciding either with selected vertices of the outer one or with centres of its faces. Most of these problems are adaptations of those in *Elements*, Book 15.[15] Apart from a final short series of problems on the sphere, and a few miscellaneous problems, the fourth section of Piero's *Libellus* considers bodies

[14] The six solids Piero rediscovered are those now known as the truncated dodecahedron, the truncated icosahedron, the truncated cube, the cuboctahedron, the truncated octahedron and the truncated tetrahedron. The names are due to Johannes Kepler (1571–1630) who described, and illustrated, all thirteen Archimedean solids in *Harmonices mundi libri V* (Linz, 1619), Book II, reprinted in *KGW* 6. Kepler may have known Piero's work through its appearance, in Italian translation, in Pacioli's *De divina proportione* (Venice, 1509), see below. I am happy to accept the implication that Kepler had a good visual imagination since the gift seems indispensible for his work on the orbit of Mars, in which one has to hold the changing configurations of Sun, Earth and planet in the mind's eye.

[15] The ascription of Books 14 and 15 to Euclid began to be questioned in the Renaissance. They are now believed to be later additions to the *Elements*. However, they were included in all the versions of the *Elements* Piero was likely to know, such as the translation by Campanus.

which, while they can be inscribed in a sphere so that their vertices all touch it, nevertheless have faces of more than one shape. Five of the bodies concerned are Archimedean polyhedra, one being the truncated tetrahedron, already described in the *Trattato*.

Neither Piero's *Trattato d'abaco* nor his *Libellus de quinque corporibus regularibus* was printed under his name in the Renaissance. However, most of the *Trattato*'s problems found their way into Luca Pacioli's hugely successful textbook *Summa de arithmetica* . . . (Venice, 1494) and an Italian version of the *Libellus* appeared as the second part of Pacioli's *De divina proportione* (Venice, 1509). The work includes illustrations of polyhedra drawn, as Pacioli tells us in the preface, 'by the divine left hand of my friend Leonardo of Florence'.[16] Vasari, who lived in a time when intellectual property rights were beginning to be recognized as important – and who, as a friend of Michelangelo (1475–1564), was no friend to Leonardo da Vinci (1452–1519) – clearly regarded Pacioli as having stolen Piero's work. This is probably unfair to Pacioli. For instance, as we have already noted, almost all the problems in the *Trattato* can in fact be traced to previous sources. As the modern editor of the work makes clear, all Piero can really be credited with, in regard to the algebraic part of the treatise, is a degree of clarity and order in his exposition.[17]

'On perspective for painting'

This judgement echoes the terms in which a competent mathematician of the sixteenth century, Egnazio Danti (1536–1586), wrote of Piero's treatise on perspective in 1583. Danti expresses his regret that no text on perspective has come down to us from the Ancients, and then says

> But in our own time, among those who have left something of note in this Art, the earliest, and one who wrote with the best method and form, was Messer Pietro della Francesca del Borgo Sansepolcro, from whom we have today three books in manuscript, most excellently illustrated.[18]

As Danti went on to note, some of Piero's treatise had by then been published, with acknowledgement to the original author, in Daniele Barbaro's *La Pratica della perspettiva* (Venice, 1569).[19]

Despite its Latin title, Piero's treatise on perspective was written in the vernacular, though a Latin version is known.[20] There can be little doubt that the work was intended to be instructional, addressed to an apprentice: the reader is called 'tu' and is

[16] A modern edition of *De divina proportione* (Milan, 1956) includes 'facsimiles', i.e. very good colour photographic reproductions, of Leonardo's beautiful coloured drawings, which appeared in Pacioli's book in the form of woodcuts.

[17] See Piero della Francesca, ed. G. Arrighi, *op. cit.* (note 13).

[18] G. Barozzi da Vignola (1507–73), *Le Due regole della prospettiva pratica*, ed. and with commentaries by E. Danti (Rome, 1583), Preface (by Danti).

[19] For an account of Danti's history of perspective, see J. V. Field, 'Giovanni Battista Benedetti on the mathematics of linear perspective', *J. Warburg Courtauld Inst.*, 1985, **48**: 71–99.

[20] One copy is in the British Library, London, Add MSS 10,366.

given detailed instructions on how to draw the diagrams. Almost throughout, the style is exactly that of the vernacular textbooks used in abacus schools, and of Piero's own *Trattato d'abaco*. That is, Piero proceeds by series of worked examples. A similar style was employed in the training of apprentices in artists' workshops, where they were taught to draw by being made to copy series of drawings. Each workshop seems to have had a manual of such drawings. The manuals were, of course, not intended for publication, but substantial traces of them survive in elements found repeated in picture after picture produced by particular workshops. Thus, although Danti may well be correct in identifying Piero's perspective treatise as the first of its kind (and no earlier example is known to survive) the work is not without clear antecedents.

The descendants of *De prospectiva pingendi* were legion. All subsequent treatises on perspective addressed to painters follow more or less faithfully the pattern laid down by Piero. Yet *De prospectiva pingendi* did not find its way into print. The reason seems to have been that artists found the treatise too difficult. It is probably significant that Daniele Barbaro's publication of part of the work dates from the second half of the sixteenth century, by which time the general level of mathematical education was much higher than it had been in Piero's time, and was in a book which is addressed not to artists but to their employers or patrons. We must thus be doubly cautious in reading Piero's treatise. First, as in reading any instructional work, particularly one that considers itself to be elementary, we must bear in mind that the author is not necessarily giving a full exposition, but may be intending merely to lay down useful rules. Second, we must remember that Piero's treatise apparently did not prove useful to artists in its original form, that is, it is probably not a reliable guide to the practice of painters other than its author and, contrariwise, looking at the work of other artists may not help us to understand Piero's text.

Although all three of Piero's mathematical treatises are in the style associated with the books used to teach practical mathematics in abacus schools, it seems that the only one of the three that was in fact intended to be used for teaching was the treatise on perspective. It is thus somewhat ironic that the most noticeable difference between *De prospectiva pingendi* and the two other mathematical works should be one which gives it a superficial resemblance to learned works on geometry: because the treatise is intended to teach an apprentice to draw, Piero proposes the problems in purely geometrical terms. For instance, the nineteenth problem in Book I is

> In the square plane shown in perspective, draw an equilateral hexagon.

In the *Trattato* and in the *Libellus*, geometrical problems are almost always given in a numerical form, as in this problem from the former, relating to one of the Archimedean solids

> There is a sphere and its diameter is 6; we want to put in it a body with eight faces, four triangular and four hexagonal. I want to know its side.[21]

The logical structure of *De prospectiva pingendi* resembles that of Piero's treatise on the five regular solids. Both works start with plane figures, pass on to relatively simple

[21] Piero della Francesca, ed. G. Arrighi *op. cit.* (note 13), 230. The answer is given in the form 'the square root of 6 and 6 elevenths'.

three-dimensional figures and then to more elaborate bodies.[22] One point of difference between the works has already been noted, namely that the perspective treatise does not use numerical examples and therefore gives the impression of being more concerned with pure geometry of the kind found in the *Elements*. The other difference tends to reinforce this misleading impression: the perspective treatise is much more self-contained, and, in the manner of the *Elements*, tends to refer back to its own earlier propositions. The explanation for this difference lies partly in the subject matter of the treatise – the work on perspective having no predecessors to refer to – and probably also partly in the social context in which the works would be used – the work on the regular bodies being addressed to educated and relatively affluent readers who would have access to a collection of mathematical texts, while the work on perspective would be the indigent apprentice's only textbook. It was presumably because abacus books were addressed to this latter kind of readership that they are so often mere compendia of examples assembled from earlier books of the same kind. Thus the self-containedness of Piero's work on perspective should be seen as an indication of its social closeness to the abacus book as much as its intellectual closeness to the *Elements*. Which said, one must nevertheless admit that Piero has to some extent modelled his work on that of Euclid.

Unlike Euclid, however, Piero begins his treatise with a discursive introduction. He explains, for instance, that what follows is not an account of the whole of painting but only of a part of it.[23] Moreover, each of the three books of the treatise has a short introduction summarizing its contents.

In Book I, after some preliminaries, which are clearly derived from Euclid's *Optics* but do not go into enough detail to allow one to decide whether he is using Euclid's text or Theon's recension, Piero deals with plane figures, beginning with a square, then subdividing it, then drawing various regular polygons within it. Several of the patterns Piero constructs are possible designs for elaborate tiled floors of the kind found in many contemporary paintings, for instance, Piero's own *Flagellation* (Fig. 5.2). In Book II some of the plane figures of Book I become ground plans of simple three-dimensional objects, such as the shaft of a fluted column (idealized as a polygonal prism), a house (idealized as a cube) and a polygonal well-head. The work is not, however, unrealistically simple. For instance, the house is not shown merely as a plain cube. Openings are made for doors and windows, and the thickness of the house wall is visible in the openings.[24] In Book III, Piero deals with more complicated shapes, such as the bases and capitals of columns, and human heads seen from various angles. His method is to choose series of points which define the shape, and then to find the perspective image of each of these points.

[22] Any reader who is inclined to regard this progression as so natural as to be inevitable should look at the treatment of the regular solids in, say, Albrecht Dürer, *Underweysung der Messung mit dem Zirkel und Richtscheyt* (Nuremberg, 1525) or Wentzel Jamnitzer, *Perspectiva corporum regularium* (Nuremberg, 1568).
[23] Piero della Francesca, ed. G. Nicco Fasola *op. cit.* (note 13), 63. This is the first line of Piero's text.
[24] Piero della Francesca, *ibid.*, Book II, proposition 9, 113–8.

Each example begins with a heading, corresponding to the statement of what it is required to prove at the beginning of each proposition in, say, Euclid's *Elements*. There follow extremely detailed instructions on how to draw the diagram, first the original 'perfect' figure and then its 'degraded' form as seen in perspective. For the huge majority of Piero's propositions, these drawing instructions cannot really be said to constitute a proof of the correctness of the result to which they lead. Piero does not tell the reader how any of the steps is related to the underlying theory of perspective. That is, there is no equivalent to Euclid's referring the reader back to the earlier results being used in each new proof. However, in much of the treatise the examples can be seen as forming logical sequences, for instance in the increasingly complicated subdivision of the square in Book I. The pattern of exposition by means of worked examples is thus like that in many treatises on algebra. The examples have, of course, been chosen both to be typical of the kind met with in real life by the prospective practitioner and to be as realistic as possible. The result, in *De prospectiva pingendi*, can often be long series of drawing instructions whose repetitiousness is mind-numbing. Moreover, despite these detailed instructions, Piero seems to have intended the text of each proposition to be accompanied by an appropriate figure. These were almost certainly to be drawn by someone other than the scribe who wrote the text, and in some manuscripts of the work the series of drawings is incomplete. None the less, the drawings must be taken as integral to the text, and it is therefore of interest that in the manuscript whose drawings are of the highest quality, and therefore presumably closest to Piero's original illustrations (perhaps even direct copies of them), many of the figures appear to correspond closely with elements found in Piero's surviving paintings.[25]

Shafts and capitals of columns are perhaps too standard for comparison to be worthwhile, and although the house already mentioned agrees in almost every detail with one of those shown on the right in the scene of the raising of the young man from the dead by the True Cross in the fresco cycle at Arezzo (Fig. 5.4), it may still be regarded as too banal to be treated as evidence. However, the head of the man holding the Cross shown in the same fresco scene provides a much more striking example. In his treatise, Piero first provides front and side elevations of a head, together with several sections (Fig. 5.5). Then he numbers appropriate series of points, finds the perspective image of each point and draws the complete perspective image of the head. This process is carried out first for a head seen straight on, but the next image is for a head seen a little from underneath (Fig. 5.6) and the similarity to the head shown in the fresco is very marked. Piero goes on to show a head seen more sharply from below, and one seen both from below and slightly to one side (Fig. 5.7). These views correspond rather closely to those of the heads of the sleeping soldiers in the *Resurrection* (Fig. 5.3).

Since our reading of the spatial arrangement in the *Resurrection* depends strongly on the human figures concerned, it is not unreasonable to suppose that Piero should

[25] The manuscript concerned is now in Parma (Biblioteca Palatina MS no. 1576).

Figure 5.4. Piero della Francesca, *The Story of the True Cross* (detail), fresco, height of register 356 cm, probably painted 1456–60, Arezzo, San Francesco. (Photograph Alinari.)

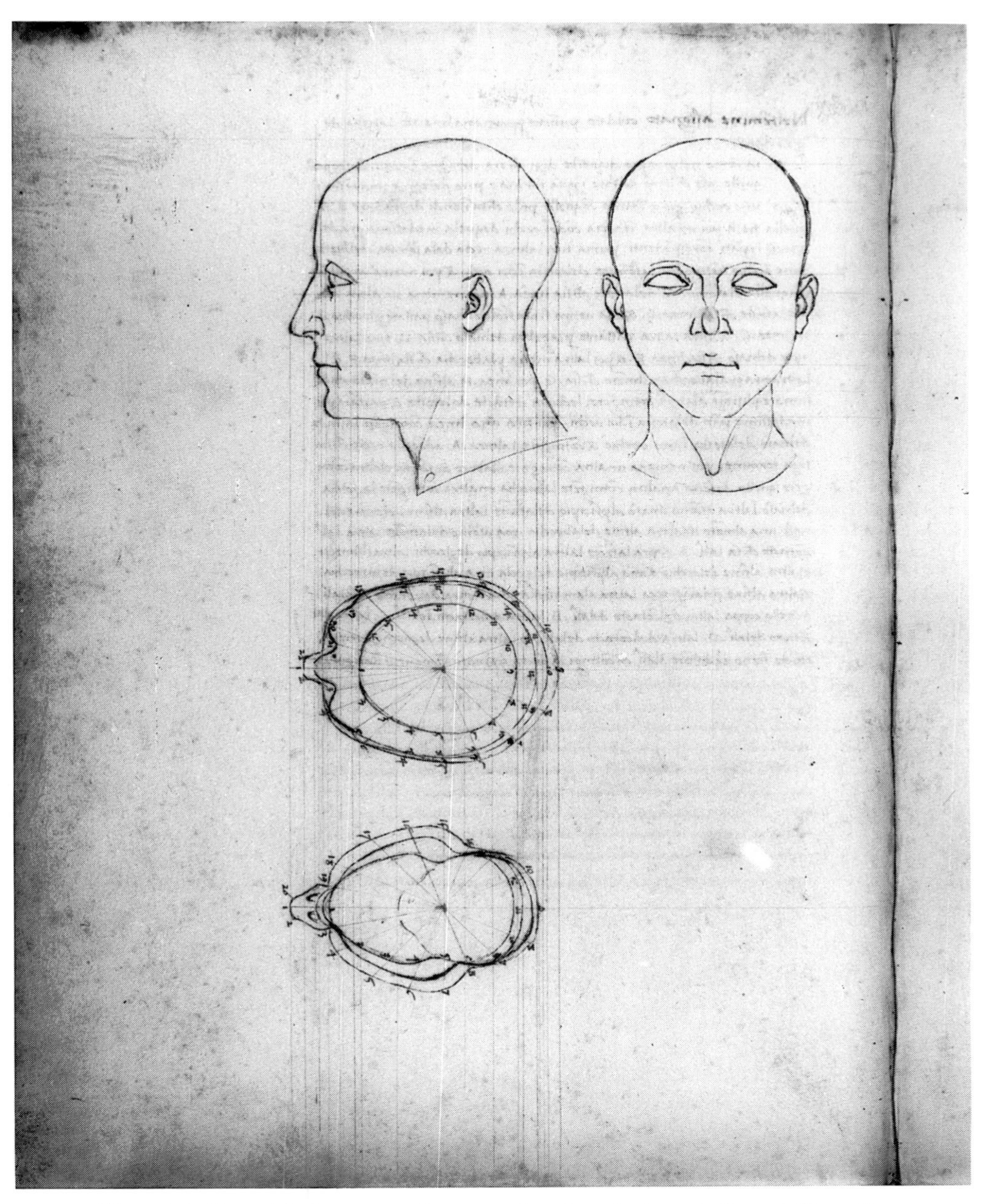

Figure 5.5. Piero della Francesca, *De prospectiva pingendi*, Parma MS no. 1576, f. 59 *verso*.

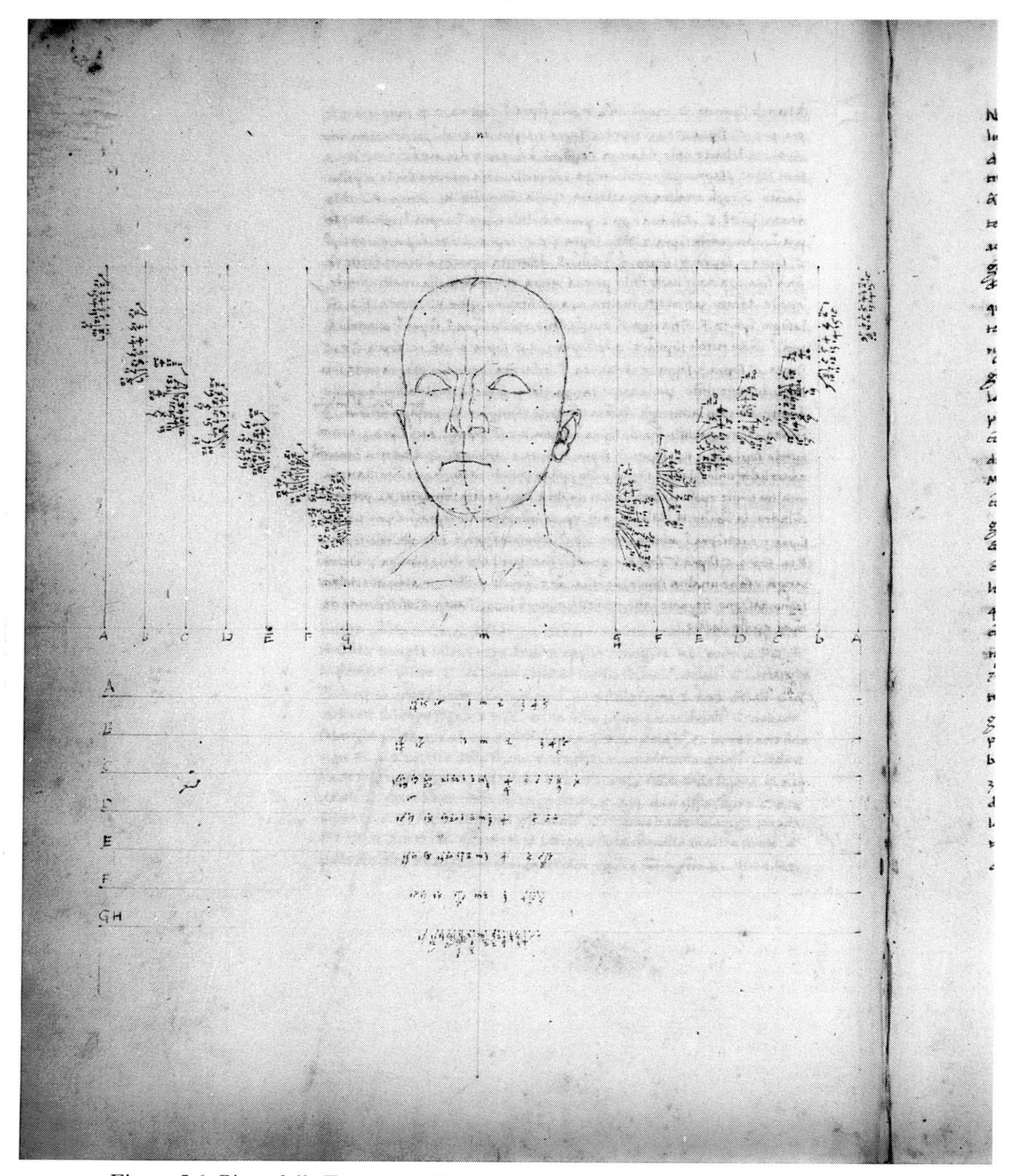

Figure 5.6. Piero della Francesca, *De prospectiva pingendi*, Parma MS no. 1576, f. 66 *verso*.

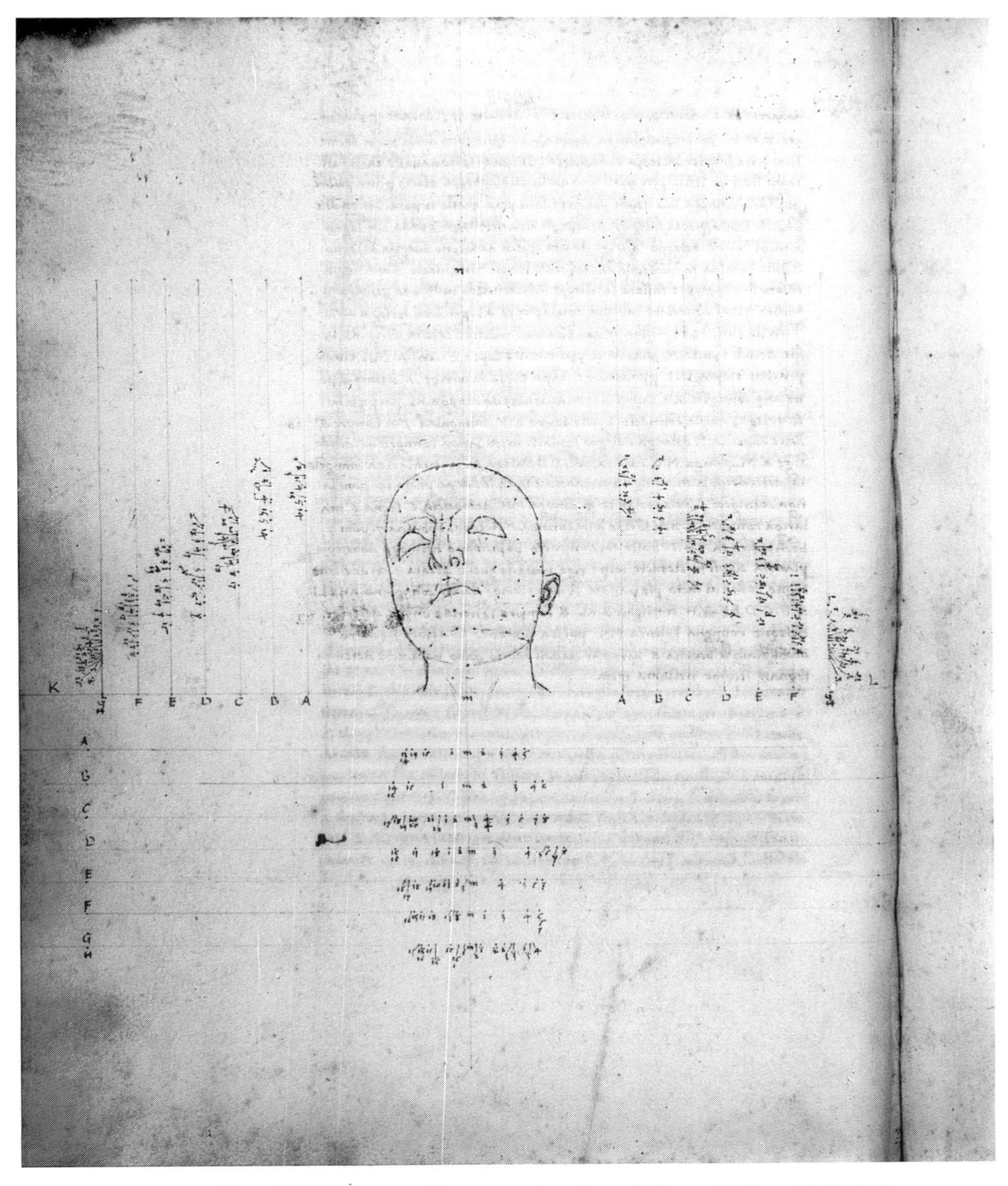

Figure 5.7. Piero della Francesca, *De prospectiva pingendi*, Parma MS no. 1576, f. 76 *verso*.

have taken care to draw the sprawling soldiers in accurate perspective, as a guarantee that the figures would carry conviction. Thus, to read the piece from Piero's own point of departure, the overall design of the fresco is a consequence of his decision to use his mathematical skill in the figures.

This degree of reliance on elaborate mathematical exercises is not usual in Piero's painting. For instance, in the *Baptism* (Fig. 5.1), painted early in Piero's career, our sense of space depends partly on the rendering of colour and the flow of light and partly on the relative sizes of various repeated elements shown in the landscape, such as trees and human figures. The mathematics involved is a matter of series of proportionalities.[26] Moreover, in both the *Flagellation* (Fig. 5.2) and the fresco showing the raising of the man from the dead by the True Cross (Fig. 5.4), the sense of space depends very strongly on mathematically simple elements. In the *Flagellation* we have the classical architecture with its rectilinear repeated elements and regular tiling patterns, together with human figures of appropriate related sizes. In the fresco from the cycle of the *Story of the True Cross*, the space within which the action takes place is defined by the house on the right and the classicizing façade of the building behind the figures. The shape of the group of figures round the bier is defined by the foreshortened image of the Cross. Thus the perspective illusion in both of these pictures essentially depends upon the relatively simple mathematics of Books I and II of *De prospectiva pingendi* – though, curiously enough, Piero's worked examples do not include showing a cross in perspective (perhaps because the shape is so simple that it can easily be developed from one of the pavement patterns in Book I?).

Despite this reliance upon simple elements, in which Piero's use of perspective resembles that of most of his contemporaries, there can be little doubt that Piero did carry out detailed perspective constructions for more complicated elements. The heads of the figures kneeling round the bier (Fig. 5.4) show the characteristic pattern of small black dots obtained when a drawing is transferred to the plaster by the method known as 'spolvere', in which a series of small holes is made round the outlines and the drawing is transferred by patting with a pad filled with powdered charcoal. The conical headdress worn by the Empress Helena, which appears in several pictures in the series, also shows signs of the use of 'spolvere'. Moreover, the elaborate turban of the spectator turning his back to us in the *Flagellation* carries 'spolvere' marks as well. Here Piero's concern is presumably not with perspective but merely with accurate transfer of a preliminary drawing – possibly even one worked up from a sketch of a real turban. The size of the turban in the finished picture is about 35 mm across.[27] Faced with this degree of concern for pictorial detail in a context of mathematical illusionism one can hardly avoid concluding that if Piero's pictures look 'right' the explanation is that they are indeed mathematically correct. Art hides art,

[26] This method of construction can be used to achieve perspective effects much more elaborate than those in Piero's *Baptism*. See M. J. Kemp and A. Massing, 'Paolo Uccello's "Hunt in the forest" ', *Burlington Mag.*, 1991, **133**: 164–78.

[27] See Lavin *op. cit.* (note 8).

but we have in fact every reason to suppose that the mathematical techniques Piero describes in his treatise on perspective were actually used in his pictures. Indeed, the illustrations in *De prospectiva pingendi* may reflect preliminary drawings made for particular paintings.

Mathematical theory

The connections that can be shown to exist between Piero's treatise and his practice place the treatise very firmly within the practical tradition. None the less, *De prospectiva pingendi* does show traces of Piero's wider mathematical interests and, in particular, an interest in the mathematical foundations of perspective construction.

The introduction to Book I is followed by definitions of a point and of a line. Piero cites the correct geometrical definitions and then says that since such entities can be seen only by the intellect, while he intends to draw visible diagrams, what he will mean by the terms will be different.

> So I shall say a point is something as tiny as it is possible for the eye to take in; the line I shall say is extension from one point to another, whose thickness is of the same nature as that of the point.[28]

After defining a 'surface', that is an area, and giving some examples of the shapes an area may have, Piero proceeds to some optical preliminaries

> Each quantity presents itself as subtending an angle at the eye. . . .
> All segments seen subtending the same angle, even if they are in different positions, present themselves to the eye as equal. . . .[29]

The first few such preliminaries are specifically optical and are either taken directly from Euclid (or Theon) or slightly adapted. There follow some results which omit references to the eye, and then some which refer to the division of areas by transversals of various kinds (and are clearly preliminaries to the division of an area into the pavement patterns we find later in Book I). For some of the propositions Piero gives an abstract geometrical proof and then adds 'Alternatively, do it with numbers . . .' and goes on to give a numerical equivalent.[30] This second form of proof suggests that the reader may be expected to check the result for himself by drawing an accurate diagram. As we shall see, Piero seems to have had this form of proof in mind in some later sections of his treatise.

The first proposition directly related to perspective construction appears in the twelfth section of Book I and is followed by the fundamental propostion that introduces the method of construction Piero uses in the rest of Book I

> [XIII] To draw a square in perspective
> As in the preceding, let .DC. be a line divided in the point .B. and let .BF. be drawn perpendicular [to it] . . .[31]

[28] Piero della Francesca, *De prospectiva pingendi* Book I, *ed. cit.* (note 13), 65.
[29] Piero della Francesca, *ibid.*, 66.
[30] See, for example, Piero della Francesca, *ibid.*, I.[V], 68.
[31] Piero della Francesca, *ibid.*, 76.

In the Parma manuscript, the proposition covers a folio page. The only way to follow the proof is to follow the drawing instructions step by step. If one does so, the resulting diagram turns out to be considerably more complicated than the one supplied in the Parma manuscript (and in most of the others). Indeed, if one tries to follow the proof from the diagram supplied one finds that lines to which Piero refers are missing. In any case, one is faced with the uncomfortable fact that Piero gives the same letter to two different points, and does so for three separate pairs of points. As a result, parts of the proof are ambiguous. Closer inspection reveals that when a statement is ambiguous *both* versions are in fact true, and that the ambiguity has allowed Piero to shorten his proof. Fifteenth-century mathematicians may have been adept at dealing with two pairs of similar triangles at the same time, but it is rather hard to believe that the apprentice to whom most of the treatise appears to be addressed would have made much headway with this section of the work.[32] Piero presumably put it in because he recognized the theorem as being fundamental.

There are two other places in *De prospectiva pingendi* where Piero again gives something more like a general theorem than a worked example. The first is at the end of Book I and the second is at the end of Book II. In both propositions Piero is seeking to defend the use of perspective construction against the charge that it leads to an absurdity and therefore cannot be legitimate. The proposition at the end of the first Book puts the matter explicitly, saying that some people claim that perspective cannot be a 'true science' (*vera scientia*) because it can make a foreshortened edge appear longer than it is in reality. Piero shows that this problem of distortion is chimerical: mathematically, it can only occur if the angle the picture subtends at the eye is more than a right angle and this case is physically impossible since the eye cannot actually take in more than a right angle.[33] Piero's proof, which I have discussed at length elsewhere, is mathematically incorrect – inevitably so, since the theorem he is trying to prove is in fact untrue.[34] His 'proof' is, however, adequate in the sense that if one drew an accurate diagram the result would almost certainly appear to be correct, and the actual procedure he recommends to avoid 'distortion' would achieve its end. It is, also, of interest that a crucial part of the proof depends upon perspective construction – 'perspective for painting', *prospectiva pingendi* – being regarded as an extension of the science of vision, *perspectiva* proper. This accords well with Piero's apparently rather scientific attitude to the correct rendering of appearances in his pictures.

The awkwardness that Piero is concerned to clear up at the end of Book II is the rather similar example of 'distortion' in the perspective rendering of a row of columns placed parallel to the picture plane. Piero shows, with some sleight of hand in ignoring inconvenient sines and cosines, that the columns at the edge of the picture will not appear thicker than those near the centre. The theorem, unlike that at the end of Book I, is in fact true, but the proof is not mathematically correct. Piero has shown a

[32] A twentieth-century art historian has even gone so far as to claim that Piero's proof is incorrect, see J. Elkins, 'Piero della Francesca and the Renaissance proof of linear perspective', *Art. Bull.*, 1987, **69** (2): 220–30.
[33] Piero della Francesca, *De prospectiva pingendi*, I.[XXX], *ed. cit.* (note 13), 96–9.
[34] See Field, *op. cit.* (note 2).

diagram which does not correspond exactly to what is required and permits a simpler proof. Given his known mathematical competence, we must presume he did this knowingly. It is not clear whether his prospective readers were expected to forgive the adjustment or fail to notice it. This is, of course, the same ambiguity that faces us in relation to Piero's occasional departures from correct perspective in his pictures.

Conclusion

De prospectiva pingendi was written as a contribution to the practical tradition, to be a workshop manual to train an apprentice to draw a useful selection of objects in correct perspective, but it also shows signs of its author's interest in the learned tradition. As in the unusual amount of geometry in the *Trattato*, and in the original nature of the *Libellus* in being concerned almost entirely with a subject otherwise treated only by Euclid, we find in the treatise on perspective, though in more explicit form, evidence of Piero's concern with mathematical matters for their own sake.

What we find cannot easily be related to the *quadrivium* as a whole. We have traces of interest in geometry, and some arithmetic, and in the 'mixed science' of *perspectiva* (seen as belonging to natural philosophy but as capable of being studied in mathematical terms). From the point of view of someone with university training in mathematics, Piero was making non-standard applications of arithmetic in algebra and of geometry in perspective construction instead of the standard ones in the mathematical sciences of Music and Astronomy. In this sense, Piero was not a 'scholar'. Moreover, it is not certain that he read Latin. None the less, with historians' hindsight, it is clear that in an important sense Piero should be seen as a scholar: he took an interest in the theory of his craft and wrote a treatise designed to pass his skill on to succeeding generations.

Since, in the event, the treatise fulfilled its purpose only indirectly through the works of others, Piero's credentials as a scholar are in need of a little defence. His credentials as a craftsman are, however, established beyond question. Indeed, it is Piero's high visibility as a painter that makes him such a useful case of the scholar-craftsman. Painting is a craft that has received long years of scholarly attention from critics and historians. In reintegrating the image of Piero – in the same sort of way that modern scholarship has reintegrated our understanding of the multifarious activities of Leonardo[35] – it is the historians of mathematics and of the mathematical sciences whose contribution is visibly deficient.

Remedying this deficiency is of interest not only as aiding understanding of the intellectual map of an important historical period but also in the more specific matter of understanding the part played by mathematics in the development of natural philosophy. The mathematics of the practical tradition shows many of the characteristics that historians of science associate with the mathematics imported into natural

[35] See, for example, M. J. Kemp, *Leonardo, the Marvellous Works of Nature and Man* (London, 1981).

philosophy from the late sixteenth century onwards. For instance, there is a total absence of discussion of the philosophical problems that Aristotelian philosophers generally saw as besetting the use of mathematics. No prospective merchant seems to have worried out loud whether the ducats in the real world behaved in the same way as those in his abacus book. This silence may or may not cover a reflection as profound as Galileo's saying in the *Saggiatore* (Rome, 1623) that philosophy is written in the book of the Universe in the language of mathematics, but the effect on practice is the same, as Galileo himself seems to have noticed.[36]

Another element in which the practical tradition may have contributed to the kind of mathematics used in pursuit of goals recognized as proper to natural philosophy is the deliberate use of approximation. Both the method of indivisibles and its descendant the infinitesimal calculus require the mathematician to accept an answer as merely 'good enough'. The counterpart of this in natural philosophy is Kepler's acceptance that, as he wrote to a fellow astronomer in 1605:

> the most true path of the planet [Mars] is an ellipse, . . . or certainly so close to an ellipse that the difference is insensible.[37]

In the published work, the result is given in the cruder form now known as Kepler's First Law:

> So the ellipse is the path of the planet.[38]

As in Galileo's innovative experimental researches on free fall, carried out at about the same time, we have here, but in the context of a well-established observational science, the triumph of a simple mathematical model, paid for by some use of approximation.

The admissibility of approximation in craft activities is so obvious that it passes without comment in any period, but the Renaissance in fact provides rather striking examples of the use of approximation in two crafts recognized as having mathematical foundations: painting and music. In the fifteenth century painters' use of perspective, which claimed to give a scientifically correct rendering of appearances, was generally far from scientific in practice. It seems to have been realized from very early on that correct mathematics was not necessary for a convincing illusion.[39] Since perspective construction was seen as linked with the mathematical part of the science of vision, one might perhaps have expected signs of this tension to appear in the learned world. However, in the event it seems that painters' techniques did not begin to attract attention from competent mathematicians until the middle of the sixteenth century, and the long-term results, for the practice of perspective, were mainly felt in

[36] G. Galilei, *Il Saggiatore* (Rome, 1623), 25; trans. in *The Controversy of the Comets of 1618*, tr. S. Drake and C. D. O'Malley (Philadelphia, 1960), 183–4. The reference to commercial arithmetic is in G. Galilei, *Dialogo sopra i due massimi sistemi del mondo* (Florence, 1632), 202; trans. as *Dialogue concerning the two chief world systems*, tr. S. Drake (Berkeley, 1953), 207–8.

[37] Kepler to Fabricius, 11 Oct 1605, *KGW* **15**, 249, letter 358, ll. 390–2.

[38] J. Kepler, *Astronomia nova* . . . (Heidelberg, 1609), 285. Reprinted *KGW* **3**, 366, l.4. As the historical record shows, for astronomers the proof of the truth of Kepler's first two laws lay not in his calculations in the *Astronomia nova* but in the long-term accuracy of prediction achieved by his *Rudolphine Tables* (Ulm, 1627), *KGW* **10**.

[39] A good example is provided by what is certainly one of the earliest surviving paintings to use mathematical perspective, namely Masaccio's *Trinity* fresco, in Santa Maria Novella, Florence, probably painted about 1426. See J. V. Field, R. Lunardi and T. B. Settle, 'The perspective scheme of Masaccio's *Trinity* fresco', *Nuncius*, 1988, **4** (2): 31–118.

theoretical arguments that most painters could allow to pass them by. By the later seventeenth century, perspective had become a part of mathematics, for mathematicians, while painters borrowed appropriate bits of mathematics and got on with illusionism.[40]

The mathematical problems that arose during the Renaissance in the theory and practice of music were altogether more serious. Music was one of the mathematical sciences of the *quadrivium* and its foundations were in series of ratios of numbers, that is in arithmetic. The increasing use of polyphony, which was a socially acceptable way of displaying the skill of large choirs and hence the wealth of a prince of the Church or a nobleman with a private chapel, gave rise to questions about the acceptability of new 'musical' ratios. Tensions between theory and practice were also brought out by the increased use of pre-tuned instruments to accompany singers. As Walker has shown, by the mid sixteenth century the mismatch between theory and practice had become sufficiently intricate to accommodate florid controversy in the style of the arguments over the solution to cubic equations.[41] Prominent among the combattants was Galileo's father Vincenzo Galilei (*c.* 1520–1591). In view of the industrial scale of Galileo research it is perhaps surprising that this area of his intellectual background has been so little studied.[42]

The importation of exact mathematical techniques into painting (where they were practised inexactly) and the unavoidable use of approximate mathematics in the practices associated with the allegedly exact science of music theory (which did not recover its former coherence until the eighteenth century) present features of mathematical thought that in some ways resemble those that characterize work associated with the Scientific Revolution. Many histories of the period have shown ways in which changes in the crafts can be linked to the social changes associated with the humanist Renaissance.[43] Thus, tracing intellectual connections between the craft mathematics of the Renaissance and the use of mathematics in the Scientific Revolution opens up the possibility of writing a history of the period that takes intellectual developments through from Dante to Newton and is not hobbled by the imposition of categories inappropriate to the period. One good reason for being interested in the work of Piero della Francesca is that it cuts across the disciplinary boundaries of our own time, thereby warning us of their inapplicability in his.

[40] See M. J. Kemp, *The Science of Art* (New Haven and London, 1990).
[41] D. P. Walker, *Studies in Musical Science in the Late Renaissance* (London, 1978).
[42] The only discussion of it known to me is Stillman Drake's short article in his *Galileo Studies* (Ann Arbor, 1970).
[43] See for example, R. Goldthwaite, *The Building of Renaissance Florence* (Baltimore, 1980), which connects artistic developments with social and economic ones.

6

Johannes Hevelius and the visual language of astronomy

MARY G. WINKLER AND ALBERT VAN HELDEN

To historians of the seventeenth century, Johannes Hevelius (1611–1689) (Fig. 6.1) is remembered chiefly for constructing very long but useless telescopes and being on the wrong side of an argument with Robert Hooke and John Flamsteed about the usefulness of telescopic sights.[1] He is seen as a methodical but plodding astronomer for whom hard work was a virtue but who never contributed an original idea to science.[2] History is not generous to losers, and as a result Hevelius has not been the beneficiary of a modern biography and his correspondence lies largely ignored in archives. We want to argue that Hevelius's greatest contribution to astronomy was crucially important for the astronomy of his day and was accepted so quickly that it was taken for granted by his successors and has, therefore, been invisible to historians up to now. Hevelius's great contribution is to be found in his first book, *Selenographia*, published in 1647 (Fig. 6.2), but we have to look carefully to find it.

In 1660 a little tract entitled *Epistola de mundi*, written by a certain Jacobus Coccaeus, appeared in Amsterdam.[3] In it, Coccaeus proposed an entirely new system

[1] Eugene Fairfield McPike, *Hevelius, Flamsteed and Halley: Three Contemporary Astronomers and their Mutual Relations* (London, 1937).

[2] In his article on Hevelius in the *DSB* John North writes: 'He does not belong to the highest rank of theoretical astronomers, although he was the doyen of mid-seventeenth-century astronomers. His character might well be judged from the sentiments expressed on his engraved title pages, two of which stand out: "Not by words but by deeds" and "I prefer the unaided eye."' (VI: 364). Such judgements are ultimately based on those of Hevelius's influential contemporaries. After visiting Hevelius in Gdańsk and making measurements with him, Edmond Halley initially expressed admiration for Hevelius and his instruments, but a few years later, in 1686, he wrote to William Molyneux that Hevelius's insistence on the superiority of naked-eye measuring instruments caused him to have to 'vindicate the truth from the aspersions of an old peevish gentleman, who would not have it believed that it is possible to do better than he has done'. See Eugene Fairfield McPike, ed., *Correspondence and Papers of Edmond Halley* (London, 1932), 60.

[3] *Epistola de mundi, quae circumferentur systematis et novo alio illis certiore dialogismum paradoxem complex: auctore Jacobo Coccaeo* (Amsterdam, 1660). Dutch edition: *Brief, over de t'samenstellinghen des wereldt welcke in swangh gaen, ende over een nieuwe stellingh sekerder as de selve, een onghemeene bedenckinghe behelsende, door Jacobus Coccaeus: ende vertaelt door I.K.V.W.* (Haarlem, 1660). Biographical details about Jacobus Coccaeus are very scant. He is usually described as a philosopher, physician, and mathematician who flourished in Amsterdam or Haarlem in the middle of the seventeenth century. The editors of Huygens's *Oeuvres Complètes* state that he was born in Bremen in 1615 and studied at Leiden in 1633, living in Haarlem thereafter. Their information about his parents would indicate that he

of the world in which the fixed stars were centred on the Earth and the planets on a point halfway between the Earth and Sun. The necessity for this rearrangement of the Universe stemmed, according to Coccaeus, from the fact that the observations of the phases of Mercury published by Hevelius in his *Selenographia* did not support any of the known world systems.[4]

Galileo had observed the phases of Venus late in 1610, and his observations had quickly been verified by others.[5] That Mercury would show similar phases was accepted by all astronomers but Mercury's smallness, its proximity to the Sun, and the poor optical qualities of telescopes had prevented astronomers from definitively verifying the planet's phases until the late 1630s.[6] Hevelius was one of the first to publish visual evidence for this phenomenon. In *Selenographia* he showed a sequence of three decreasing phases of Mercury observed in November and December 1644 and three increasing phases observed in May 1645.[7] Hevelius's seeming success was the result of stopping the aperture of his telescope down to the size of a large pea for observations of the fixed stars and Mercury.[8] His procedure, however, produced not real but spurious discs, and the phases of Mercury he thought he saw were products of his telescope. As Coccaeus correctly pointed out, in late 1644 the phases should have been increasing while in the spring of 1645 they were decreasing – the exact reverse of what Hevelius showed.[9] Why, however, did Coccaeus not simply reject Hevelius's observations rather than go through all the trouble of redesigning the Universe?

The answer to this question must be found in Hevelius's authority. In his tract, Coccaeus referred to the Polish astronomer as 'the sharp-seeing Hevelius', 'the Prussian Lynx', and 'the star of Danzig'.[10] His admiration for Hevelius as an observer was unbounded. If Hevelius showed a heavenly phenomenon, then that was the way that phenomenon really appeared. And Coccaeus was by no means the only one who thought of Hevelius in this manner. In the years following 1647, Hevelius was the acknowledged authority on telescopic astronomy,[11] and we need to investigate whence this authority derived.

was the brother of Johannes Coccaeus (1603–1669), who was at this time professor of theology at the University of Leiden. See *OCCH,* **III**: 66.

[4] Coccaeus, *Brief*, *op. cit.* (note 3), 8–32.

[5] Galileo to Benedetto Castelli and Galileo to Christopher Clavius, 30 December 1610, *OGG*, **X**: 499–504. See also the mathematicians of the Collegio Romano to Robert Cardinal Bellarmine, 24 April 1611, *ibid.*, **XI**: 93.

[6] Johannes Baptista Zupo had observed phases of Mercury with telescopes made by Francesco Fontana, beginning in 1639. These observations were published by Fontana in *Novae coelestium, terrestriumque rerum observationes* (Naples, 1646), section 5. Hevelius did not see Fontana's book before he published *Selenographia* in 1647. For better visual representations of Zupo's observations, see Giovanni Baptista Riccioli, *Almagestum novum*, (2 vols., Bologna, 1651), **I**: 484.

[7] Johannes Hevelius, *Selenographia: sive Lunae Descriptio* (Gdańsk, 1647), between 70 and 71.

[8] *Ibid.*, pp. 37, 74. [9] Coccaeus, *Brief*, *op. cit.* (note 3), 12–13. [10] *Ibid.*, 10, 12, 14, 23.

[11] At its meeting of 21 February 1663 (new style), the Royal Society instructed its secretary Henry Oldenburg to write 'in his own name a letter to Monsieur Hevelius of Dantzick, to assure him of the esteem, which the society had of his merits, of which he had given such demonstrations to the learned world in the books published by him'. See Thomas Birch, *The History of the Royal Society of London*, (4 vols., London, 1756–7; reprinted Brussels, 1967–8), **I**: 194. In his letter of 28 February, Oldenburg wrote to Hevelius that the recent publication of one of his books 'provided an occasion for your merits in the republic of letters to be praised to the skies in that assembly of the Muses [i.e., the

As we have argued elsewhere, the telescope caused the need for a visual language in astronomy.[12] Words and diagrams no longer sufficed to convey information about heavenly bodies. In *Sidereus Nuncius* (1610) Galileo made a start in this process, but we can hardly call his renditions of the Moon in that book naturalistic representations of our neighbour. The engravings show a distorted view designed to illustrate a verbal argument: reality was still conveyed by words. With the exception of his efforts in his second sunspot letter, Galileo did not pursue the development of a visual language further. It was not until the 1640s that the visual dimension began to take its place next to the verbal in astronomical accounts. But these efforts were not immediately successful. The first attempted picture book of telescopic astronomy was *Novae coelestium terrestriumque rerum observationes* published in 1646 by the telescope maker Francesco Fontana of Naples. It was described by Evangelista Torricelli (admittedly a rival in telescope making) as a book 'of foolishness observed, or rather dreamed, by Fontana, full of insane things, that is absurdities, fictions, affronts, and a thousand similar outrages'.[13]

Now Fontana's book does, indeed, show that an observer must have some sense of which part of the information reaching the eye comes from the heavens and which part is an artefact of one's telescope. But the problem went deeper than that. Torricelli, the Grand Duke's mathematician, and Fontana, a telescope maker, were not equals in the prestige hierarchy of the Italian patronage system. How could a person such as Fontana claim authority and objectivity for his observations? The reader of the book needed assurances that Fontana's instruments did not deceive, that he had made his own observations, and that nothing was lost or added in the process of translating the observation into the printed engraving. In short, the reader needed to be assured that he was, so to speak, there at the observation, that he was a 'virtual witness'. These were the questions with which Hevelius had been dealing for some time. The reader of *Selenographia*, which appeared a year after Fontana's book, got an entirely different impression of the quality of the pictures of heavenly bodies.

Selenographia was not merely a book about the appearance of the Moon; it also reviewed the state of the art regarding all heavenly phenomena accessible with the telescope. Its 548 pages contain a large number of engravings as well as numerous diagrams. We find accounts of all the planets, the moons of Jupiter, and sunspots, besides his exhaustive treatment of the Moon. The work was carefully planned and meticulously executed. Before the reader ever got to the lunar observations, he had already read a complete treatise on telescopic astronomy, including a full description of the instruments involved and a guided tour through the universe as revealed by the telescope, all this including a complete review of the literature.

Royal Society]'. See *The Correspondence of Henry Oldenburg*, ed. A. Rupert Hall and Marie Boas Hall (13 vols., Madison, London, Philadelphia, 1965–87), **II**: 27.

12 Mary G. Winkler and Albert Van Helden, 'Representing the heavens: Galileo and visual astronomy', *ISIS*, 1992, 83: 197–217.

13 Torricelli to Vincenzo Renieri, 25 May 1647, *Le opere dei discepoli di Galileo Galilei. I Carteggio 1642–1648*, ed. Paolo Galluzzi and Maurizio Torrini (Florence, 1975), 366.

Figure 6.1. Johannes Hevelius, *Selenographia* (Gdańsk, 1647). Portrait of the author on second page after title page.

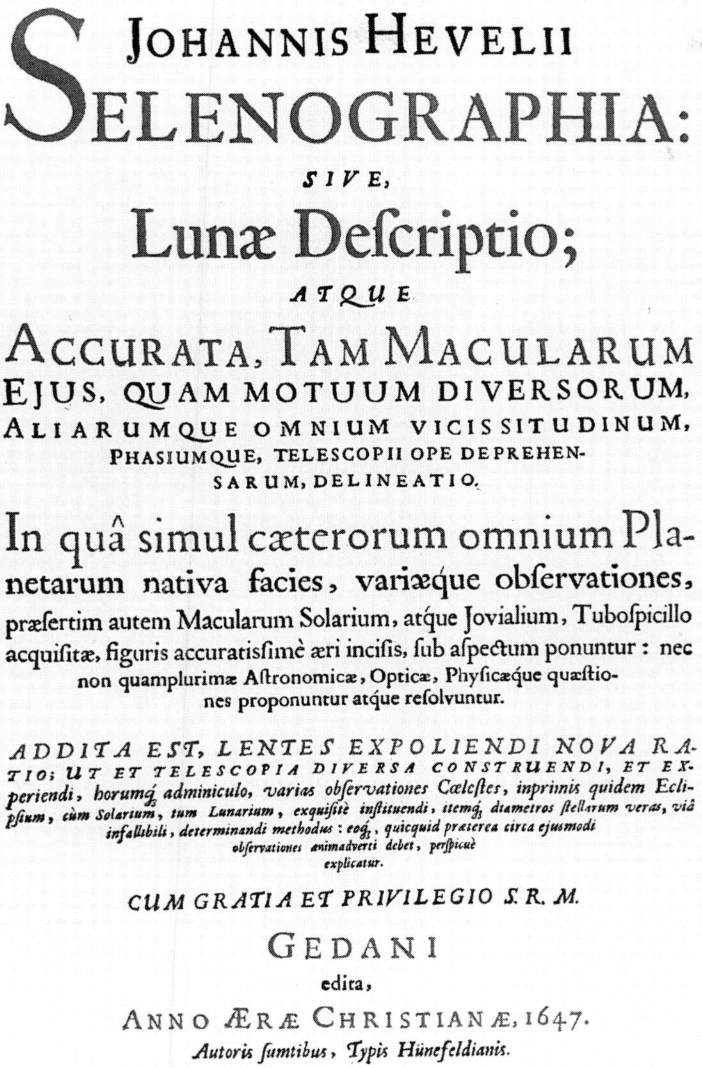

JOHANNIS HEVELII

SELENOGRAPHIA:

SIVE,

Lunæ Deſcriptio;

ATQUE

ACCURATA, TAM MACULARUM
EJUS, QUAM MOTUUM DIVERSORUM,
ALIARUMQUE OMNIUM VICISSITUDINUM,
PHASIUMQUE, TELESCOPII OPE DEPREHEN-
SARUM, DELINEATIO.

In quâ simul cæterorum omnium Pla-
netarum nativa facies, variæque obſervationes,
præſertim autem Macularum Solarium, atque Jovialium, Tuboſpicillo
acquiſitæ, figuris accuratisſimè æri inciſis, ſub aſpectum ponuntur: nec
non quamplurimæ Aſtronomicæ, Opticæ, Phyſicæque quæſtio-
nes proponuntur atque reſolvuntur.

*ADDITA EST, LENTES EXPOLIENDI NOVA RA-
TIO; UT ET TELESCOPIA DIVERSA CONSTRUENDI, ET EX-
periendi, horumq; adminiculo, varias obſervationes Cœleſtes, inprimis quidem Ecli-
pſium, cùm Solarium, tum Lunarium, exquiſitè inſtituendi, itemq; diametros ſtellarum veras, viâ
infallibili, determinandi methodus: eoq;, quicquid præterea circa ejusmodi
obſervationes animadverti debet, perſpicuè
explicatur.*

CUM GRATIA ET PRIVILEGIO S. R. M.

GEDANI

edita,

ANNO ÆRÆ CHRISTIANÆ, 1647.

Autoris ſumtibus, Typis Hünefeldianis.

Figure 6.2. J. Hevelius, *Selenographia* (Gdańsk, 1647). Title page.

Hevelius begins this treatise with a discussion of lenses. Besides lenses with spherical curvature, there are those whose surfaces are elliptical or hyperbolic. But although these latter lenses are beautifully demonstrated in formal optics, they have never been made and put into use as far as Hevelius knows. He therefore limits his discussion to lenses with spherical curvature. The more lenses are made convex in order to enlarge the image, the less their curvature must be, and it is to be noted that plano-convex lenses amplify the angle of vision twice as much as biconvex lenses of the same curvature. A plano-convex lens with a radius of curvature of twelve feet, inserted in the aperture of a camera obscura, will show a clear image at a distance of twelve feet, provided that the aperture is not too large or small. The best way to check the quality of the image is to project it onto a concave mirror or a convex lens covered with paper. Telescope lenses are made on a lathe, which Hevelius not only describes but also shows in a beautifully executed picture (Fig. 6.3). This picture is so detailed that even without the verbal description an instrument maker could copy it, but the text gives practical information about the machine and its use, impossible to show in a picture. With this visual and verbal information the reader could set up his own lens-grinding operation.

Hevelius next launches into a discussion of glass quality (which was a major problem at the time). The best glass for telescopes was Venetian glass, and it was better to use a slightly green than an overly milky glass. Venetian glass was also often better than natural rock crystal, for crystal, even if it was very clear, produced more deleterious refractions, 'something an inexperienced person will perhaps hardly believe'.[14] Further, it would be hard for persons not practised in the art to judge the quality of a lens. For this reason Hevelius gives four criteria by which to judge: the surface must be smooth, without little holes or scratches; the convex lens should be equally thin on all sides, indicating that the surfaces are symmetrical; when an object of known shape is viewed through the lens, its shape should not be distorted; and finally there should be no curvature of the field near the edges of the lens.[15] He goes on to give some more practical advice on this score.

Hevelius devotes the next chapter to the question of how to combine lenses to make telescopes. The length of the instrument depends on the curvature of the convex object-lens and to a lesser extent on that of the eye-lens. Thus, a biconvex lens whose curvature is 12 feet, when combined with a biconcave eye-lens whose curvature is $5\frac{1}{2}$ inches will produce an instrument of 5 or $5\frac{1}{2}$ feet. But one should be careful not to make the eye-lens too strong, for then images will be confused.[16] Telescopes can also be made with different combinations of lenses, including those of two convexes which show an inverted image, and those with three convexes which show an erect image.[17]

The composition of the tubes is another problem. Those of paper, favoured by Torricelli and Johannes Wiesel, the Imperial Optician, have serious drawbacks: if they are properly tight under normal circumstances they become too tight to draw out and

[14] Hevelius, *op. cit.* (note 7), 9. [15] *Ibid.*, 9–10. [16] *Ibid.*, 11–13. [17] *Ibid.*, 15.

Figure 6.3. Hevelius's lens-griding lathe, *Selenographia* (Gdańsk, 1647), pp. 6–7.

push in when the night is humid; if they are looser then dry weather makes them too loose so that they cannot be kept adjusted. Further, they are not durable but break easily, especially if they get wet (which must have been more of a problem in northern climes), and the pulling out and pushing in creates dust that fogs up the lenses. Finally, worst of all, it is impossible to align the lenses exactly in this type of tube. For this reason, sturdier tubes are needed. But sheet-metal makes them too heavy, and therefore one should make telescope tubes out of good wood that was well dried. This tube should consist of as few little tubes as possible. Thus, Hevelius made the tube for a twelve-foot telescope out of two box-shaped sections, one eight feet and the other four feet long.[18] As far as the apertures are concerned, that of the eye-lens should be not too large and not too small, but just the right size to allow the eye to travel from

[18] *Ibid.*, 15–17.

one lens to the other. The aperture of the object-lens should not exceed $1\frac{1}{2}$ inches (about $3\frac{1}{2}$ cm) in the longest telescopes.[19] In this chapter he also describes the 'helioscopium', an instrument used for projecting the Sun's image, the 'microscopium' (the one-lens 'flea-glass' or *vitrum muscarium*, as well as the instrument consisting of two convex lenses), and the 'polemoscopium' or periscope of his own invention.[20] The best way to compare the quality and magnification of telescopes is to project the Sun's image through them and carefully check the image thus produced.[21] Hevelius also gives five rules for the refraction of rays, taken from theoretical treatises on optics.[22]

For those reading the literature on telescopic astronomy, what little there was, this careful description of instruments was a first. Galileo's description was practical but extremely limited. In *Sidereus Nuncius* he had related how he had made his first three-powered instrument by fitting a convex and concave lens together and had then gone on to produce instruments up to a power of thirty. He had further shown how the magnification of an instrument might be determined. But all this was a minimum by way of introduction to a subject of unprecedented novelty. His followers fell into several categories. There were those who wrote about the theory of the instrument (Kepler, Scheiner, Descartes), and a few writers who discussed telescope-making (e.g., Girolamo Sirtori);[23] and there were those who described observations but gave no details on their telescopes (e.g., Gassendi).[24] Only Christoph Scheiner, in his massive treatise on sunspots, *Rosa Ursina* (1630), had combined a discussion of instruments with a description of observations, but his discussion had been virtually entirely theoretical. Hevelius combined all three approaches, giving the formal optical principles on which the telescope is based (citing Euclid, Witelo,[25] Maurolyco,[26] Kepler, De Dominis,[27] Scheiner, and Descartes), the configurations of lenses, the technique of lens-grinding, and the practical problems of mounting instruments. He thus demonstrated his complete grasp of the subject, from abstruse theory to the most practical of detail, and the reader was here as an apprentice to the master. Could anyone doubt that the instrument with its sophisticated mounting shown in plate F (Fig. 6.4) was an actually existing instrument built by Hevelius (and, presumably, his 'invisible technicians'),[28] and that the man behind it was the author himself? Such

[19] *Ibid.*, 17. Hevelius does not specify what particular measure of the inch he uses, and we may infer that the measurement was meant as approximate. The Gdańsk inch was 2.3 cm. There were 12 inches in a Gdańsk foot.
[20] *Ibid.*, 22–31.
[21] *Ibid.*, 17–19.
[22] *Ibid.*, 19–21.
[23] Johannes Kepler, *Dioptrice* (Augsburg, 1611) reprinted in *KGW* 4; Christoph Scheiner, *Rosa Ursina* (Bracciano, 1630); René Descartes, *Discours de la Methode Pour bien conduire sa raison, & chercher la verité dans les sciences. Plus La Dioptrique. Les Meteores. et La Geometrie. Qui sont des essais de cete Methode* (Leiden, 1637); Girolamo Sirtori, *Telescopium: sive ars perficiendi* (Frankfurt, 1618).
[24] Pierre Gassendi, *Mercurius in Sole visus et Venus invisa Parisiis mdcxxxi* (Paris, 1632).
[25] Witelo's *Perspectiva* had been printed by Frederick Risner in his *Opticae Thesaurus* (Basle, 1572).
[26] Francesco Maurolyco, *Photismi de lumine et umbra ad prospectivam radiorum incidentium facientes* (Venice, 1575).
[27] Marco Antonio de Dominis, *De radiis visus et lucis in perspectivis et iride* (Venice, 1611).
[28] For a discussion of the problem of identifying the assistants who built Robert Boyle's instruments and ran the experiments, see Steven Shapin, 'The invisible technician', *Am. Scientist*, 1989, 77: 554–63.

Figure 6.4. Hevelius at the telescope, *Selenographia* (Gdańsk, 1647), pp. 42–3

pictures were doubly instructive: they certified Hevelius's instrumentation and at the same time gave the reader important information on the instrument's construction. There was an honesty here, an openness, an effort to share with the reader the practice of telescopic astronomy.

The next problem was, what does one see when one turns a telescope to the heavens? Hevelius relates the observations of Galileo and others and states that he too has seen all these things, in fact he had improved on the observations of his predecessors. Where Galileo, Kepler, and Hortensius had never been able to see the fixed stars and Mercury as round bodies, he had succeeded. For the observation of these very small and very bright objects he stopped down his aperture to the diameter of a large pea. In general, the brighter the star or planet (e.g., Sirius or Venus) the smaller the aperture. For stars whose light is somewhat more languid, e.g., Aldebaran, one can use a slightly larger aperture. For the Moon, however, one needs a large aperture.[29] Now it is perhaps of interest, in retrospect to note that Hevelius had here stopped down his aperture so much that he was observing not real but rather spurious discs, and that he wasted a lot of time measuring the apparent diameters of these spurious discs. What is of importance for us here, however, is his detailed description of the practice of observation, something that was entirely new in this field of endeavour.

Hevelius next devotes an entire chapter to his observations of the planets and the moons of Jupiter. Previous observations are not always correct. Thus, in his *Detectio Dioptrica* of 1643, Matthias Hirzgarter had reported the figure of Saturn observed by Francesco Fontana of Naples.[30] (Hevelius had not yet seen Fontana's own book, published in 1646.) Hevelius had doubted the appearances seen by his predecessors, but with better and better telescopes he was able to verify that 'this described figure of Saturn was not a mere dream'.[31] The planet actually appeared this way. But Hevelius did quarrel with his predecessors a little, and showed a different figure (one, incidentally, which was very wrong, see Fig. 6.5). About this figure he stated: 'And with a long tube of exquisite construction I have been able to observe accurately that this is the true figure of Saturn and to consider everything properly, so that everyone who is guided by a desire to discover the truth can trust in this indefatigable observation'.[32] We can now begin to appreciate how the claim for authority is built up. Hevelius has revealed all about his instruments, thus giving the reader the tools with which to check his observations. The reader has become a participant in instrument building and is now, through detailed pictures engraved by Hevelius, himself becoming a witness in the actual observations. He can imagine himself as the observer in plate F, seeing through the instrument shown there the appearances Hevelius so faithfully depicted.

When it came to the satellites of Jupiter, Hevelius addressed himself to a recent argument. In 1643 Antonius Maria Schyrlaeus de Rheita had published a little tract in which he claimed to have observed not four but nine satellites of Jupiter, six of Saturn, and 'some' of Mars. This claim was supported by some but dismissed by Gassendi.[33] The arguments revolved around visual information that was not well presented by any of the participants in this controversy. Hevelius had been observing Jupiter at the same time and he now showed detailed maps of the appropriate regions of Aquarius and Pisces (Fig. 6.6), which could be verified by anyone, and demonstrated that Rheita had mistaken fixed stars for satellites of Jupiter. What was clinching in this argument was Hevelius's use of visual evidence. Rheita had shown only the little telescopic stars that he claimed to be satellites; Hevelius mapped the region and superimposed Rheita's diagram on that. An observer could point his telescope to that region of the sky and verify that the only objects missing were Jupiter and the four moons discovered by Galileo; the five new satellites claimed by Rheita were still there in the

[29] Hevelius, *op. cit.* (note 7), 36–8.

[30] Matthias Hirzgarter, *Detectio dioptrica* (Frankfurt, 1643), 20–2.

[31] Hevelius, *op. cit.* (note 7), 43.

[32] *Ibid.*, 43–4.

[33] A. M. Schyrlaeus de Rheita, *Novem stellae circum Iovem, circa Saturnum sex, circa Martem non-nullae, a Antonio Rheita detectae & satellitibus adiudicatae. De primis (& si mavelis de universis) Petri Gassendi Iudicum. Ioannis Caramuel Lobkowitz eiusdem Iudicij censura* (Louvain, 1643).

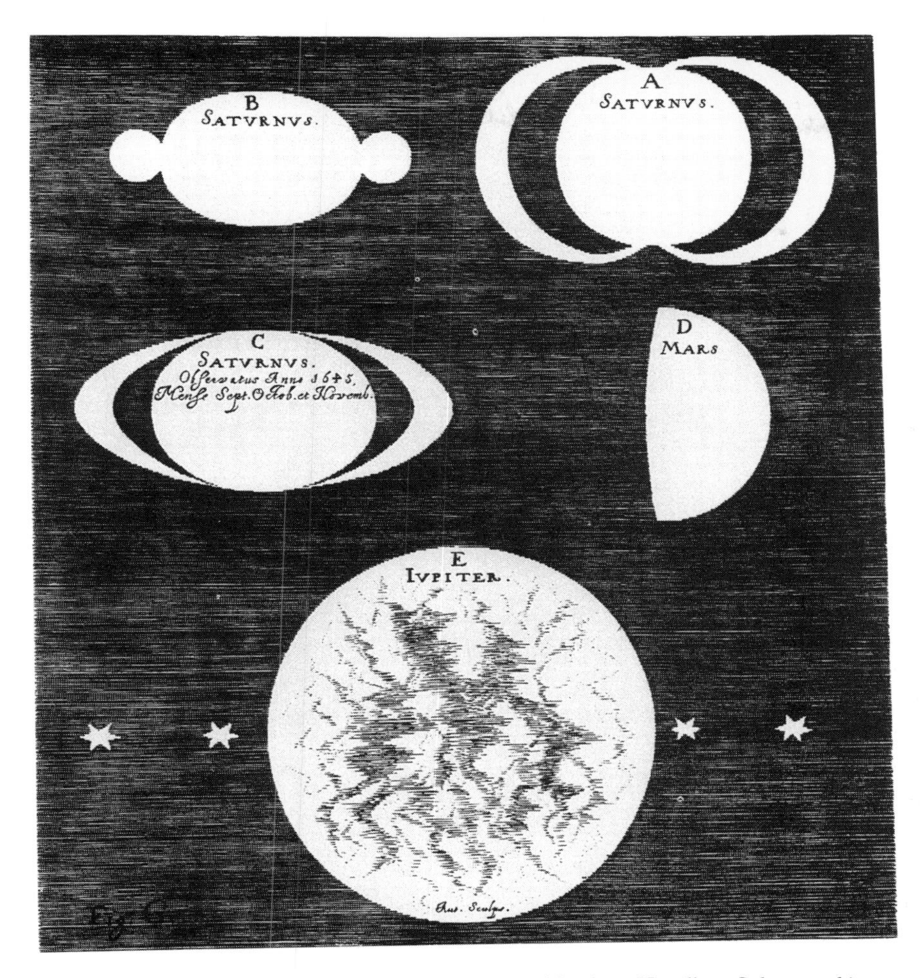

Figure 6.5. The appearances of Saturn, Mars, and Jupiter. Hevelius, *Selenographia* (Gdańsk, 1647), pp. 44–5.

same position they had occupied in 1643: Jupiter had left them behind. After Hevelius's intervention, learned men stopped paying attention to Rheita's claim.

It is only after this tour of the telescopic universe, and a chapter on sunspots, that Hevelius finally arrives at the Moon, the subject that will occupy him for the remainder of the book. After dealing with the various arguments about the roughness of the lunar surface, he turns his attention to our neighbour's telescopic appearance. Here he differs with Galileo, whose figures, published in *Sidereus Nuncius* show rather distorted views of the Moon because they were meant as illustrations of his verbal arguments about the Moon's nature.[34] Not realizing that Galileo's purpose was

[34] Galileo Galilei, *Sidereus Nuncius* (Venice, 1610), 8r, 9v–10v. These engravings are faithfully reproduced in *OGG*, III: 63, 66–7, and *Sidereus Nuncius or the Sidereal Messenger*, tr. Albert Van Helden (Chicago, 1989), 41, 44–6. For an analysis of Galileo's lunar illustrations, see Winkler and Van Helden *op. cit.* (note 12).

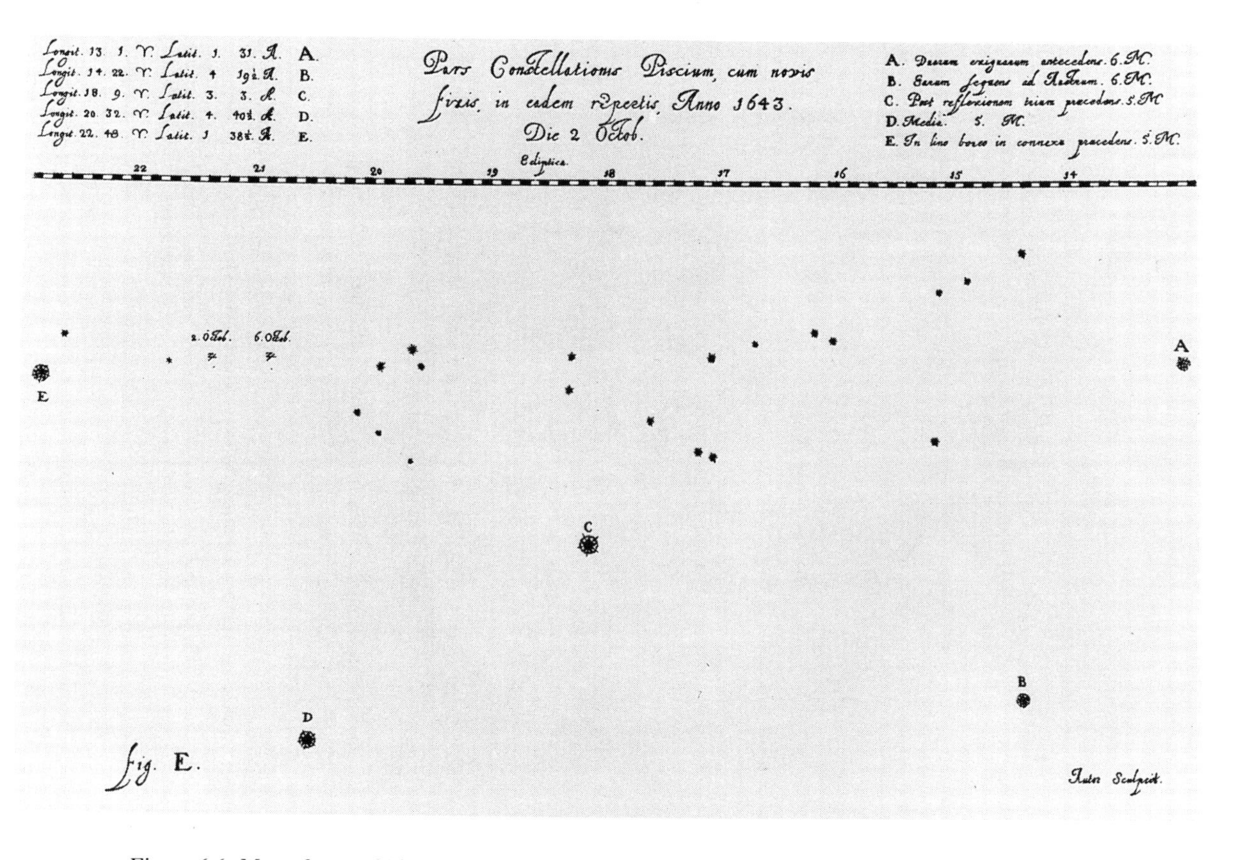

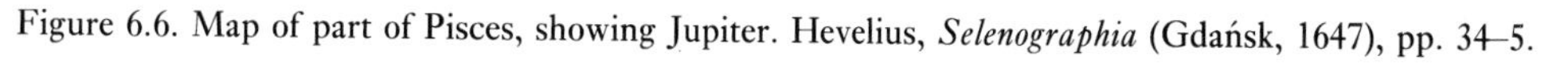

Figure 6.6. Map of part of Pisces, showing Jupiter. Hevelius, *Selenographia* (Gdańsk, 1647), pp. 34–5.

different from his own, Hevelius concluded that Galileo's instruments were inferior, that he did not take sufficient care, or that he could not draw.[35] Clearly, it is assumed that observers of the Moon would proceed as Hevelius did, for his was the right way.

As the title page of *Selenographia* (Fig. 6.2) announces, Hevelius's purpose was to show the Moon, as well as the planets, in their natural figures. His aim was to *picture* the heavens, but there were few antecedents for this. As his discussion up to this point had shown, other people's depictions of the Moon and planets were unreliable. How could he convince his readers that his were better? Hevelius's approach was to build a seamless environment. Although we do not know to what extent he used workmen to help him, we do know that he was personally involved in the fabrication of his instruments. Perhaps he ground lenses with his own hand, but it is to be doubted that he machined the fittings for his machines and instruments himself. At any rate, we can say that he and his workmen built the instruments, and that from there he did everything himself.[36] Hevelius set out to represent every phase of the Moon. This was no easy task. Only someone who had tried to render at least one phase accurately without any previous example could appreciate the imagination and memory necessary and also the patience and dedication. As shown by the example of seeing one's own face in the mirror, it is extremely difficult to hold the image in one's mind, 'for while the eyes linger on the mirror it appears to them that they have embraced everything accurately, but when the eyes are removed from the mirror they sense crudely and hardly easily the images seen to be in the mind'.[37] The same happens when we try to draw the Moon. While our eye is at the eyepiece, we know exactly what we want to draw and how to do it,

> but at the very point of time when we sketch the figure and start a picture we find out quickly that most, if not all, escapes our memory. It is therefore necessary that every little point is observed ten times or more before the proper place of some spot, its figure and its form, can truly correctly and accurately be reproduced on the paper.[38]

And even then, during subsequent observations we find that we have made mistakes and have been deluded in some appearance. Patience, dedication, and an excellent memory are therefore essential. Note also that Hevelius was using a powerful Galilean or Dutch telescope, which would only show a part of the Moon at one time, so that the topographical relations between lunar features was problematical.

But this was only the beginning. One must also be uncommonly skilled in drawing and picturing. Without these skills the observations will be useless. No matter how

[35] Hevelius, *op. cit.* (note 7), 205.
[36] We know little about who, if anyone, assisted Hevelius in his astronomical observations during this period. Later, his second wife, Elizabeth Koopman, whom he married in 1663, observed with him, became an astronomer in her own right, and took over much of the overall operation of the observatory. After his death she edited and published his final works. For an illustration of Koopman and Hevelius observing together, see Johannes Hevelius *Machinae coelestis pars prior* (Gdańsk, 1673), 222–3, plate M.
[37] Hevelius, *op. cit.* (note 7), 209.
[38] *Ibid.*, 209–10.

many words we use, we can not make someone else grasp exactly what we have seen. If the brush of the painter strays it may take as many as ten corrections before the picture is adequate. The work is therefore extremely slow. Each phase occurs only once per month, and no clear night can be skipped. Even so, clouds often make observation impossible, so that it may be years before all the phases have been observed and depicted. Does the artist therefore spend every night with the observer, or is he sometimes called from his bed? But if so, he must live with the observer, and this brings great inconvenience to both. Further, how much of the cost of correcting should the artist have to bear? Obviously, the task will go much smoother and the results will be more accurate if the observer has learned also to use his hands. The work must therefore be undertaken by an observer who is also a draughtsman, or else it will be poorly done or remain unfinished.[39]

Hevelius was such a draughtsman. But drawing the pictures was only part of the task. These pictures had to be engraved so that they could be printed. The problem was similar to that presented by hiring a painter. Even a very skilled engraver could hardly understand what Hevelius had in mind. He would 'transgress in the representation, now in this way and now in another, or neglect the proportion, or do something else that could not easily be perfected or put right'.[40] Hevelius therefore engraved every observation he made in copper with his own hand, a work that took him almost as long as the observations themselves.[41] All this toil has as its purpose to present a 'history of the Moon', as part of the universal philosophy espoused by Francis Bacon.[42]

Still, there remained the printing process, which Hevelius could not do himself. As he explicitly states on the title page, however, he paid for the printing, and this means that he retained some control over the process. Although a fair number of small verbal errors escaped the proofing process,[43] the engravings escaped this fate: only the uneven shrinkage of the paper on which the plates were printed prevented total perfection in the reproduction. In discussing his lunar representations, Hevelius gives their original diameters on the copper plates (in fractions of Paris, Rhineland, and Gdańsk feet!) and then tells the reader that the paper shrunk ten parts in width to six parts in length as it dried after the impression had been made.[44]

Selenographia was thus very different from previous books on telescopic astronomy. The detailed description presented the reader with a wealth of detail about the instruments, their construction and use, the observations, and the methods of

[39] *Ibid.*, 210–12. [40] *Ibid.*, 218.

[41] The frontispiece and the author's portrait are signed by others. Plates A (optical lathe), B (periscope), F (the observer at the telescope), L, L* (projection chamber with equipment), and Q (the full Moon shown in the likeness of a terrestrial map) are not signed. The style of all except plate Q suggests that they may have been engraved by Hevelius, but we cannot be certain. The putti and scrolls in the plates of the full Moon (P, Q, and R) appear to have been done by Adolph Boij, the artist of the frontispiece.

[42] Hevelius, *op. cit.* (note 7), 219. [43] See the list of errata, *ibid.*, 563.

[44] *Ibid.*, 214.

recording and transmitting the observations. Other observers had sometimes cited witnesses to certify their results. Hevelius does not cite witnesses; instead he makes the reader, in the term used by Shapin and Schaffer, a virtual witness.[45] He took the reader, as it were, by the hand, led him through his workshop and showed him how his instruments were made, and then showed him the observations. How could the reader be sure that what was on the page was really what Hevelius observed? The visual information passed from Hevelius's eye through his brain to his hand, and by his own drawing and engraving directly to the page without the interposition of any agent. Hevelius stressed this seamless process, from observation to the printing of the engravings, in order to certify his illustrations: his pictures actually represented what he saw. In verbal communication there were certain safeguards that tended to preserve the accuracy of information. Accepted spelling and rules of grammar and syntax, besides rhetorical fashions, provided a strong force conserving the original meaning of the text in copying; and the printing process allowed for proof-reading before final printing. Through long usage, astronomers were familiar with these safeguards. In communicating visual information, however, other factors come into play: copying pictures requires understanding of the visual language of the original, and, as Ernst Gombrich has demonstrated, pictures may lose their intended message through an ignorant and misguided copyist.[46] One needs only to compare woodcuts of the 1610 Frankfurt edition of *Sidereus Nuncius* with the engravings of the original Venice edition to see this process of deterioration. Hevelius solved this problem by copying from himself, and made visual representation in astronomy acceptable and respectable. He showed his fellow astronomers how it should be done.

But there was another factor. If astronomy was to become a visual science, astronomers had to find or create a visual language to convey the new things they had seen. Describing a new sensory experience is always difficult, but in using pictures as direct translations of what the eye has seen there are more difficulties than may be apparent. Besides the artistic skill of the observer (always a variable), there is the whole problem of artistic style. Astronomy needed to find its own 'style' if the representations of the heavens were to be believed.

This is not a trivial problem, as a look at Galileo's representations of the Moon will illustrate. There are two issues here. One is the relationship between picture and verbal text. The other is related to the technology of printing. Modern selenographers have been somewhat mystified by Galileo's illustrations. Measuring the relative positions of lunar features and trying to identify the exaggerated crater ('spot', as Galileo called it) almost invariably lead them to conclude that Galileo's lunar pictures have little to do with our Moon.[47] Only some features can actually be identified. But

[45] Steven Shapin and Simon Schaffer, *Leviathan and the Air-Pump: Hobbes, Boyle and the Experimental Life*, (Princeton, 1985), 60–5.

[46] E. H. Gombrich, *Art and Illusion: A Study in the Psychology of Pictorial Representation* (New York, 1960), 74–7. See also Samuel Y. Edgerton, 'The renaissance artist as quantifier', in *The Perception of Pictures*, ed. Margaret A. Hagen (2 vols., New York, 1980), **I**: 179–212, pp. 202–7.

[47] J. Classen, 'The first maps of the Moon', *Sky Telescope*, 1969, **37**: 82–3, p. 82; Zdeněk Kopal, *The Moon* (Dordrecht, 1969), 225; Zdeněk Kopal and Robert W. Carder, *Mapping the Moon: Past and Present* (Dordrecht, 1974), 4.

after thus dismissing the pictures, these same selenographers frequently feel the need to remark that Galileo's verbal descriptions surrounding them are curiously and surprisingly accurate.[48] This impression should not surprise us at all. The pictures in *Sidereus Nuncius* were not meant to stand on their own feet. They made sense only when placed in the context of verbal exposition; they derived their meaning from that text. In conformity with long practice, the visual was subordinate to the verbal. In the world of learning, accuracy could reside in words or numbers but not in pictures. Moreover, the printed pictures were not skilfully executed. Lacking art, they fail to convey the information that Galileo had included in his pen and wash drawings.[49] Small wonder therefore that the few lunar illustrations that we find in the works of Galileo's contemporaries were little more than diagrams, generic moons, not our Moon.[50] In the first decades of telescopic astronomy there was no accepted visual language, and therefore pictures could not be independent of verbal texts, could not independently convey information and meaning.

The artists had, in the meantime, been developing just such a visual language. One needs only to open the machine books of Besson and Ramelli to see that here meaning could reside in the visual without any verbal text: in fact, the verbal texts are predicated on the pictures. Moreover, especially in northern Europe, artists were becoming very inventive in developing printing and engraving techniques to fit the concerns and spatial experiments of Renaissance art. In fact, the interests of northern artists contributed a great deal to the development of the visual language Hevelius chose for his astronomical work.

Shifting the discussion of visual language to northern Europe and focusing on the concerns of northern artists is very illuminating. Here, the questions and problems were different from what they were in Italy. Svetlana Alpers finds a marked difference between the art of northern Europe and Italy in the sixteenth and seventeenth centuries. Northern art, she asserts, is an 'art of describing' rather than an 'art of narrative'.[51] The difference had important implications not only for theorists but for anyone attempting to represent natural phenomena pictorially. Alpers argues that unlike Italian art, northern art does not offer easy verbal access. Nor is it an art that expresses elaborate optical theories, a fact that led Italian artists and theoreticians to somewhat denigrate it, as Michelangelo did, for being 'without reason or art'.[52]

According to Alpers, this northern lack of theory as well as an un-Italian concern with 'nature', produced a split between a craft-based northern art and the treatise-driven, speculative Italian art. In sum, there was a difference (always mitigated by the

[48] Classen, *op. cit.* (note 47), 82; Ewen A. Whittaker, 'Galileo's lunar observations and the dating of the composition of "Sidereus Nuncius" ', *J. Hist. Astr.*, 1978, **9**: 155–69, p. 158.

[49] A visual language had to be invented for printing as well. Samuel Edgerton suggests that engravers and woodblock cutters did not even have the requisite skills to do scientific illustrations until after the second decade of the sixteenth century. It required skill and an inventiveness like that of Albrecht Dürer to translate artists' experiments in perspective and chiaroscuro into a linear system. See Samuel Y. Edgerton, Jr., *The Heritage of Giotto's Geometry* (Ithaca, 1991), 148.

[50] Winkler and Van Helden *op. cit.* (note 12).

[51] Svetlana Alpers, *The Art of Describing: Dutch Art in the Seventeenth Century* (Chicago, 1983), xxi.

[52] *Ibid.*

constant exchange of artistic innovations and developments across the Alps) between the particularistic, descriptive tendencies of northern art and the humanistic idealizing concerns of the Italians.

This difference is relevant to a discussion of Hevelius and the development of a visual language for astronomy. Hevelius's native artistic milieu was essentially German art with its tenacious craft tradition, especially the tradition of goldsmiths and other metal workers who specialized in a sophisticated linear style.[53] He thus found a tradition for printing descriptive pictures ready to hand. A second quality of northern art is also significant. Northern art, since the Van Eycks, specialized in a kind of naturalism that combines empiricism, a trust in individual perception, with an interest in, and empathy with, the particular.[54] Northern Europeans were accustomed to seeing pictures that contained these elements.

Finally, Hevelius – unlike Galileo – was *hors de combat* when it came to debates about the liberal versus the mechanical arts, or questions of professional hierarchy: they were, literally, academic to him. As a magistrate, a wealthy man, his position in Gdańsk was recognized and secure. Nor would he have found his lack of training in *disegno* an embarrassment. He felt himself perfectly capable of translating what he had seen into a picture – and he assumed his readers (viewers?) would accept his empirical, linear, naturalistic style, especially since that was generally the northern mode of depicting nature.

The religious tradition within which Hevelius grew up resonated with this artistic tradition. During the Protestant Reformation, when ecclesiastical reformers turned their attention to the visual arts, some northern artists found a deeper rationale for their interest in the details of the natural world. Protestant theology on Divine Providence gave the study of nature a new religious legitimacy. Luther, for example, found a rational progression from observation and contemplation of nature, to an opening of the heart through wonder, to faith. The ability to wonder at nature is a gift of grace and an aid to salvation:

> Nothing – even raising the dead – is comparable to the wonderful work of producing a bird out of water. We do not wonder at these things because through our daily association we have lost our wonderment. But if anyone believes in them and regards them more attentively, he is compelled to wonder at them and his wonderment gradually strengthens his faith.[55]

Svetlana Alpers argues that northern art demonstrates that meaning is to be found in what the eye can see.[56] For Protestant illustrators the meaning was the revelation of the working of Providence. There are hints of Providential theology in some of

[53] Alexander Nagel, writing on a recent exhibition of German Renaissance art, stresses that German art is characterized by a 'propensity to carve, producing a particular kind of willful, form-searching line'. Never in the history of art, he asserts, have pen, burin, knife been used so forcefully as in the German Renaissance. See Nagel, 'How German is it?' *J. Art*, December 1990, 4: 50. See also Martin Kemp, *The Science of Art* (New Haven, 1990), 62.

[54] Kemp, *op. cit.* (note 53) 109; Nagel, *op. cit.* (note 53), 50.

[55] *D. Martin Luthers Werke*. Kritische Gesamtausgabe (Weimar, 1883–), **62**, 36–7. See *Luther's Works*, vol. 1, ed. Jaroslav Pelikan (St Louis, 1958), 49.

[56] Alpers, *op. cit.*, (note 51), xxiv.

Figure 6.7. Hevelius, *Selenographia* (Gdańsk, 1647). Frontispiece by Adolph Boij.

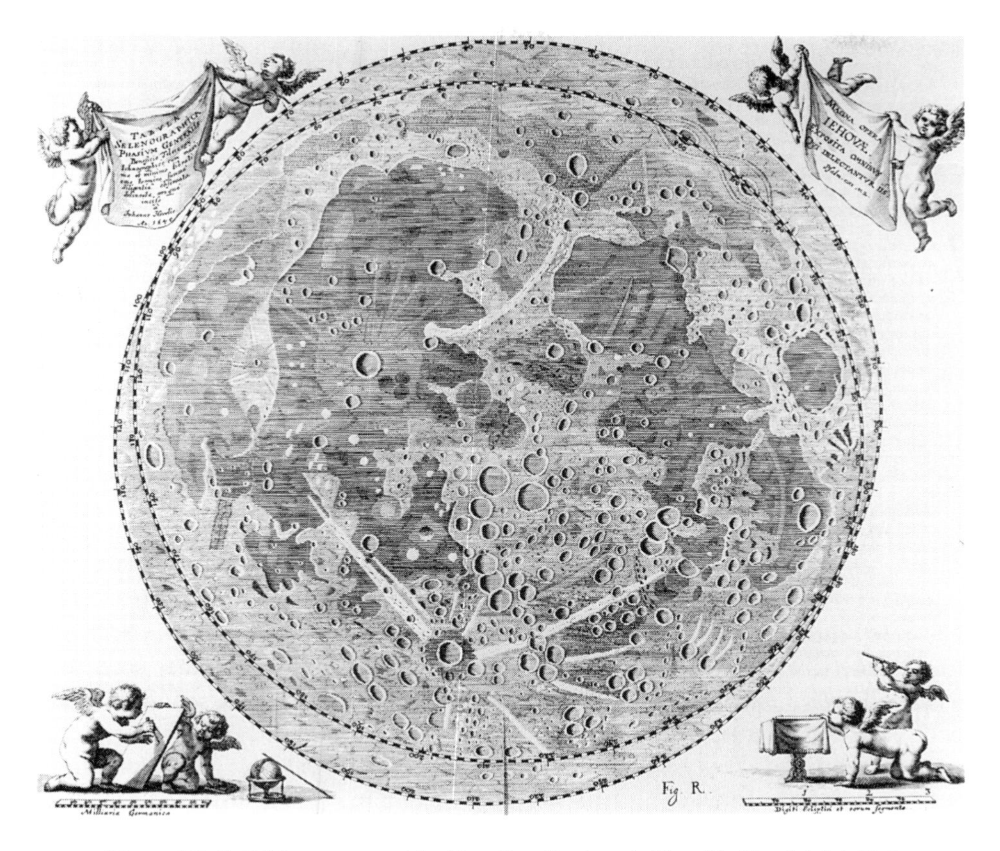

Figure 6.8. Full Moon engraved by Hevelius. Putti probably added by Adolph Boij. *Selenographia* (Gdańsk, 1647), pp. 262–3.

Hevelius's engravings. His plan makes little room for fanciful illustration because the focus is on visual description. At a few key places, however, Hevelius allows for an iconographical programme, executed by a 'professional'. In these instances – the frontispiece and in two illustrations of the full Moon – human figures and putti adorn the text. On the frontispiece (Fig. 6.7), Contemplatio, a star on her brow, emerges from a break in the clouds riding an eagle and holding a telescope. To her right is a Hevelius Moon, to her left, the Sun. Beneath her two putti unfurl a scroll with the text of *Isaiah* 40:26: 'Attollire in sublime oculos vestros, et videre qui creaverit ista', or (in the King James version) 'Lift up your eyes on high and behold who hath created these things'. Inside the book, two putti fly above Hevelius's visual description of the Moon (Fig. 6.8, top right), proclaiming these words from Psalm 111: 'Magna opera IEHOVAE exposita omnibus qui delectantur iis', or 'The works of the Lord are great, sought out of all that have pleasure therein'. In other words, Hevelius has sought out the works of the Lord, has lifted up his eyes to behold who created these celestial things. The scriptural quotations imply the meaning Hevelius found in what the eye can see: he looked at the heavens to see their Creator, and that gave weight to his endeavours as a witness and as an illustrator.

Selenographia was a model of how to go about communicating astronomical information visually. The principle had to be that the observer represented what he saw on paper as faithfully as possible, without intermediaries, distortion, or embellishment, and then to make sure that what he had drawn was passed on to the reader as faithfully as possible through the engraving and printing process. Others could not live up to this ideal method because they did not have the entire range of skills. No one after Hevelius in the seventeenth century could build his own instruments, observe, draw, and engrave. In his *Micrographia* of 1665 Robert Hooke blamed errors on the 'graver'.[57]

In the second half of the seventeenth century pictures of heavenly phenomena observed through the telescope became common, and these pictures spoke for themselves. In fact, they sometimes contained information that was not explained until much later. Campani's[58] and Hooke's[59] observations in the 1660s of what turned out to be the C ring (or crêpe ring) of Saturn (discovered in 1848) are good examples of this. Hevelius's authority flowed from his copious use of visual evidence whose reliability he was able to impress upon the reader. Yet, by the end of the century *Selenographia* was already obsolete. As others discovered that some of his observations were in error, the influence of Hevelius's work in telescopic astronomy waned, and by the time of his death it had been superseded. His method of visual communication had, however, by then become an accepted part of astronomy and science in general. Indeed, it had become so common that this achievement of Hevelius had become invisible.

[57] Robert Hooke, *Micrographia* (London, 1665; Brussels, 1966), e.g., 181–2. For a discussion of Hooke's approach to visual representation, see Michael Aaron Dennis, 'Graphic understanding: instruments and interpretation in Robert Hooke's *Micrographia*', *Sci. Context*, 1989, 3: 309–64.

[58] Giuseppe Campani, *Ragguaglio di due Nuove Osservazioni* (Rome, 1664), 18. See also *OCCH*, V: 118.

[59] 'A late Observation about Saturn made by [Robert Hooke]', *Phil. Trans.*, 1666 (no. 14), 1: 246–7, plate. See also A. F. O'D. Alexander, *The Planet Saturn* (London, 1962), 108–9.

7

Mathematical sciences and military technology: the Ordnance Office in the reign of Charles II

FRANCES WILLMOTH

The Ordnance Office was, to quote a seventeenth-century writer,

> always . . . of great account and importance, as being the only standing and grand magazine of the principal preparatives, habiliaments, utensils and instruments of war, as well by sea as by land, for the defence and safety of the kingdom.[1]

It was also a long-established centre for practical mathematics, since mathematics was recognized as having numerous military applications; some of the principal Ordnance Officers, especially in the later part of the century, are known to have also taken an interest in wider aspects of the mathematical sciences, and thus to have shared in some of the concerns of the early Royal Society. Rupert Hall, in his *Ballistics in the Seventeenth Century* (1952), acknowledged these contacts briefly while arguing that there was virtually no connection between the analysis of the paths of projectiles carried out by academic mathematicians and the actual rule-of-thumb practices of gunners.[2] The effect of much recent work by others, however, has been to draw attention to the existence of a continuum of social and intellectual relationships by which the abstract theoretical speculations of mathematicians and natural philosophers were linked, directly and indirectly, with the practical skills of craftsmen.[3] In this continuum Ordnance officers had the opportunity to occupy an important, perhaps uniquely influential, place. It is therefore worth examining the Office's activities in some detail: looking first at its official duties and the character of its staff,

The following abbreviations are used in this paper. RGO, Royal Greenwich Observatory papers in the Cambridge University Library; PRO WO, War Office Papers in the Public Record Office (Kew Branch).

[1] Edward Chamberlayne (1616–1703), *The Present State of England* (1st part, London, 1669, 7th edn, 1673; 2nd part, London, 1671, 4th edn, 1673), part 2, p. 224, quoted by H. C. Tomlinson, *Guns and Government* (London, 1979), before the Preface. Details that follow here come from Tomlinson unless otherwise stated.

[2] A. R. Hall, *Ballistics in the Seventeenth Century* (Cambridge, 1952); similar arguments appear in summary form in Hall's 'Gunnery, science and the Royal Society', in John G. Burke ed., *The Uses of Science in the Age of Newton* (Berkeley and Los Angeles, 1983), 111–41.

[3] For example in work by J. A. Bennett: especially 'The mechanics' philosophy and the mechanical philosophy', *Hist. Sci.*, 1986, **24**: 1–28, and 'Robert Hooke as mechanic and natural philosopher', *Notes Rec. Roy. Soc. Lond.*, 1980, **35**: 33–48.

and then at its dealings with the outside world, particularly with the Royal Society. This in turn reveals something about the Royal Society's attitude to technological improvements.

The Ordnance Office

A form of organization created in the time of Henry VIII remained the basis for the Ordnance Office's operations in the seventeenth century, but they nevertheless underwent frequent rapid expansions in size and scope. The Office had by that time ceased to be directly engaged in the manufacture of armaments, but from its headquarters in the Tower of London (Fig. 7.1) supplied firearms, ammunition, gunpowder and other weapons to infantry and cavalry, and heavier guns, of iron or bronze, and a great variety of related stores, to fortresses, garrisons, armies in the field, and the ships of the Royal Navy. When an artillery train was equipped, the technical and administrative staff to accompany it had also to be provided. During the 1660s and 1670s arming the Navy was the chief concern because the Second and Third Dutch Wars (of 1665–67 and 1672–74) were fought almost entirely at sea, but demands of other kinds also increased: colonial expansion brought new commitments and stores were held at a growing number of 'outports', at home and overseas. In 1667 the Ordnance Office was given the task of constructing and maintaining all fortifications in the kingdom, in 1670 it took on the care of the old Armoury housed in the Tower, and from 1682 it was required to provide training for the gunners at all other garrisons. Thus during the second half of the century the Ordnance's total workforce (mostly clerks, storekeepers, artificers, and labourers) increased from around 175 to over 400; an annual expenditure of about £12,000 in the early 1660s mounted to a peak of over £70,000 in the 1670s and settled at about £50,000 in the 1680s. Adept financial management and sound credit helped sustain this level despite underfunding, and the most recent detailed study of the subject indicates that most of the demands made upon the Office at this period were adequately met.[4]

As any government relied heavily upon the efficiency and loyalty of Ordnance personnel, the Office continued to be of immense political importance. At the same time the Tower served as a fortress, housing a garrison to keep watch over the City of London, and accommodated the Royal Mint (which paralleled the Ordnance Office in its demand for personnel with technical skills, such as metalworkers, assayers and engineers).[5] The head of the Ordnance Office, the Master General, was usually a prominent political figure and influential as a military adviser to the Crown. Between 1664 and 1670 this post was entrusted to commissioners, but one of them, Sir Thomas Chicheley, then continued as sole Master until 1679. The Master or commissioners made general policy decisions and received instructions from the Crown or Privy

[4] Tomlinson, *op. cit.* (note 1), Conclusion, esp. 219–22.
[5] John Craig, *The Mint* (Cambridge, 1953), is the most detailed account.

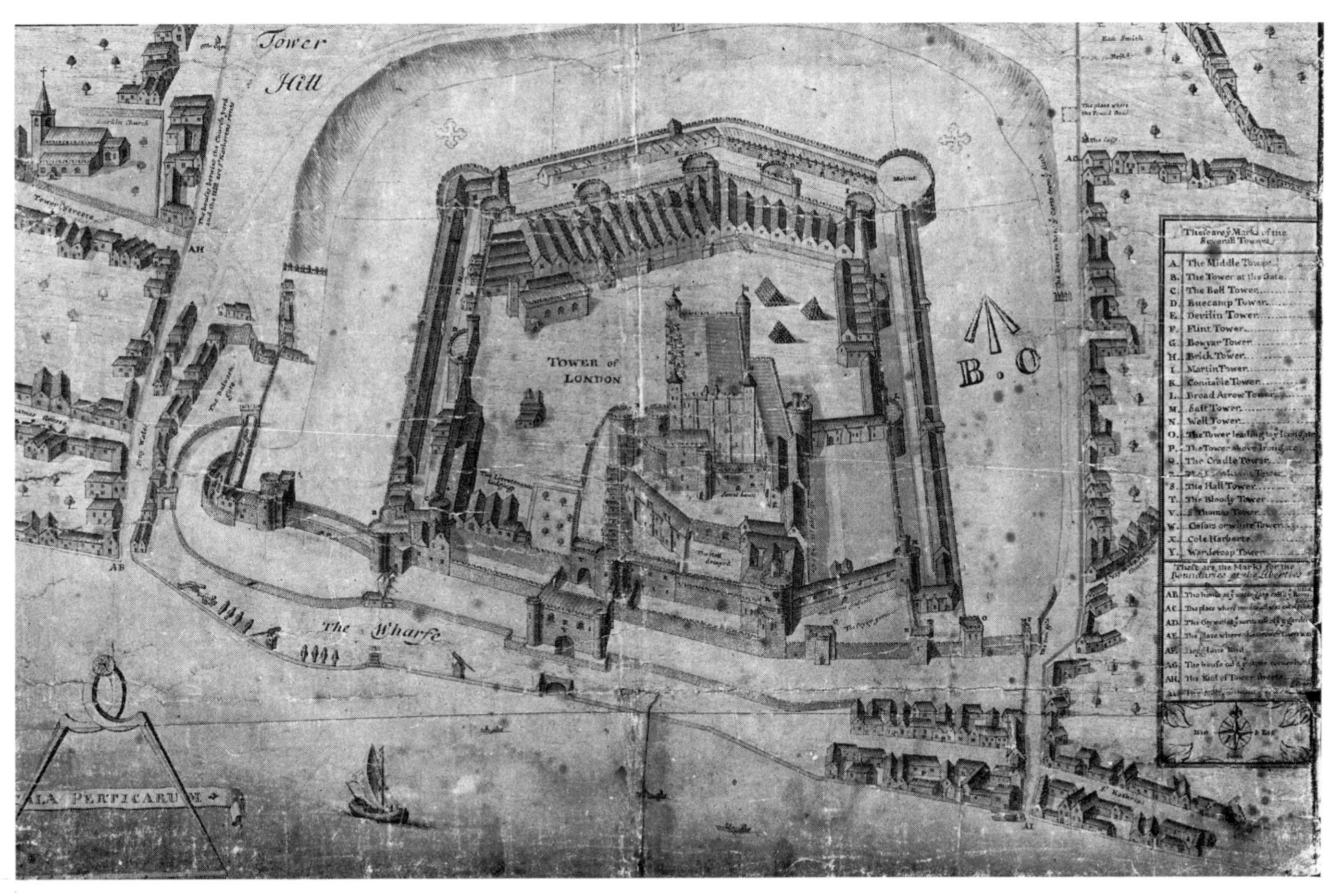

Figure 7.1. An extract from the earliest known detailed plan of the Tower of London, drawn by William Hayward and John Gascoyne in 1597; it survives only in this eighteenth-century copy. (Reproduced by permission of the Museum of London.) In the sixteenth century the Ordnance Office occupied buildings in the north-west corner of the site, near the Tower Chapel; in the next century new offices were built for it in the complex south of the White Tower. The official residences of the Clerk, Surveyor and Storekeeper were in the block adjoining the Constable Tower, on the north-eastern side of the central courtyard. Note the cannon-balls stacked up in that area, and gun-barrels laid out along the Wharf.

Council via a Secretary of State, or from the Admiralty or the Navy Board, for transmission to a board of principal officers who were collectively responsible for the day-to-day conduct of business. The leader of this Board was the Lieutenant-General, often a courtier, who had oversight of estimates and contracts. Next to him in rank were the Clerk, the Storekeeper, and the Surveyor-General, who were all of roughly equal standing and senior to the Clerk of Deliveries and the Treasurer, the last a post created in 1670. All these and a few other senior officers, notably the Master Gunner and Chief Engineer, were Crown nominees, appointed by Letters Patent under the Great Seal; at this period they were often beneficiaries of patronage exercised by the Duke of York.

Each principal officer had specific responsibilities: the Clerk for administration and keeping the main series of records, the Storekeeper for the safe custody of stores and the Clerk of Deliveries for issuing them. The Surveyor assessed the quality and quantity of goods on their initial receipt or return, supervised the Proofmaster's testing of guns and ammunition, and surveyed land and building work when fortifications were constructed; he thus had to be a competent practical mathematician, skilled in the measuring techniques of what was then termed 'gauging' as well as in geometrical land surveying. The others could in theory get away with having only sufficient numeracy to keep accurate accounts (or the slightly higher level needed for fiddling them), but in practice all of them acquired some familiarity with the more technical aspects of the Office's activities. This was partly because of the way in which work was shared in busy periods and partly because, in order to reduce opportunities for corruption, it was laid down that many procedures were only to be carried out by officers jointly and many decisions could be made only by the whole Board (three members making up a quorum).

Amongst the Office's employees, mathematical skills were essential for the engineers, the Firemaster (who dealt with gunpowder and explosive devices) and the Master Gunner, and were recognized as useful, if not absolutely necessary, for the ordinary gunners and 'fireworkers'. The most prestigious of all these were the engineers, whose main function was to design, build and demolish fortifications. They nominally belonged to the military side of the service and had the rank of army officers under the direct command of the Master General, though at this period the distinction between civil and military sections was unimportant and did not prevent the exchange of personnel between the different specialist groups. The numbers of engineers employed grew during the second half of the century and a second and third in command were appointed to assist the Chief Engineer; according to the comprehensive set of Instructions for the conduct of the Office drawn up in 1683, none would be appointed unless 'well skilled in all parts of the mathematics . . . and . . . perfect in architecture civil and military'.[6] At its highest levels military engineering was so skilled a trade that it had international currency and it was not unusual for men of foreign origin to enter English service: the Chief Engineer Sir Bernard De Gomme was

[6] Tomlinson, *op. cit.* (note 1), 57.

Dutch, his assistant and successor Martin Beckmann was Swedish, and so was the Chief Firemaster Ernst Heinrich De Ruis. No convenient list of the full complement of engineers is available, but one of fireworkers suggests that foreigners were also recruited at the lower levels; the supply of English candidates with equivalent experience grew only slowly despite occasional attempts to boost it by sending Ordnance employees abroad to secure training at royal expense.[7]

Gunners were not drawn from so far afield, but nor were they as uneducated or unskilled as writers hoping to sell books of instruction sometimes asserted.[8] Gunnery was still a respectable employment for a gentleman's impecunious younger sons, and a listing of 1664 shows that many of its exponents came from skilled trades: several were associated with shipping or shipbuilding, one was a master smith in the Office, more were among its clerks, and one, Joseph Hone, was a mathematical instrument maker with a shop on Tower Wharf.[9] The Master Gunner was responsible for training, and from 1683 at least it was intended that a regular programme of instruction should be provided.

Mathematics and mathematicians

Given all this, and given that military applications had long been accorded an important place in the mainstream of practical mathematics, it is no surprise to find that the Ordnance Office was accustomed to employing mathematicians who applied their skills to a wide range of interests. This habit was established by the late sixteenth century when the scholar, mathematical practitioner and engineer Thomas Bedwell was Storekeeper; in the early seventeenth century the Storekeeper at Deptford was John Wells, a former seaman, writer on dialling, and friend of Henry Briggs, Edmund Gunter and Henry Gellibrand.[10] His contemporary John Reynolds, an instrument-maker in Tower Street, became Master Gunner of England (and his son of the same name Assistant Master of the Mint); one of Reynolds' pupils, Robert Norton, was engineer at the Tower from 1627, a mathematical practitioner and the translator of Stevin's *Disme (The Art of Tenths)*.[11] William Oughtred's pupil and opponent Richard Delamain petitioned for an engineer's place after being employed to measure fortifications; Thomas Rudd, Chief Engineer in the 1630s and for the Royalists in the 1640s, later published a practical geometry and a Euclid; and in the 1650s one of the gunners was Thomas Rice, a pupil of Samuel Foster's and friend of John Collins,

[7] *Ibid.*, 237–8.

[8] A. R. Hall in 'Gunnery, science and the Royal Society', *op. cit.* (note 2), 112–14, adopts a bleak view based upon printed evidence and anecdotes from a later period; material presented in ch. 2 of his *Ballistics op. cit.* (note 2), suggests the true picture was more complex.

[9] Tomlinson, *op. cit.* (note 1), 49; for Hone see also E. G. R. Taylor, *Mathematical Practitioners in Tudor and Stuart England* (Cambridge, 1954), 263.

[10] For Bedwell, see S. Johnston, 'Mathematical practitioners and instruments in Elizabethan England', *Ann. Sci.* 1991, **48**: 319–44; for Wells, Taylor, *op. cit.* (note 9), 199.

[11] *Ibid.*, 186, 194; Norton *ibid.*, 189.

who described him as an expert maker of dials.[12] This is by no means a complete list.

It seems possible, too, that the Office played a part in sustaining a practical mechanical tradition as well as a mathematical one. The best known evidence of this concerns the Ordnance workshop at Vauxhall in the 1630s and 1640s, where ingenious contrivances of many kinds were constructed by the Dutch mechanic Caspar Kalthof for his patron Edward Somerset (Earl of Glamorgan, better known under his later title of Marquis of Worcester).[13] The Marquis's *Century of . . . Inventions*, written in 1655 and published in 1663, reflects the activities of the establishment as a whole in that more than a quarter of the inventions it describes had a strictly military purpose, and it seems particularly appropriate that one of the means of 'driving up water by fire' (since claimed as a primitive steam-pump) was made out of a broken cannon.[14] However, by the time the book was compiled the Marquis had been ousted from the Vauxhall workshop, which was run down and sold off in the 1650s; any of its activities that were carried on were presumably accommodated at the Tower until a similar establishment, with facilities for proving guns and a fireworkers' 'laboratory' but no independent patron, was set up at Woolwich in 1681.[15]

In the two decades following the Restoration the Ordnance's principal officers included several who were known for the breadth of their interests in mathematics and the mathematical sciences. These were the Clerk, Edward Sherburne, who had succeeded his father in the office in 1641, secured reinstatement in 1660 and received a knighthood in 1683; Jonas Moore, knighted in 1673, who was Assistant Surveyor from 1665 and Surveyor from 1669 until his death ten years later (Fig. 7.2); and (Captain) George Wharton, who was chief clerk to the Lieutenant-General from 1660 and Treasurer from 1670 until 1681. Sherburne had an established reputation as a poet specializing in publishing translations of classical pieces into English verse; amongst these was *The Sphere of Marcus Manilius* (1675) (Fig. 7.3), which he related to the concerns of modern astronomers by means of extensive annotations and supplemented with a long 'Astronomical Appendix' including 'A Catalogue of the most Eminent Astronomers, Ancient and Modern'.[16] That the author had been interested in astronomy since the 1650s is indicated by the inclusion of his own observations of a comet, and by a letter he wrote from France at that time seeking information about a solar eclipse on behalf of the Duke of Orleans' professor of mathematics.[17] Sherburne's common-place books, now in the Sloane collection, contain not only the

[12] *Ibid.*, 201; for Rudd *ibid.*, 357–58 and *DNB*; for Rice Taylor, *op. cit.* (note 9), 222, and John Collins, *Geometricall Dyalling* (London, 1659), preface.

[13] Arthur MacGregor, 'A magazin of all manner of inventions', *J. Hist. Coll.*, 1989, **1**: 207–12, and sources there cited.

[14] Worcester, *A Century of the Names and Scantlings of such Inventions, As at present I can call to mind to have tried and perfected* . . . (London, 1663), Invention 68, pp. 46–7.

[15] A. Forbes, *A History of the Army Ordnance Services* (3 vols., London, 1929), **1**: 106; Tomlinson, *op. cit.* (note 1), 120–2, also mentions a laboratory at Greenwich from 1683.

[16] For biography see introduction to F. J. Van Beeck, ed., *The Poems and Translations of Sir Edward Sherburne* (Assen, 1959), Charles Davies Sherborn, *A History of the Family of Sherborn* (London, 1901), 82–6, and *DNB*.

[17] Letter Bodleian MS. Ashmole 243 f. 292, printed in C. H. Josten, ed., *Elias Ashmole* (5 vols., Oxford, 1966), **2**: 688–9.

Figure 7.2. The frontispiece to Sir Jonas Moore, *A New System of Mathematicks* (London, 1681) by N. Yeates. (Reproduced by permission of the Royal Greenwich Observatory.) Little is known about this engraver, except that a few years later he worked on Thomas Phillip's views of Tangier before and after the destruction of the Great Mole in 1684. The figures at the top of the frontispiece are presumably Geometry on the right (or perhaps Geography or Cosmography?) and Astronomy on the left: this is a New Astronomy since her globe-stand is empty and she is about to write upon a blank sheet. For details of the work itself, see my *Sir Jonas Moore*, *op. cit.* (note 20), chapter 5.

draft of *Manilius* and notes on classical themes, but extensive extracts from works on astronomy, geometry, perspective, fortification, the finding of longitude by astronomical methods, optics, and natural philosophy.[18] The catalogue of his library embraces an equally wide spread of subjects.[19]

Moore had derived his livelihood from mathematical practice from the 1640s onwards and had become quite well known as a teacher of mathematics and as Surveyor to the fifth Earl of Bedford's fen drainage company.[20] His *Arithmetick* of 1650 included an unusually extensive algebra section, reflecting William Oughtred's influence; appended to the 1660 edition were *Contemplationes Geometricae In Two Treatises*, namely *A New Contemplation Geometrical upon the Ellipsis* and a short piece on *Conical Sections* derived from Claude Mydorge by way of Oughtred. The *New Contemplation* included a reprint of the brief *Resolutio Triplex*, first published in 1658 as a reply to a challenge set by 'Jean de Montfert'; to issue this was to enter into competition with other problem-solvers including the Savilian Professor of Geometry, John Wallis and the Gresham Professor of Astronomy, Christopher Wren. All these publications, then, helped establish Moore's claim to competence in theoretical matters, and some friends, at least, seem to have been convinced: John Aubrey boldly characterized him as 'a good mathematician and a good fellowe'.[21] After the Restoration Moore enjoyed the patronage of the Earl of Sandwich and the Duke of York, and may first have come into contact with the Ordnance Office when employed to survey and help plan a Mole (a fortified harbour of extremely ambitious proportions) for the newly-acquired (1662) town of Tangier.

Following his appointment as Surveyor-General of the Ordnance in 1669, Moore continued to produce mathematical books. His *Modern Fortification: or, Elements of Military Architecture* (1673) was a concise but comprehensive manual containing preliminary information about geometry, surveying, and the measures of various countries, and complete instructions for planning, building, besieging and defending

[18] British Library MS. Sloane 824, 829, 832, 837. [19] *Ibid.*, MS. Sloane 857.

[20] For his career see my *Sir Jonas Moore: practical mathematics and Restoration science* (Woodbridge, 1993), a revised version of my Ph.D. dissertation, 'Sir Jonas Moore (1617–79): practical mathematician and patron of science' (Cambridge, 1990).

[21] Andrew Clark, ed., *'Brief Lives' . . . set down by John Aubrey* . . . (2 vols., Oxford, 1898), 2: 80. Only the first epithet was a compliment, as elsewhere Aubrey notes that 'to be habitually a good fellow' is 'to drink every day wine with company, which, though not to drunkennesse, spoiles the braine' (*ibid.*, 1: 350).

Figure 7.3. The frontispiece to Edward Sherburne, *The Sphere of Marcus Manilius made an English Poem: with Annotations and an Astronomical Appendix* (London, 1675). (Reproduced by permission of the Syndics of Cambridge University Library.) The signature of the engraver, Wenceslaus Hollar, appears at the foot below the armillary sphere. The classical figures depicted are: Urania, the Muse of Astronomy, who 'shows both the ways of the heavens and the stars'; Pan, representing 'the Universe of Nature', with a starry cloak and beneath it scenery reminiscent of one of John Speed's maps; and Hermes/Mercury, the god of eloquence, 'Interpreter of the Universe'. The work concludes by exhorting readers to support moves to found an English observatory in emulation of the Scots.

Figure 7.3. For caption see opposite

all kinds of fortifications according to the instructions laid down by a variety of continental authorities. The following year saw the first appearance of the *Mathematical Compendium*, which was to be the most often reprinted of all Moore's works. It was edited by his servant Nicholas Stephenson and there can be no doubt that it was, as its title page stated, collected out of Moore's papers. The selection must have been made with his approval. A fairly accurate impression of the content is given by the subtitle:

> Useful Practices in Arithmetick, Geometry, and Astronomy, Geography and Navigation, Embattelling, and Quartering of Armies, Fortification and Gunnery, Gauging and Dyalling. Explaining the Logarithms, with new Indices; Nepair's Rods or Bones; making of Movements [for clocks], and the Application of Pendulums: with the Projection of the Sphere for an Universal Dyal, &c.

It may be added that it began in almanac style with a diary, times of high water at London Bridge, information about tides, newly calculated by royal command, and a perpetual calendar with subsidiary tables. The second edition of 1681 was dedicated to the then Lieutenant of the Ordnance, George Legge, who was said to have perfected himself 'in all the parts of the Mathematicks' and applied them to 'Navigation, Fortification, the Art of War' and gunnery. This is insufficient to prove that Legge should be included in a list of Ordnance officials with mathematical interests but other evidence confirms at least that he took an interest in mathematical navigation.[22]

Sir George Wharton, like Moore, had once been associated with the mathematics teacher William Milburne, and he had gone on to serve as the leading royalist astrologer throughout the Civil War and Interregnum.[23] After 1660 he continued to produce almanacs as well as taking on extra Ordnance business (as keeper of saltpetre from 1663, and briefly clerk of deliveries); both occupations were profitable and at his death in 1681 he left a fortune of over £6,000.[24] His eldest surviving son, Sir Polycarpus, became a gunpowder contractor and married Sherburne's niece.[25] Though Sherburne later lost his post and complained of poverty, both he and Wharton could well afford to join Moore in making benefactions: a minor but typical example appears in the catalogue of the library founded by a London vintner at Hawkeshead Grammar School (Lancashire) in the late 1670s to which Moore donated a 'Juvenal & Persius', Sherburne a copy of his *Sphere of Marcus Manilius*, and Wharton Claude Milliet de Chales's three-volume *Cursus seu Mundus Mathematicus*.[26]

There are several ways in which the outlook and interests of the principal officers may have influenced the long-term character of the institution they served. The 'Instructions for the Government of Our Office of Ordnance' approved by the King in

[22] Edwin Chappell, ed., *The Tangier Papers of Samuel Pepys*, Navy Records Society, 73 (1935), 126–8.
[23] Biographies in Edward Sherburne, *The Sphere of Marcus Manilius* (London, 1675); *Athenae Oxonienses* 3rd edition (5 vols., London, 1813–30), 4: cols 5–9; *DNB*.
[24] Tomlinson, *op. cit.* (note 1), 84.
[25] *DNB*.
[26] Richard Copley Christie, *The Old Church and School Libraries of Lancashire* (Chetham Society New Series 7, Manchester, 1885).

1683 were drawn up by the then Master, George Legge (Lord Dartmouth), or according to Sherburne by Sherburne, and formed the basis of the Office's organization until the middle of the nineteenth century. These should probably be seen primarily as a codification of established practice, but the care with which procedures were laid down and the stress on qualifications and training appear to have been comparatively new.[27] At the same time, the skills of the workforce may have been enhanced through the patronage which all the principal officers had opportunity to exercise. The right of appointment to most posts belonged to the Master, but he often simply confirmed nominations submitted to him by members of the Board.[28] The result, during the 1670s at least, seems to have been that men with mathematical ability were favoured. Thus in 1673 a gunner's place was bestowed upon Michael Dary, a mathematician of sufficient standing to have corresponded with James Gregory and Isaac Newton (often through John Collins, who had previously found him employment in the Farthing Office).[29] He fulfilled his duties as a gunner conscientiously enough, by Collins' account, and later dedicated his *Interest Epitomized* to Moore, Sherburne, Wharton, and an official of the Mint, James Hoare, who had jointly funded its publication.[30] Then there was Nicholas Stephenson, Moore's editor, who served as a 6d-per-day gunner from 1674 and second clerk to the Surveyor from December 1673 to at least September 1678, while continuing, as clerks not uncommonly did, to act as a personal servant to his principal officer.[31] His mathematical competence is indicated not only by his association with the *Compendium* but also by his producing the *Royal Almanack* for Moore in subsequent years (1675–78). It might also be appropriate to include here Jonas Moore junior, who was 'bred in the Office' as a clerk and a gunner and sent on a tour to study the fortifications of Europe before being granted the reversion of his father's post; he was to prove unsatisfactory as Surveyor, but more because of his arrogance, and consequent inability to work harmoniously with his colleagues, than through technical incompetence.[32]

The Ordnance Office and the Royal Society

All this is sufficient to show that Ordnance personnel had both practical mathematical concerns and wider mathematical interests; we must now consider whether this had

[27] This document is discussed in Forbes, *op. cit.* (note 15) vol. 1, ch. 5, 96; and in Tomlinson, *op. cit.* (note 1), 16–17.
[28] Tomlinson, *op. cit.* (note 1), 73–5.
[29] See Collins's correspondence with Dary and others in Stephen J. Rigaud, ed., *Correspondence of Scientific Men of the Seventeenth Century* (2 vols., Oxford, 1891); and H. W. Turnbull, *James Gregory Tercentenary Memorial Volume* (London, 1939), 327 and see index.
[30] Michael Dary, *Interest Epitomized, Both Compound and Simple* with *A Short Appendix for the Solution of Adfected Equations in Numbers by Approachment* (London, 1677).
[31] PRO (Kew) WO 54/32; Tomlinson, *op. cit.* (note 1), 227, 48.
[32] PRO (Kew) WO 54/27 *et seq.*; travels WO 54/31 and *Calendar of State Papers, Domestic 1671–72*, 261. He and not his father translated Tomaso Moretii's *General Treatise of Artillery* (London, 1683) (described in Hall, *Ballistics op. cit.* (note 2), 50); its inadequacies cannot necessarily be taken to reflect especially badly upon either of them, since it was published only after both their deaths.

any signficance beyond the walls of the Tower. The Office seems to have had some links with the Royal Society from the early 1660s, through such Fellows as William Winde, a military officer and student of civil and military architecture who helped fortify Gravesend and Tilbury in the late 1660s, and Thomas Povey, who served as Treasurer to the Duke of York and to the Committee for Tangier.[33] It thus came about that Moore (long before becoming a Fellow of the Royal Society himself) carried a sea-sounding device of Robert Hooke's on his trip to survey Tangier in 1663, supplied information about the use of gunpowder to quarry stone, and promoted Ralph Greatorex's diving engine in the hope that a version could be developed for use in the building of the Tangier Mole.[34] By 1670 Moore had become a close friend of Hooke's. He participated fairly actively in the Society's business after his election in December 1674, entertaining meetings with appropriate anecdotes and donating the foot of a sea-fowl and a West Indian hornets' nest to the Society's repository.[35] He made slightly more substantial contributions to the serious business of the Society when questions arose that were also of interest in Ordnance terms: once, for instance, he asked the Society's advice on the Office's behalf about methods of analysing the composition of gunpowder; occasionally he joined in discussions about the path of projectiles.[36] But there are no signs of his being attracted to experimental natural philosophy for its own sake.

This is an appropriate point at which to mention the ballistic experiments carried out in September 1674 to test the assertion in Robert Anderson's recently published *Genuine Use and Effects of the Gunne* that the trajectory of a bullet was a parabola and not affected by air resistance, which could be regarded as negligible.[37] Hall describes these experiments as undertaken by members of the Royal Society and thus considers them as evidence of the Society's (albeit limited) interest in ballistics.[38] Actually, as John Collins' report to James Gregory makes clear, they were carried out by and for the Ordnance Office, Anderson having 'published printed challenges against the gunners of the Tower, which hath caused the Master of the Ordnance, Sir Thomas Chicheley, Sir Jonas Moore, the Lord Brouncker and the gunners to go many times a shooting with mortar pieces at Blackheath'.[39] The recently appointed gunner Michael Dary was in all probability a leading instigator of the enterprise, since he had been carrying on a personal rivalry with Anderson for some years.[40] Brouncker was present in his capacity as a leading member of the Navy Board. Moore as an officer of the Ordnance (he was not yet a Fellow of the Royal Society); between them they drew Hooke into the discussions commented on in his diary, but on at least one occasion he

[33] Michael Hunter, *The Royal Society and its Fellows, 1660–1700* (Chalfont St Giles, 1982), 180 (F113), 168 (F47). Several of Povey's relatives held Ordnance posts.

[34] Thomas Birch, *History of the Royal Society of London*, 4 vols. (London, 1756–7), 1: 259, 307; *ibid.*, 330; *ibid.*, 370, 425, 433.

[35] *Ibid.*, 3: 385.

[36] *Ibid.*, 3: 402, 464–5, 474.

[37] For the content of this work see Hall, *Ballistics op. cit.* (note 2), 120–1.

[38] *Ibid.*, 125; Hall, 'Gunnery, Science and the Royal Society', *op. cit.* (note 2), 124–5, recognizes the professional military interests of some participants but still views them primarily as Fellows of the Royal Society.

[39] Turnbull, *op. cit.* (note 29), 286; see also 82–4.

[40] *Ibid.*, 73; the preface to Dary's *Miscellanies* (London, 1669) claimed that the work was a response to Anderson's attack on Dary's *Art of Practical Gauging*, ed. John Newton (London, 1669).

relied upon their account rather than attending the trials in person.[41] The results were in any case inconclusive; later experiments confirming the general conviction that Anderson was wrong were reported to Collins by John Flamsteed in 1677 and to the Royal Society by Moore in 1678.[42] In this first instance, however, the Society took no direct interest in the matter and may indeed have been in no fit state to do so, as in the autumn of 1674 its fortunes were at such a low ebb that its continued existence was said to be in doubt.[43]

Nevertheless there was enough life left in the Royal Society for Moore to be elected a Fellow shortly afterwards. This clearly happened because both he and the Society rather suddenly found it useful: he had already discussed with Hooke the feasibility of setting up an observatory at Chelsea College, which the Society owned, and was in the process of negotiating about the site.[44] As it turned out, though, the move proved abortive because Chelsea was found unsuitable, and it was only by making use of his official position and contacts at Court that Moore finally achieved his aim.[45] The building of the Royal Observatory at Greenwich was authorized in 1675 and completed in the following year.

The Royal Observatory at Greenwich

The Royal Observatory was thus very much Moore's personal project: he instigated the plan, paid for clocks and instruments, and, most significantly of all, persuaded his Ordnance Office employers and colleagues to lend their support.[46] The main structure of the large sextant was made by Tower smiths who worked at the Ordnance forge. This was situated amongst forges belonging to the Royal Mint, in buildings on the North side of the Tower, and was more usually employed in repairing weapons and armour. The celebratory prints of the Observatory commissioned by Moore were engraved by Francis Place from drawings made by Robert Thacker, the Ordnance's sketcher of fortifications.[47] The Office was also directly involved in a number of ways: providing funding and materials for the building, which was situated near the gunnery practice ground in Greenwich Park, and in the longer term accepting responsibility for maintaining it, for paying the astronomer's salary, and providing him with an assistant.

The relationship between the Royal Society and the Royal Observatory was an uneasy one from the start, as the Society was eager to see speedy results that would meet the needs of some of its own members and demonstrate the ability of the English to match French achievements; Moore's attempts to persuade or compel the astronomer, John Flamsteed, to supply the required data proved counter-pro-

[41] Henry W. Robinson and Walter Adams, eds., *The Diary of Robert Hooke* (London, 1935), 122 (23.9.1674).
[42] Rigaud, ed., *op. cit.* (note 29), 2: 172–4.
[43] Letter from Moore to Flamsteed (10.10.1674), RGO 1/37, f. 77.
[44] *Ibid.*; Birch, *op. cit.* (note 34), 3: 139. [45] *Ibid.*, 3: 161.
[46] For full details of instruments see D. Howse, *Greenwich Observatory: Vol. 3. Buildings and Instruments* (London, 1975).
[47] Reproduced in D. Howse, ed., *Francis Place and the Early History of The Greenwich Observatory* (New York, 1975).

ductive.[48] Wrangles of a similar kind continued in later years. The Society began to get the upper hand in 1712 when the appointment of a Board of Visitors to the Observatory was secured, but this did not by any means solve the problem.

To the Ordnance Office, in contrast, the project seemed worthwhile because of the Observatory's practical purpose as stated in the initial warrants: it was to develop astronomical methods of finding longitude at sea.[49] This was, of course, a matter of great commercial interest but could also be acutely important in naval warfare. The whole project was closely related, through Moore and others, to the development of the Royal Mathematical School at Christ's Hospital, where boys were to be trained up for the Royal Navy; had a workable astronomical method of finding longitude ever been produced, the skills taught there would have enabled it to be implemented.[50] Flamsteed himself was briefly required to instruct selected Hospital boys (although he swiftly repudiated this obligation on Moore's death); the names of Ordnance personnel appear regularly in his list of pupils and assistants over many years.[51] His participation in ballistic experiments in 1677, presumably at Moore's request, has already been noted. In the longer term, after Moore's death, it was the continuing goodwill of the principal Ordnance officers that enabled Flamsteed to retain his post and even to enjoy a degree of independence.

Conclusion: varieties of mathematics

The picture that emerges from all this is of the Ordnance Office as the institutional embodiment of an unfailing confidence in the power of mathematics to solve practical problems. This is directly expressed in the verse eulogy published after Moore's death in 1679 by one of his protégés, almost certainly Nicholas Stephenson.[52] The 135-line tribute (in rhyming couplets) does not mention Moore's membership of the Royal Society but concentrates on his practical achievements, attributing them specifically to his command of a mathematics that was effective in practice because firmly grounded in a theoretical understanding that produced perfect certainty. In contrast to the 'tedious scrutinie' proffered by the 'Dull Philosopher',

> Demonstration rules those noble Arts
> That so renown'd Renowned Moores great Parts
> The Man whose Genius mounted to the Skie
> And fetch'd from thence Infallibilitie.

[48] *Ibid.*, 454, 458; RGO 1/36, ff. 54–5, 59.

[49] Warrant in Francis Baily, *An Account of the Rev. John Flamsteed* (London, 1835; reprinted 1966), 111–12; from RGO 1/40, f. 60.

[50] E. H. Pearce, *Annals of Christ's Hospital* (London, 1908).

[51] Baily, *op. cit.* (note 49), 115, RGO 1/36, f. 58; Guildhall Library, London, MS. 12806/7, ff. 166, 171; RGO 1/15, ff. 165–6.

[52] Anon., *To the Memory Of my most Honoured Friend Sir Jonas Moore Knight, Late Surveyor General of His Majesties Ordnance and Armories* [London, *c.* 1679]. For identification of the author see my *Sir Jonas Moore*, *op. cit.* (note 20) 158–9.

This was essentially the view of mathematics championed a century earlier by John Dee in his *Mathematicall Præface* to the first English edition of Euclid's *Elements*. It was fairly widely accepted in England in the first half of the seventeenth century when, for example, the great teacher and figurehead William Oughtred held that mathematics must remain a unitary subject and its practice firmly based in theory. Echoes of this kind of approach may perhaps be traceable among those early members of the Royal Society whose thinking is sometimes characterized (following Frances Yates) as 'Vitruvian'.[53]

By the late 1670s, however, as the polemical tone of the eulogy indicates, the standpoint it represented could only appear controversial. The source of this change is to be found in debates about the proper nature and bounds of mathematics, which took place from around the middle of the century onwards.[54] It is impossible to do justice to their complex history in such a brief sketch as this, but to acknowledge their existence and the extent to which they raised doubts about the status and effectiveness of practice may at least help throw some light on the Royal Society's attitudes. While it was to be expected that the Ordnance Office would consistently defend the reputation of practical mathematics, there was no such unanimity to be found amongst those Fellows of the Society who had mathematical interests. These made up quite a small proportion of the membership: as Shapin's work on Boyle has shown, experimentalists might find reason to deny that mathematics had any relevance to their concerns, and many Fellows knew too little of the subject to feel qualified to comment on mathematical issues at all.[55] There was thus no incentive for the Society as a whole to become involved in mathematical controversies, and it seems to have generally distanced itself from them.

This is not to suggest that the Society ceased to draw upon an inheritance from the sphere of practical mathematics. Hooke in particular certainly continued to exploit it. But if such moves appeared increasingly problematic, this might go some way towards explaining how a former confidence in the application of knowledge to practical ends came to be partially undermined in the later decades of the century. A loss of confidence in practical mathematics and in the utility of knowledge in general might also help to explain why the Society's contributions to technological advance were at this stage less substantial than had once been hoped.

[53] Vitruvian influences were first highlighted in Frances A. Yates, *The Theatre of the World* (London, 1969), especially in ch. 2, which concerns Dee.
[54] For further details see my *op. cit.* (note 20), especially the introduction to the book.
[55] Steven Shapin, 'Robert Boyle and mathematics: reality, representation, and experimental practice', *Sci. Context*, 1988, **2**: 23–58.

8

Between *ars* and *philosophia naturalis*: reflections on the historiography of early modern mechanics

ALAN GABBEY

Many years ago, when I was a beginning research student working on what I then called Descartes's 'dynamical' thought, I came across a reference to a work of Tycho Brahe called *Astronomiae instauratae mechanica* [*The mechanics of the renewed astronomy*]. So off I went in innocent enthusiasm to consult the Dreyer edition, only to discover that this *Mechanica* was devoted entirely to the description and construction of quadrants, sextants, armillary spheres, astrolabes and other instruments, with plans of Uraniborg thrown in for good measure. There was nothing on the topics I half expected to find there, and which might have provided additional background material for my thesis. So I left Brahe and his astronomical apparatus and got back to the great men who did have something to say about bodies falling, colliding, and revolving here below and in the heavens above. Though puzzled, I was not unduly worried, at least *pro tem*, since Brahe's *Mechanica* made no appearance in the histories of mechanics that were my main reading at that time. Neither has it been studied in any book or article on 'the history of mechanics' that has appeared since, and its calculated omission as 'background' was not deemed a weakness by the examiners of my doctoral thesis.

Yet it remained a puzzle that a work with *mechanica* in the title, and by a major figure in 'The Scientific Revolution', had escaped the attention of all historians of mechanics. Something was wrong somewhere, either in my historical understanding of mechanics, or in the prevailing historiography of the discipline, or in both. As I soon discovered, there was nothing wrong with Brahe's understanding of what he was doing in his *Mechanica*, where he assigned astronomical instrumentation to 'a mechanical part of art'.[1] That is, astronomical instruments are not natural objects (as Galileo discovered to his cost), like stones, smoke, waterfalls, planets, or apples.

Neither are levers, wedges, pulleys, screws, and windlasses. They too belong to 'a mechanical part of art', and their unnatural purposes are to move heavy objects away

[1] 'Appendix de architectonicis structuris astronomicis observationibus accommodis', in Tycho Brahe, *Astronomiae instauratae mechanica*, (Uraniborg, 1598), *TBDOO*, 5: 134.

from the centre of the earth, or to force bodies to behave or act on each other in ways they would never do if left alone, and all to secure perceived practical benefits for humanity. Left to their own natural devices, so to speak, bodies, animate as well as inanimate, never do enough to satisfy our practical needs. Hence the early emergence of the practical or mechanical arts, and of the distinction between them and theoretical or speculative disciplines, particularly (in this context) the distinction between mechanics and natural philosophy. Since antiquity, the construction and operation of machines and devices that re-arrange things *contra naturam* and for human ends had been the concern of mechanics *qua* practical art. On the other hand, the theoretical discipline physics did not share the concerns of mechanics, since physics did not deal with artificial things *qua* artificial. Traditional Peripatetic definitions of *physica* (equivalently *philosophia naturalis* and *physiologia*) typically included the phrase 'the science [*scientia*] of natural bodies, *in so far as they are natural*'.

However, already in Pseudo-Aristotle's *Mechanica*,[2] as in Book 8 of the *Collections* of Pappus of Alexandria, mechanics was also a theoretical discipline in that it dealt mathematically with problems arising out of the construction and use of machines. With the rediscovery of the *Mechanica* in the Renaissance, mechanics became a developed *scientia media*, like optics, operating 'midway' between mathematics and physics: the treatment was mathematical, dealing with bare quantity separated from natural body, while the subjects treated were physical and sensible. As Pseudo-Aristotle expressed it, mechanical problems 'are not altogether identical with physical problems, nor are they entirely separate from them, but they have a share in both mathematical and physical speculations, for the method is demonstrated by mathematics, but the pratical application belongs to physics.'[3] Mechanics was therefore 'subalternated' to both mathematics and physics, though in the Renaissance and early seventeenth century there was considerable argument as to which of the two parent sciences was the principal one.[4] These disciplinary distinctions explain a significant difference between treatises on mechanics in the early modern period and Peripatetic manuals on natural philosophy. The treatises on mechanics were written by mathematicians from different walks of life, not by textbook authors working within the Peripatetic tradition. Conversely, very few (if any) Peripatetic writers on natural philosophy published anything on mechanics.[5]

[2] Aristotle, *Minor Works*, trans. W. S. Hett, Loeb Classical Library, vol. 14 (London and Cambridge, 1980, 1st edn. 1936), 331. The influential *Mechanica* was a collection of thirty-five mechanical problems and puzzles, written possibly by Aristotle's pupil Strato, and first published in 1497. See further Paul Lawrence Rose and Stillman Drake, 'The Pseudo-Aristotelian *Questions of Mechanics* in Renaissance culture', *Stud. Ren.*, 1971, **18**: 65–104. François de Gandt, 'Les *Mécaniques* attribuées à Aristote et le renouveau de la science des machines au XVIe siècle', *Etudes Phil.*, 1986: 391–405.

[3] Aristotle, *Minor Works*, 331.

[4] The story is quite complicated. See W. R. Laird, 'The scope of Renaissance mechanics', *Osiris*, 1986, **2**: 43–68, and W. R. Laird, 'Giuseppe Moletti's *Dialogue on Mechanics* (1576)', *Ren. Quart.*, 1987, **40**: 209–23; William A. Wallace, chapter on 'Traditional natural philosophy' in *The Cambridge History of Renaissance Philosophy*, eds. Charles B. Schmitt, Quentin Skinner, Eckhard Kessler, assoc. ed. Jill Kraye (Cambridge and New York, 1988), 201–35, and William A. Wallace, *Galileo and his Sources: the Heritage of the Collegio Romano in Galileo's Science* (Princeton, 1984), especially pp. 202–16.

[5] On the mathematicians' mechanical treatises, see Stillman Drake and I. E. Drabkin (eds.), *Mechanics in Sixteenth-Century Italy. Selections from Tartaglia, Benedetti, Guido Ubaldo, & Galileo* (Madison, 1969). Not all the

These points will be familiar to students of ancient, medieval or Renaisance natural philosophy and mechanics, yet it seems to be almost exclusively in these areas that serious attention has been paid to the implications of the disciplinary distinctions between the mechanical arts, mechanics as a *scientia media*, and natural philosophy. These issues have rarely been mentioned in the history of mechanics and dynamics in the century or so *after* Galileo. The five simple machines, and developments based on them, are not invariably allotted full-scale treatment in histories of post-Galilean 'mechanics', nor do those histories tell their story against the background of the disciplinary distinctions I wish to emphasize. Instead, histories of 'mechanics' typically include such topics as Galileo's work on accelerated motion and his law of fall, Descartes's three laws of nature and his analyses of collision and circular motion, Huygens's collision theory and studies on circular motion, or Newton's three laws of motion and their celestial applications. Yet these are all exercises *de motu*, and their respective aims are not to address problems *ex mechanica*, though their authors might heuristically import principles and techniques from statics. The theory of machines is wholly absent from Descartes's *summa* of the principles of natural philosophy, and to find his account of the five simple machines one must consult his correspondence with Mersenne and Constantijn Huygens. Newton wrote no *mechanicorum liber*, and the theory of machines surfaces only in passing in his significantly titled *Philosophiae naturalis principia mathematica*.[6] These facts should cause no surprise, since the purpose in both Descartes's and Newton's *Principia* was to establish laws of *nature* and to demonstrate their explanatory value within broad domains of *natural* phenomena.

Similar considerations apply to the disciplinary division we now call 'dynamics', whose Greek ancestor (δύναμις = power, potentiality) meant something wholly unmechanical before Leibniz hijacked it, to Newton's displeasure, as a highfalutin' Grecian label to pin on his own contribution to the problem of coupling forces and

mathematicians wrote exclusively on mechanics. Galileo, like Benedetti, was as much a natural philosopher as he was the author of *Le meccaniche* (a manuscript treatise written *c.* 1600). Appealing in May 1610 to Belisario Vinta, Secretary of State to Cosimo II de' Medici, for a position at the Tuscan court, Galileo made a point of requiring that 'as to the title of my position, I desire that in addition to the title of "mathematician" His Highness will annex that of "philosopher"; for I may claim to have studied more years in [natural] philosophy than months in pure mathematics.' *Discoveries and Opinions of Galileo*, ed. and trans. Stillman Drake (New York, 1957), 64.

[6] The manuscript entitled 'The Elements of Mechanicks', written about or after the mid-1690s, shows that Newton (at least post-*Principia*) took 'mechanics' to comprise both the theory of machines and the theoretical content of the *Principia*. The text covers a wide range of topics, including the three laws of motion, the law of fall and projectile motion, centrifugal and centripetal forces, Kepler's laws and the inverse square central force, resistance in a medium, hydrostatics, vibratory and pendular motions, the principle of 'virtual work', and at the end of the manuscript, the traditional problem of moving a given weight with a given force, which had appeared near the end of the Scholium to the Laws of Motion in the *Principia*. Cambridge University Library, MS Add. 4005, fols. 23–5. *Unpublished Scientific Papers of Isaac Newton: A Selection from the Portsmouth Collection in the University Library, Cambridge*, ed. A. R. Hall and M. B. Hall (Cambridge, 1962), 165, 168. *Sir Isaac Newton's Mathematical Principles of Natural Philosophy and his System of the World*, trans. Andrew Motte, ed. Florian Cajori (Berkeley, Los Angeles, 1947), 27. *Isaac Newton's Philosophiae naturalis principia mathematica*, (3rd edn, London, 1726) with variant readings, eds. A. Koyré, I. B. Cohen, Anne Whitman, (2 vols., Cambridge, 1972), 1: 72. On the complicated question of the title of Newton's treatise of 1687, see Andrew Cunningham, 'How the *Principia* got its name; or, taking natural philosophy seriously', *Hist. Sci.*, 1991, **29**: 377–92; and my 'Newton's *Mathematical Principles of Natural Philosophy*: a treatise on "mechanics"?' in Peter Harman and Alan Shapiro (eds.), *The Investigation of Difficult Things: Essays on Newton and the History of the Exact Sciences*, (Cambridge, 1992), 305–22.

motions and sending them out to work in mathematical dress.[7] It is true that long before Leibniz there was a clear understanding, for example in the Merton School, of the distinction between the causes of motion and its geometrical description in terms of time and distance, but this does not translate into an understanding of a distinction between 'dynamics' and 'kinematics'.[8] So we find everywhere in the literature analyses of the 'dynamics' or 'dynamical thought' of Aristotle, Galileo, Descartes, or Newton, without it ever being asked if these are appropriate terms to use to describe the contents of the texts under examination.

Most histories of mechanics have been written as though these were peripheral matters, as though the modern sense of 'mechanics' as a prolegomenon to physics, and of 'dynamics' as an unproblematic subdivision of mechanics, can be employed indiscriminately in the early modern period and before. A telling example from one of the pioneering histories of mechanics is the following selective quotation Mach gives of Pseudo-Aristotle's opening remark in the *Mechanica*:

> If a thing take place whereof the cause be not apparent, even though it be in accordance with nature, it appears wonderful . . . Such as the instances in which small things overcome great things, small weights heavy weights, and incidentally all the problems that go by the name of 'mechanical'.[9]

However, the original full text reads (in another translation):

> Remarkable things occur in accordance with nature [κατὰ φύσιν], the cause of which is unknown, and others occur contrary to nature [παρὰ φύσιν], which are produced by skill for the benefit of mankind. For in many cases nature produces effects against our advantage; for nature always acts consistently and simply, but our advantage changes in many ways. So when we have to produce an effect contrary to nature, we are at a loss, because of the difficulty, and require skill. Therefore we call that part of skill which assists such difficulties, a device [μηχανήν]. For it is true, as the poet Antiphon wrote, that 'we by skill gain mastery over things in which we are conquered by nature'. Of this kind are those in which the less master the greater, and things possessing little weight move heavy weights, and all similar devices which we term mechanical problems.[10]

For Mach the historian, mechanics was a branch of physics, 'at once the oldest and the simplest of which is therefore treated as introductory to other departments of this science [i.e. physics].'[11] Mach was fully aware of the mechanical arts and their early origin, he emphasized 'a distinction between mechanical experience and mechanical science',[12] and his *Science of Mechanics* includes as a matter of course descriptions of

[7] See F. E. Peters, *Greek Philosophical Terms: A Historical Lexicon* (New York and London, 1967), 42–5, and my 'Force and inertia in the seventeenth century: Descartes and Newton', in *Descartes: Philosophy, Mathematics and Physics*, ed. Stephen Gaukroger (Brighton, 1980), 230–320: pp. 242–3.

[8] Marshall Clagett, *The Science of Mechanics in the Middle Ages* (Madison, 1959), 205.

[9] Ernst Mach, *The Science of Mechanics: A Critical and Historical Account of its Development* (1st German edition, 1883) trans. by Thomas J. McCormack, new introduction by Karl Menger, sixth (American) edition [1960], with revisions through the ninth German edition [1933] (La Salle, Illinois, 1974), 13. The ellipsis is in Mach's text.

[10] Aristotle, *Minor Works*, 331.

[11] Mach, *op. cit.* (note 9), 1.

[12] *Ibid.*

machines of special theoretical interest. Yet his conception of mechanics seems to have prevented him from recognizing the older conception of mechanics as *contra*-natural in its applications and therefore to that extent 'in opposition to' physics as *natural* philosophy.

Again, only Part I (Medieval Statics) of Clagett's misleadingly titled classic study of *The Science of Mechanics in the Middle Ages* deals with mechanics in the appropriate disciplinary sense. In Parts II and III (Medieval Kinematics and Dynamics, respectively) he presents and analyses texts that address principally questions *de motu* and its causes. In Westfall's misleadingly subtitled *Force in Newton's Physics: The Science of Dynamics in the Seventeenth Century* the terms 'mechanics' and 'dynamics' and their cognates often function in inappropriate contexts. For example, Westfall discusses aspects of Hobbes's 'mechanics', concluding that 'in the development of the science of mechanics, Hobbes did not play a role of any significance'.[13] If this is true, part of the reason will be that strictly speaking Hobbes did not write anything on 'mechanics', though he wrote a great deal on motion, motive powers and *conatus*, and on their explanatory roles in natural philosophy. In a piece published as recently as 1990, I examined the case of 'mechanics' in the Scientific Revolution, and talked about the contributions to 'mechanics' of Newton and others, without once recognizing the issues I now raise in this paper.[14]

I do not mean that an historiography sensitive to taxonomic distributions of disciplines should entail the wholesale superannuation of the findings produced by historiographies that have other priorities. The histories of mechanics and dynamics can still be profitably written and read with principal 'internalist' reference to conceptual innovations and transformations. Mach's *Science of Mechanics* remains a very useful survey containing material that cannot readily be found elsewhere, Clagett's is a wonderful volume that has not been superseded, and Westfall's *Force in Newton's Physics* is the best general account of the subjects indicated in both title and subtitle. If we use the names 'mechanics' or 'dynamics' to describe Newton's laws of motion or Descartes's collision theory, or if we label as 'dynamics' Aristotle's *Physics*, bk. 7, chap. 4,[15] we describe them incorrectly or anachronistically, but that should not preclude us from doing valuable work on the texts themselves.

The danger lies elsewhere. Inattention to disciplinary taxonomies weakens interest in historical situations in which disciplinary distinctions were the issue, and distorts our appreciation of situations where such distinctions were latent factors. It prevents us from recognizing historical problems whose solutions depend on taxonomic considerations, whether in the disciplines themselves, in relevant professions, or in class or other institutional structures. It diverts us from seeing the historical

[13] Richard S. Westfall, *Force in Newton's Physics: The Science of Dynamics in the Seventeenth Century* (London and New York, 1971), 113–14. I must make it clear that similarly misleading or inappropriate uses of 'dynamics' and 'mechanics' are to be found in the work of other historians in this area, such as Dugas, Duhem, Cohen, Herivel, and myself.

[14] Alan Gabbey, 'The case of mechanics: one revolution or many?', in *Reappraisals of the Scientific Revolution*, ed. David C. Lindberg and Robert S. Westman (Cambridge, 1990), 493–528. I make amends to some extent in the essay mentioned in note 6.

[15] S. Sambursky, *The Physical World of the Greeks*, trans. Merton Dagut (London, 1963), 92–4.

significance of certain features of texts, or of scientists' methods and assumptions, that might otherwise be passed over as merely intriguing anomalies or as instances of an idiosyncratic reluctance to meet some historian's Whiggish expectations. It serves to restrict the topical range that ought to characterize a properly comprehensive history of mechanics (or dynamics). A few examples will do more than pages of historiographical theorizing.

Transformations in the nature of mechanics: Barrow to Euler

As I indicated in note 4, Renaissance discussions on the nature and status of mechanics have been a focus of interest in the work of Wallace, Laird, and others, but in the post-Galilean period there were also significant discussions on the subject that have not received anything like the same degree of attention. The 'revolutionary' transformations in mechanics and the science of motion during the latter period were accompanied by a shared awareness that the nature of mechanics as a discipline was undergoing complex and profound changes. Writers on mechanics often included in their treatises or other works a *status quaestionis* setting out both their and others' varying positions on how the discipline and its aims should be viewed.[16]

In the second of his *lectiones mathematicae* (1664–1666), Isaac Barrow discussed at great length 'the particular division of the mathematical sciences'. He reviewed the contributions of Geminus and more recent writers such as Biancani, Hérigone, and Guldin. The latter divided the mathematical sciences initially into 'two pure and primary parts' (geometry and arithmetic) and 'four mixed and subaltern' parts (optics, mechanics, astronomy and music), which they and others again subdivided in various ways. But 'the manifold division and subdivision of things into their minutest parts rather tends to breed confusion, than to remove it', so Barrow advocated a streamlined taxonomy in which all 'mixed mathematics' (that is, all mixed sciences) become integral parts of physics, on the grounds

> that there is no part of this [physics] which does not imply quantity, or to which geometrical theorems may not be applied, and consequently which is not some way dependent on geometry ... magnitude is the common affection of all physical things, it is interwoven in the nature of bodies, blended with all corporeal accidents, and well-nigh bears the principal part in the production of every natural effect.[17]

[16] There is no such *status quaestionis* in Descartes's writings and comments on mechanics, though there are more interesting reasons why his position is difficult to determine. See further my 'Descartes's physics and Descartes's mechanics: chicken and egg?' in *Essays on the Philosophy and Science of René Descartes*, ed. Stephen Voss (Oxford, 1993), 310–23.

[17] Isaac Barrow, *The usefulness of mathematical learning explained and demonstrated: being mathematical lectures read in the publick schools at the University of Cambridge ... To which is prefixed, the Oratorical Preface of our learned author, spoke before the University on his being elected Lucasian Professor of the Mathematics* (London, 1734), 21–3. The Latin *Lectiones mathematicae XXIII* were first published in London in 1683.

Consequently, all aspects of mechanics, both theoretical and practical, are parts of physics, which effectively becomes the mathematical treatment of corporeal magnitudes moving under forces of whatever kind.

Wallis and Pardies broadly shared Barrow's conception of mechanics as the mathematical science of motion. The first definition in John Wallis' *Mechanica: sive, de motu, tractatus geometricus* (first published 1670–71) is of mechanics itself, which Wallis, in keeping with the title of his treatise, defines simply as 'the geometry of motion'.[18] He explains that he will not use *mechanica* in the sense of mechanical art, or to refer to the use of mechanical instruments to do geometry. Rather, he takes mechanics 'to be the part of geometry that deals with [local] motion, and investigates, apodictically and using geometrical reasoning, the force with which such and such a motion takes place'.[19] Similarly, in the preface to his *Discours du mouvement local* (1670, 2nd edition 1674), Pardies writes that 'everything that is produced by human industry or by natural causes comes about only through motion, so that it is not possible to penetrate the secrets of physics, or in the arts to succeed in invention and practice, without the help of mechanics, that is to say, without a knowledge of the laws of motion'.[20] As for Newton, his meditation on mechanics in the preface to the *Principia* is well known, but whether and in what sense the *Principia* is a treatise in 'mechanics' is not an easy matter to determine, which in itself is a sign of the transitional nature of this superlative exercise in the mathematization of natural philosophy.[21]

Others took more traditional views than those championed by Barrow and Wallis. Six years after the publication of *Principia*, and apparently unaware of its existence or of the work of writers like Barrow and Wallis, Philippe de La Hire began his mechanics course in the Collège de France with a prolegomenon on the nature of the subject he was about to teach:

> There is a major controversy among mathematicians concerning the definition of mechanics. There are those who understand by the term 'mechanics' the explanation of motions that are violent and create admirable effects; so some define mechanics as the science from which can be taken the causes and principles for the many technical arts [*artes sellulariae*]. For others, mechanics is the art of forcing bodies as much as possible so that they are moved

[18] Wallis's 'geometria de motu' is not easy to translate. 'The geometry of motion' is not incorrect, but is misleadingly kinematical. The intended meaning is something like 'geometry in so far as it relates to motion', which at least does not exclude the all-important *force*–motion relation. (Except where otherwise indicated, all translations are my own.)

[19] J. Wallis, *Mechanica*, Part I, chap. 1 ('De motu generalia'). J. Wallis, *Opera mathematica*, 3 vols. (Oxford, 1693–1699), 1: 575. In the commentary on Definition I Wallis includes a note on the etymology of the term *mechanica*.

[20] Ignace-Gaston Pardies, *Discours du mouvement local. Avec des remarques sur le mouvement de la lumiere . . . Seconde Edition* (Paris, 1674), 1–2.

[21] I go into this question in some detail in my *op. cit.* (note 6). Cunningham, *op. cit.* (note 6), concentrates on the 'natural philosophy' of Newton's title, and argues for the questionable claim that natural philosophy in Newton's day was and always had been 'a discipline and subject-area whose *role and point* was the study of God's creation and God's attributes.' (*ibid.*, 388, italics in text). Natural philosophy, and indeed any other discipline, was necessarily a study of God's creation in the sense that God was held to be the Author of all things, but that in itself does not make for a disciplinary identification. Moreover, in the Peripatetic tradition (and not only there) the study of God's attributes was standardly a matter for metaphysics and (in a different way) for theology, not for natural philosophy. Johannes Magirus explains that as pure act and immaterial forms, spirits, including God, do not have a *natura* (the *per se* principle of motion and rest). In particular, 'since God is above Nature, He cannot be part of the subject of Physics.' *Physiologiae Peripateticae libri sex cum commentariis* (Cambridge, 1642), 8.

> against gravity. Now I say that mechanics is the art of increasing to infinity a specified motive power.
>
> This art is called admirable, not undeservedly, since nothing is more alien to the laws of nature, nothing seems more incredible, than a smaller power overcoming a larger power, or a weight of one pound balancing a hundred or a thousand pounds, or even overcoming it. So we have no doubt in affirming that if there were a fixed and immobile point away from the centre of the earth, we will be able to move, with even the least motive power, the terrestrial globe from its place.[22]

La Hire restated his 'Archimedean' conception of mechanics in the opening pages of his *Traité de mécanique*, published in 1695. There would be nothing more admirable than the operations of machines, writes La Hire, were it not that people see them every day without paying attention to them or to the law according to which they operate and on which the whole of mechanics depends. That law is the law of equilibrium of equal and opposed mechanical powers, from which it can be demonstrated geometrically that small forces can produce very large effects. 'That is why we say that mechanics is a science which shows how the effort of a power can be increased indefinitely [*à l'infini*].'[23]

Varignon in a sense coupled the traditional and the new understanding of mechanics by explicitly subsuming the former under the latter. He begins his *Nouvelle mécanique, ou statique* (1725) with the remark:

> Mechanics in general is the science of motion, of its cause and of its effects; in short, of everything that relates to it. Consequently it is also the science of the properties and uses of machines or devices fitted to facilitate motion . . . It is this latter part of mechanics that is in question here, that is to say, we are dealing only with . . . elementary machines, and some others normally regarded as composed of them. This part of mechanics is properly called statics, but since most authors let it have the general name of mechanics, I thought I ought also to call it by this name, so as not to speak differently from them . . .[24]

Euler's 1758 paper on the mechanical knowledge of bodies is perhaps the best indicator of the extent of the transformation in the conception of mechanics since the days of Galileo and Descartes. He begins with an account of a threefold way of knowing bodies: the geometrical, the mechanical, and the physical. Geometrical knowledge of bodies involves knowing only about their extension and shape.

> The mechanical knowledge of bodies considers them in so far as they are matter, without paying attention to the qualities with which the matter is

[22] Philippe de La Hire, 'In regio collegio post vacationes anni 1693. Mechanica'. Paris, Bibliothèque Nationale, Fonds français 12271, fol. 290r–338r: fol. 290r. Autograph.

[23] Philippe de La Hire, 'Traité de mécanique, ou l'on explique tout ce qui est nécessaire dans la pratique des arts', in *Mémoires de l'Académie Royale des Sciences*, 10 volumes (Paris, 1729–1732), vol. 9 (1730): 1–340, pp. 1–2. This Academy edition is a reprint of the 1695 edition, which appeared under the longer title *Traité de mécanique, ou l'on explique tout ce qui est nécessaire dans la pratique des arts, & les propriétés des corps pesants lesquelles ont un plus grand usage dans la physique* (Paris).

[24] Pierre Varignon, *Nouvelle mécanique, ou statique, dont le projet fut donné en M.D.LXXXVII. Ouvrage posthume* . . . (2 vols., Paris, 1725), **1**: 1–2.

> endowed; and here it is a question of knowing not only the quanitity of matter with which each body is composed, which we call its mass, but also the way in which the matter is distributed throughout the extension of the body. This knowledge is absolutely necessary when the motion of bodies is in question; and that is the reason it is called mechanical.

And the physical knowledge of bodies comprises all their other properties and qualities, which are the subject proper of Physics.[25]

In the light of even this highly selective sequence of texts, it is unwise to assume that everyone in the century or so before Euler who wrote treatises on 'mechanics', or who employed the term, had exactly the same enterprise in mind, or that their participation in that enterprise was independent of contemporary debates on the nature of the discipline itself.

Instruments 'philosophical' and 'mathematical'

The distinction between mechanics and physics is present as a latent, though unsuspected, factor in some important issues discussed by Deborah Jean Warner in her recent essay review of a group of books on scientific instrumentation. She notes the probable first appearance of the designation 'philosophicall apparatus' in a letter of 1649 from Hartlib to Boyle, and highlights the explicit distinction Nehemiah Grew drew between philosophical and mathematical instruments in his *Catalogue & Description of the Natural and Artificial Rarities belonging to the Royal Society* (1681). There was a category 'Of Instruments relating to Natural Philosophy', and a category of practical instruments, 'which, following long tradition, he [Grew] identified as "Things relating to the Mathematicks" '. However, to explain this new categorization, Warner claims that Grew's distinction 'reflects a clear understanding of the power of words to influence perception', and that the epithet 'philosophical'

> implied authority and prestige, and the liberal tradition open to the gentility. The verbal distinction between philosophical and mathematical obscured the connections between men of science . . . and the aspiring tradesmen . . . It implied that the observations, measurements and experiments of natural philosophers were made in a search for truth, and thus differed from the observations, measurements and experiments which mathematicians and mechanics made for merely practical purposes . . .[26]

On this account the story is little more than a sociological case history. However, the philosophical–mathematical distinction was clearly more than verbal, since it originated in the real distinction between natural philosophy and the mechanical arts, between the natural and the artificial, to go no further than the title of Grew's catalogue. If the new 'philosophical' instruments of the Scientific Revolution were used to 'search for truth', that was because natural philosophy had never had any

[25] L. Euler, 'Recherches sur la connoissance mécanique des corps' (1758, published 1765). *Opera Omnia*, vol. 2.8, 178.
[26] Deborah Jean Warner, 'What is a scientific instrument, when did it become one, and why?' *Brit. J. Hist. Sci.*, 1990, **23**: 83–93, pp. 83, 84.

other purpose. If the mathematical instruments were used for 'merely practical purposes', that was because the mechanical arts had never had any other kind of purpose. As Bennett explains, writing on a closely related theme,

> The mathematical instruments were problem-solving tools, not investigative tools. They represented practical techniques, applied within some area of the mathematical sciences, aimed at some definite goal – the solution of some technical or practical mathematical problem, like finding latitude or drawing a map. They were not expected to throw up new truths about the natural world. They were, in other words, mathematical not natural philsophical instruments.[27]

Yet for Warner the distinction in the case in question is an instance of successful 'social manipulation', the instrument of manipulation being the above supposed 'terminological ploy', which succeeded in creating a contrast between historians' extensive studies of 'the controversies of natural philosophers with scholastic philosophers' and their neglect of the efforts of the same natural philosophers 'to distinguish themselves from other empirical investigators'.[28] If these natural philosophers did aim for social exclusivity, they were exploiting a much older distinction that had nothing to do with social manipulation, but had everything to do with the distinction between human manipulation of nature and speculative inquiry *de rerum natura*.

Two puzzles for the philosopher of technology

Disciplinary taxonomies alert us to the existence of certain anomalies and puzzles that arise not from within mechanics *per se*, but from views of its nature and purpose held by those who contributed to the discipline, or from their opinions on related matters. La Hire's admiration for mechanics on the grounds that 'nothing is more alien to the laws of nature, nothing seems more incredible, than a smaller power overcoming a larger power' (quoted above) was universally shared in the Renaissance and early modern period. In his influential *Mechanicorum liber* (1577), Guido Ubaldo proclaimed that mechanics, the most noble of the arts, 'not only crowns and perfects geometry . . . but also holds control of the realm of nature . . . since it operates against nature or rather in rivalry with the laws of nature, [it] surely deserves our highest admiration'.[29] Indeed, but how is it *possible* for mechanics to act against nature and its laws? If mechanics is the mathematical science of what machines can do *against* nature, what is the historical and philosophical significance of the canonic view in the early modern period of the laws of mechanics as the fundamental laws *of* nature? How can Nature act against Nature?

This intriguing puzzle in the 'philosophy of technology' seems not to have worried Guido Ubaldo and his successors in the mechanical tradition, nor as far as I am aware

[27] J. A. Bennett, 'The mechanics' philosophy and the mechanical philosophy', *Hist. Sci.*, 1986, **24**: 1–28, p. 2.
[28] Warner, *op. cit.* (note 26), 84.
[29] Drake and Drabkin, *op. cit.* (note 5), 241.

has it bothered any historian of early modern science or philosophy since. One seventeenth-century exception in the Peripatetic tradition (and there were others) was Giovanni Battista Giattini, who in his questions on Aristotle's *Physics* (1653) wondered at one point 'Whether Art can do something that Nature cannot do'. He answered in the negative,

> because art does not operate except by the application of naturally productive causes [*causas naturaliter factivas*], and so unless there is a natural force in the causes applied, the application will certainly produce nothing: nevertheless it is true that such an application cannot happen in practice [*moraliter*], and on that account we say commonly that art does something above nature [*supra naturam*], because it applies causes that would not be applied in the natural run of things.[30]

This does not solve the problem, of course, but Giattini's *quæstio* and *responsio* raise issues that we have still not come to grips with. On the other hand, and more positively, they clarify (for example) what Descartes was getting at in a letter to Constantijn Huygens of March 1638, six months after he had sent him a small treatise on mechanics (i.e. the five simple machines). In response to Huygens's wish that he get down to writing a full-scale version of his mechanics, Descartes wrote:

> I still cannot see any likelihood of publishing my *Monde*, at least not for a long time; and without that, neither would I be able to finish the Mechanics you wrote to me about, for it depends entirely on my *Monde*, principally in what relates to the speed of motions. And it requires having explained what the laws of nature are, and how she acts in the ordinary way, before one can really explain how she can be applied to effects to which she is not accustomed.[31]

There is something illogical about the idea of Nature *not* acting in an ordinary way and being *un*accustomed, even in a manner of speaking, to effects that according to the canons of Descartes's own mechanical philosophy must ultimately instantiate the laws *of* nature that he had already presented in *Le Monde* and was to present in revised form in *Principia philosophiae*. It is not at all clear from Descartes's writings that there is a readily available resolution of the problem.

Before the early nineteenth century there were many who, understanding mechanics, believed that 'mechanical' perpetual motion was an impossibility. However, the same people believed that 'physical' perpetual motion *was* in principle a possibility.[32] The atomist Isaac Beeckman, an early protagonist of the mechanical philosophy, was

[30] *Physica P. Io. Baptistae Giattini Panormitani Societatis Jesu in Collegio Romano, ter olim, Philosophiae nunc sacrae Theologiae Professoris* ... (Rome, 1653), 149–50. Although the title page does not say so, p. 1 of the main text reveals that Giattini's *Physica* is in fact a set of 'In VIII libros Aristotelis De physico auditu quaestiones'.

[31] *Oeuvres de Descartes, publiées par Charles Adam & Paul Tannery, nouvelle présentation*, ed. P. Costabel, J. Beaude and B. Rochot (11 vols., Paris, 1964–1974), **2**: 50.

[32] I am concerned here only with those who were *au fait* with the principles of mechanics (however defined) and their application. Many cranks and some rulers also believed in perpetual motion machines, right through to the twentieth century – but that is another aspect of the story. I give a full account of what follows in 'The mechanical philosophy and its problems: mechanical explanations, impenetrability, and perpetual motion', in *Change and Progress in Modern Science: Papers Related to and Arising from the 4th International Congress on History & Philosophy of Science, Blacksburg, Virginia, November 1982*, ed. J. C. Pitt (Dordrecht, 1985), 9–84, pp. 38–84. To avoid footnote clutter, I refer the reader to this study for most of the references to the primary sources cited below.

adamant that in no mechanical device could the principle of work be abrogated. Yet his *Journal* contains descriptions of other-than-mechanical perpetual motion machines, including the one that Cornelius van Drebbel presented to James I of England on entering his service in 1605. Beeckman did not reject such machines out of hand. Indeed he supported attempts to devise one machine using the tides as power source. Descartes's statement of the work principle is among the clearest of the seventeenth century, yet in his early writings he countenanced the idea of a perpetual motion device powered by magnets or by a substance (if such there be) whose weight varies with the lunar phases. To make a methodological point in *Regulae XIII*, he used the distinction between 'the natural kind' of perpetual motion, 'such as the perpetual motion of the stars or of fountains', and 'a perpetual motion produced by human industry', that is 'by art'. But there is no explicit indication, here or elsewhere, that Descartes thought the former kind of perpetual motion an absolute impossibility.

Perhaps the most striking example is Huygens, who demonstrated Torricelli's Principle by arguing that if it were false, then one could build a mechanical perpetual motion, an evident absurdity. Nonetheless, Huygens's comment on the two perpetual motion machines built in 1659 by Becher (of phlogiston fame) is revealing:

> one of the machines ... is purely mechanical and the other physico-mechanical. As for the latter, the thing does not seem to me absolutely impossible, but as for the other I confess it passes belief.[33]

Much later Huygens wrote to Leibniz (12 January 1693) to say that anyone who thought mechanical perpetual motion possible was simply ignorant, but in the case of 'physico-mechanical' perpetual motion, 'it still seems there is some hope, such as by using the loadstone'. Newton too, writing in the late 1660s, was sufficiently hopeful to conjecture that gravitational or magnetic particles might be harnessed to turn specially prepared paddle-wheels perpetually. Even Leibniz, who publicly trumpeted the impossibility of mechanical perpetual motion as the knockdown argument against the Cartesian measure of motive force, drafted a couple of manuscript pages in which he describes a perpetual motion device comprising cylindrical bellows alternately rising and sinking in water. And in the mid-1690s Leibniz admired an apparatus for producing perpetual motion which depended on the differential between the specific gravities of two miscible liquids, the inventor of the apparatus being the 23-year-old Johann Bernoulli, who was to formulate the first formal statement of the principle of virtual work.[34]

These case histories, and many others on the theme of perpetual motion, are quite unintelligible unless we clearly understand the distinction between contra-natural mechanical actions and natural physical processes, between the mechanical and the physical. In each case the distinction is either explicitly present, taken for granted, or seen as itself an issue. However, according to the mechanical philosophy, of which major advocates figure in the cases just mentioned, the Universe is a machine, 'an

[33] Huygens to Pierre Carcavy, 26 February 1660: *OCCH*, 3, 28. Becher had already visited Huygens in January to tell him about his machines, which by that date had been in perpetual operation for six months.

[34] Bernoulli stated his principle of 'virtual speeds' in the letter to Pierre Varignon of 26 January 1717: Varignon, *op. cit.* (note 24), **2**: 175–6.

αὐτόματον or self-moving engine', as Boyle put it in the *Origin of Forms and Qualities* (1666). In short, a perpetual motion machine. So does it contravene the law against mechanical perpetual motion by being the only exception, the unique mechanical perpetual motion machine, maintained in motion by God? If so, what happens to the distinction between the mechanical and the physical? Is the Universe perhaps a physical perpetual motion machine, the divinely powered source of man-made physical machines? If so – and if the mechanical–physical distinction still holds – what becomes of the mechanists' cosmic machine?

Concluding remarks

Disciplinary taxonomies enable us to widen the topical extension of early modern mechanics by calling attention to mechanical problems that contemporary authors saw as legitimate challenges, yet which are accorded patchy representation in most histories.[35] More generally, these taxonomies provide proper frameworks within which the histories of mechanics and dynamics should be written. Without these frameworks, the historiography of these subjects will lack indispensible tools for exploring such topics as the relationship between and evolution of statics and dynamics in the early modern period, the influential role outside statics and the theory of machines of the law of the lever, the emergence of the modern senses of 'mechanics' and 'dynamics', or the evolving relations between mechanics (and dynamics) and physics. Indeed, we might well conclude from a survey of the conceptions of mechanics found in Guido Ubaldo and his contemporaries, in Barrow, Wallis, Pardies, Newton, La Hire and Varignon, who all saw a need for a re-examination of the nature of the discipline in the face of change, and from Euler's bland, businesslike statement in the mid eighteenth century, that an adequate history of early modern mechanics remains to be written.

[35] One indicative text among many is Philippe de La Hire's *Traité*, *op. cit.* (note 23), which I discussed above in another context. In addition to the five simple machines, centres of gravity, geostatics, motion on inclined planes, projectiles and water-jets, La Hire analyses Roberval's Balance (1669), percussion (including centres of percussion and centres of oscillation), a water-raising wheel with minimal friction, catenary suspensions, the resistance of suspended solids, the optimum shape of windmill arms, problems in building construction, the contractive force of moist rope, and hypotheses that might explain, on mechanical principles, the extraordinary manoeuvres of the tongue of the chameleon and green woodpecker. For further consideration of the chameleon, see Freudenthal's paper in this volume.

9

The conscience of Robert Boyle: functionalism, 'dysfunctionalism' and the task of historical understanding

MICHAEL HUNTER

It could be said that we currently have two Robert Boyles, reflecting the differing emphases of two contrasting scholarly traditions. On the one hand, there are studies which lay predominant emphasis on the intellectual enterprise to which Boyle devoted himself in refashioning knowledge about the natural world along corpuscularian and anti-scholastic lines, his motivation being taken to be an altruistic pursuit of scientific theories which would be internally coherent and compatible with his theological and philosophical commitments. Such a view of Boyle as the searcher after truth is writ large in Marie Boas Hall's various writings on his scientific work; more recently, it has been vigorously asserted by Rose-Mary Sargent and by Timothy Shanahan, who has argued that 'intellectualist considerations have explanatory and expository value of their own' in response to those who have thrown doubt on this.[1]

Others, on the other hand, have emphasized the extent to which Boyle's intellectual goals fitted into a more or less overt programme aimed at furthering identifiable social and political ends. Thus J. R. and M. C. Jacob have argued that 'both the corpuscularianism and the experimentalism' of Boyle and other 'reforming philosophers' 'were designed to combat two threats, heresy and social insubordination, at the same time'.[2] More recently, Steven Shapin and Simon Schaffer, while agreeing that Boyle was out to enrol the new science in defence of a fundamentally conservative order, have laid more stress on a conscious attempt on Boyle's part to redefine the manner in

I am grateful to Edward B. Davis, Scott Mandebrote and John Spurr for their comments on a draft of this paper, but I am aware of the extent to which they retain reservations about its main thrust, for which I am solely responsible.

[1] Timothy Shanahan, 'God and nature in the thought of Robert Boyle', *J. Hist. Phil.*, 1988, **26**: 547–69, p. 549n. See also Marie Boas, *Robert Boyle and Seventeenth-century Chemistry* (Cambridge, 1958); M. B. Hall, *Robert Boyle on Natural Philosophy* (Bloomington, 1965); Rose-Mary Sargent, 'Scientific experiment and legal expertise: the way of experience in seventeenth-century England', *Stud. Hist. Phil. Sci.*, 1989, **20**: 19–45.

[2] J. R. and M. C. Jacob, 'The Anglican origins of modern science: the metaphysical foundations of the Whig constitution', *ISIS*, 1980, **71**: 251–67, pp. 256–7. See also J. R. Jacob, *Robert Boyle and the English Revolution* (New York, 1977); J. R. Jacob, 'Boyle's atomism and the Restoration assault on pagan naturalism', *Soc. Stud. Sci.*, 1978, **8**: 211–33.

which knowledge claims were delimited and assessed. According to this reading, Boyle consciously sought to promote a public image for himself with a political purpose, emphasizing his independence, probity and piety as a means of establishing his credentials as a polemicist, and hence helping to solve the problem of order to which conflicting knowledge claims otherwise gave rise.[3]

For all their differences, what these approaches have in common is that both are essentially rationalistic. Ulterior motives as much as overt intellectual aims are *intentional*. Yet it seems to me that both leave something out, and it is this missing dimension of Boyle that I wish to explore in this paper. I want to examine what might be described as the 'unintentional' or 'irrational' dimension to Boyle's intellectual outlook, considering the extent to which he was the subject of powerful impulses which were not necessarily 'functional' and which he did not even always have wholly under control. In doing so, I shall be teetering on the brink of psychoanalysis, but shall restrict myself to observations of a commonsensical variety: I remain agnostic about the value of an overtly Freudian or other approach to a long-dead figure like Boyle.[4]

That Boyle was 'mixed up', to adopt a fashionable phrase, was not altogether unfamiliar to his contemporaries. In his own time, the most noticeable facet of this was perhaps the stutter which he acquired as a child and which he never entirely lost for the rest of his life. The impression that this made on those who met Boyle was perhaps most strikingly illustrated by the Italian visitor, Lorenzo Magalotti, who wrote:

> He speaks French and Italian very well, but has some impediment in his speech, which is often interrupted by a kind of stammering, which seems as if he were constrained by an internal force to swallow his words again and with the words also his breath, so that he seems so near to bursting that it excites compassion in the hearer.[5]

There is also evidence that certain elements of Boyle's outlook which militated against the sober, rational view of him that has prevailed since his own time have been deliberately suppressed. An example of this is Boyle's belief in day fatality, as recorded in the autobiographical notes which he dictated to Bishop Gilbert Burnet in the last years of his life. Among other things, these frank memoranda record that it was on May Day 1642 that Boyle heard of the outbreak of the Irish Rebellion which had such a catastrophic effect on the Boyle family fortunes, and he noted of this day:

[3] Steven Shapin and Simon Schaffer, *Leviathan and the Air Pump: Hobbes, Boyle and the Experimenal Life* (Princeton, 1985). See also Steven Shapin, 'Pump and circumstance: Robert Boyle's literary technology', *Soc. Stud. Sci.*, 1984, **14**: 481–520; Shapin, 'Who was Robert Hooke?', in Michael Hunter and Simon Schaffer (eds.), *Robert Hooke: New Studies* (Woodbridge, 1989), 253–85, esp. pp. 269f.

[4] It is for this reason that I have placed the word 'dysfunctionalism' in inverted commas in my title. I have adopted it because it is the best term I can find to counter the restrictively defined 'functionalist' terms of reference of the authors discussed in the text: but I am aware that it echoes the terminology of modern psychotherapists, and I would not wish to push its direct applicability in this context too far. In fact, according to a scheme of otherworldly values, Boyle's views may have had a clear rationale, and I might be accused of imposing a dichotomy of my own on them. On the other hand, my sense is that even his casuistical advisors thought that he overdid things: see further below.

[5] W. E. Knowles Middleton (ed. and trans.), *Lorenzo Magalotti at the Court of Charles II* (Waterloo, Ontario, 1980), 135. Compare R. E. W. Maddison, *The Life of the Honourable Robert Boyle* (London, 1969), 4.

'observed often Inauspicious'.[6] Apparently in the first full life of Boyle to be written (though it was never published and is now mostly lost), its author, the scholar, William Wotton, 'disapproves of the notion & hints of the evils which may attend such a perswasion', but he felt bound to include it since he understood that these notes itemized topics which Boyle *definitely* wanted included in any life of him.[7] On the other hand, Henry Miles, who helped to collect materials for the first adequate life of Boyle that *was* published, that of Thomas Birch, felt that Wotton was wrong about the mandate from Boyle; he therefore urged Birch to omit the episode, and it was forgotten until noted by Marie Boas Hall, in one of the very few modern references to the facet of Boyle that I intend to dwell on here, who describes Boyle's belief in day fatality as 'a rather endearing superstition for such a determined rationalist as he was to become'.[8]

Boyle's religiosity

Though eighteenth-century commentators may have suppressed *this*, however, the most important facet of Boyle's make-up from the point of view that I want to stress here was one that they took very seriously – even if interpreting it selectively – and this was his deep religiosity. There are important facets of Boyle's religious outlook which are familiar, and which I am going to take for granted here. Essentially, these are the features which are at one with the intellectualist tradition that I have already described, the profound, pious adulation of God's design in the Universe and the conviction that this was well-illustrated by the findings of the new science. These themes are perhaps best expounded by R. S. Westfall in his *Science and Religion in Seventeenth-century England* (1958).

But I believe that, to understand Boyle properly, we must lay equal stress on an aspect of his religiosity at which Westfall merely glanced.[9] Boyle needs to be placed in the context of the tortured spirituality exemplified by works like John Bunyan's *Grace Abounding to the Chief of Sinners* (1666) and classically depicted by William Haller in *The Rise of Puritanism* (1938). Nearer to home, the tradition is exemplified in Boyle's sister, Mary Rich, Countess of Warwick, whose religious attitudes and activities have been the subject of a perceptive, if slightly reductionist, account by Sarah Mendelson in *The Mental World of Stuart Women* (1987). Boyle shared the deep, agonized, piety of his sister, though in him, as in her, there is evidence of a

[6] British Library Add. MS 4229, fol. 60. An adjacent note records a fall from a horse on May Day. On this text, see my 'Alchemy, magic and moralism in the thought of Robert Boyle', *Brit. J. Hist. Sci.*, 1990, **23**: 387–410, pp. 387–8.

[7] Miles to Birch, 21 October 1742, British Library Add. MS 4314 (not 4316, as stated by Boas, *op. cit.* (note 1), 11 n.4), fol. 70. On Wotton's life, see Michael Hunter, *Letters and Papers of Robert Boyle* (Bethesda, MD, 1992), Introduction.

[8] British Library Add. MS 4314, fol. 70; Boas, *op. cit.* (note 1), 11.

[9] R. S. Westfall, *Science and Religion in Seventeenth-century England* (New Haven, 1958), 40–1, see also pp.142–3.

degree of conflict between this outlook and the fashionable world with which their upbringing also brought them into contact.[10]

One text which places Boyle firmly in this tradition is his autobiographical 'Account of Philaretus during his Minority', written in the late 1640s, which chronicles his conversion experience and the religious traumas which preceded and followed it.[11] This has long been well known, but what has not been generally recognized is that the intensity of religious experience which Boyle felt at that time continued throughout his life, influencing his social role and intellectual attitudes more profoundly than has hitherto been acknowledged. This is perhaps most clearly illustrated by Boyle's lifelong concern with casuistry – in other words, with the assessment of 'cases of conscience', the resolution of difficult moral dilemmas in the context of a highly acute sense of sin. Such pastoral activity was highly regarded in religious circles in Protestant England as in Catholic Europe in the seventeenth century, with a distinctive 'reformed' strand of interpretation emerging which was codified in works like William Perkins' *Whole Treatise of the Cases of Conscience* (1606), Jeremy Taylor's *Ductor Dubitantium* (1660) and Richard Baxter's *Christian Directory* (1673).[12] It is to this tradition that Boyle belongs.

The evidence for Boyle's casuistical concerns appears to have been more profuse in the generation after his death than it is now, probably because this style of piety was becoming less popular in Anglican circles even in his own later years. By the eighteenth century, embarrassment about it had become acute, and Henry Miles, whose role in relation to Birch's *Life* has already been referred to, mentions an item of this kind – now lost – which 'a very judicious friend to whom I shewd the MS' considered 'not suited to the genius of the present age'.[13] Such distaste evidently accounts for the rarity of such material among Boyle's extant remains, but fortunately enough survives to illustrate his unremitting concern with matters of conscience; he was constantly reviewing his activities in the light of a strong sense of right and wrong, valuing the advice of churchmen as experts who could help him and others to come to informed decisions on such issues. In 1659 Boyle paid a pension to the divine, Robert Sanderson, so that he could prepare for the press his lectures on conscience, while in 1662 it was to Sanderson that Boyle applied for advice as to whether it was legitimate for him to accept a grant of impropriations from former monastic lands in Ireland which was made to him as part of the Restoration settlement. These impropriations seem to have been the subject of particular anxiety for Boyle, evidently

[10] British Library Add. MS 4229, fol. 68v, printed in Maddison, *op. cit.* (note 5), 53–4; S. H. Mendelson, *The Mental World of Stuart Women* (Brighton, 1987), ch. 2.

[11] Maddison, *op. cit.* (note 5), 2–45, esp. pp. 34–6.

[12] For modern studies of casuistry, see H. R. McAdoo, *The Structure of Caroline Moral Theology* (London, 1949); Thomas Wood, *English Casuistical Divinity during the Seventeenth century with Special Reference to Jeremy Taylor* (London, 1952); G. L. Mosse, *The Holy Pretence* (Oxford, 1957); Camille Slights, 'Ingenious piety: Anglican casuistry of the seventeenth century', *Harvard Theo. Rev.*, 1970, **63**: 409–32; and Edmund Leites (ed.), *Conscience and Casuistry in Early Modern Europe* (Cambridge, 1988).

[13] For documentation of this and the other points in this paragraph see my 'Casuistry in action: Robert Boyle's confessional interviews with Gilbert Burnet and Edward Stillingfleet, 1691', *J. Ecc. Hist.*, 1993, **44**: 80–98.

because those in possession of such rights could be regarded as guilty of the sin of sacrilege.[14]

In the following decades Boyle repeatedly consulted his friend, Thomas Barlow, Bishop of Lincoln, on matters of conscience: in fact, it was Barlow's answers to such questions which Miles's friend expressed distaste for in the 1740s and which have since disappeared. What *have* survived are Boyle's notes on two confessional interviews that he had in the last months of his life with two prominent churchmen, Gilbert Burnet, Bishop of Salisbury, and Edward Stillingfleet, Bishop of Worcester. This striking document is now among the Boyle Papers at the Royal Society, and I have published it with a commentary elsewhere.[15] Here I will summarize what can be learnt from it about the kind of anxieties that Boyle was harbouring as his death drew near; it also reveals the amount of mental effort that he invested in salving his conscience. Thus he had prepared himself for the interviews by writing a paper, though this no longer survives, and at one point in his notes on his interview with Stillingfleet, he added a poignant note: 'I wish I had put him to speake more positively & roundly about this Point'.

One of Boyle's concerns was the legitimacy of his receipt of the impropriations already referred to, and especially the question of whether the earlier casuistical advice that he had been given ought to have inspired him to more charity than it had. Boyle was also concerned as to whether he had dealt uprightly with transactions over the landed estates that he had inherited from his father, the great Earl of Cork (it is perhaps worth speculating that Boyle's acute anxiety on moral issues may have owed something to an awareness of the rather dubious legitimacy with which the family fortunes had been made in the first place).[16] In addition, Boyle asked both churchmen for their advice about the blasphemous thoughts that assailed him. Clearly he suffered agonies of religious doubt, believing that he had committed the ultimate blasphemy of the Sin against the Holy Ghost, as reported in the Gospels. The idea that a believer had committed unforgiveable blasphemy was not an uncommon one in godly circles in Boyle's period, and that such worries were not new at the end of Boyle's life is shown by earlier references of a similar kind.[17]

The bishops were inclined to try to minimize Boyle's anxieties. Entire treatises were written at this time to assure those who feared that they had committed the Sin against the Holy Ghost how unlikely it was that they were guilty of such outright apostasy, and Stillingfleet echoed such views in his advice to Boyle. He stated 'That none that are affraid of haveing committed that Sin are guilty of it; since one cannot commit it without haveing a full Intention to do it'. Both Stillingfleet and Burnet were also inclined to give a naturalistic explanation of such worries as 'depressions or weaknesses of the Animal Spirits oftentimes proceeding from the want of Nourishment or Free Air or Exercise or pleaseing Circumstances &c'. Such reactions are

[14] See esp. Keith Thomas, *Religion and the Decline of Magic* (London, 1971), 96ff.

[15] Hunter, *op. cit.* (note 13).

[16] Compare Nicholas Canny, *The Upstart Earl* (Cambridge, 1982).

[17] For this and the points in the next paragraph, see Hunter, *op. cit.* (note 13).

themselves an interesting sign of the way in which Anglican attitudes were changing in Boyle's later years. Even earlier, however, it had been a commonplace of casuistry that individual believers were prone to undue disquiet in matters of conscience – what casuists described as 'scrupulosity'[18] – and it is in this context that one should see the bishops' attempt to quieten Boyle's worries about past moral dilemmas. Indeed, it is hard to avoid the impression that they were slightly irritated by the extreme to which Boyle took his anxiety, finding it tiresome and counterproductive, and, though I am aware of the difficulties of passing judgment on matters which are by their nature inscrutable, I think that it is hard not to see this aspect of his religiosity as 'dysfunctional'.

Boyle's casuistical concerns have hitherto been largely ignored. Insofar as they have been noticed, their significance has been misconstrued, due to an attempt to link them to a supposed political programme on Boyle's part. The culprit here is J. R. Jacob, who devotes three pages to the episode in 1659 when Boyle paid Sanderson a pension to publish his lectures on conscience, but interprets it much more narrowly than is warranted: he sees Sanderson's book as 'an answer to the republicanism rampant at the time', on the grounds that theories of popular sovereignty had needed refutation in the late 1640s when the lectures were originally given, and were topical again in the aftermath of Oliver Cromwell's death.[19] Now undoubtedly the political turmoil of the Interregnum caused acute dilemmas for many, and Sanderson's casuistical advice on such issues was widely valued, perhaps particularly as to whether royalists should subscribe to the so-called Engagement Oath in 1649–50.[20] But, though Sanderson alluded to the political circumstances of the time, his lectures provided a broad treatment of the relationship between divine and human law, significant not just in its immediate context but more generally: if this had not been the case, it would be hard to understand why, in 1877, Bishop Christopher Wordsworth should have had the work reprinted for the use of theological students in his diocese of Lincoln.[21] It is the purest of speculation to suggest that Boyle's reasons for sponsoring the book stemmed from its topicality at a particular moment, rather than from an appreciation of its long-term value similar to that of Bishop Wordsworth.[22]

This is not, however, atypical of the approach of Jacob in his *Robert Boyle and the English Revolution*, one of the main expositions of the 'social functionalist' view of Boyle. For, when the arguments of this book are examined closely, it repeatedly turns out that Boyle's supposed views have been divined by a highly selective reading of his

18 McAdoo, *op. cit.* (note 12), 75, 96.

19 Jacob, *Robert Boyle*, *op. cit.* (note 2), 172. See also *ibid.*, pp. 130–2.

20 See, for instance, the manuscript copies of Sanderson's response to Anthony Ascham's views on allegiance to ejected princes and his case of conscience on the subject, published in 1649, now to be found in British Library Add MS 32093, fols. 272–5, Egerton MS 2982, fol. 254; Stowe MS 746, fols. 146–7. See also Sanderson, *Works*, ed. Jacobsen (Oxford, 1854), v. 20–36, Isaac Walton, *Lives*, ed. George Saintsbury (London, 1927), 391, and Slights, *op. cit.* (note 12), 419ff.

21 C. Wordsworth (ed.), *Bishop Sanderson's Lectures on Conscience and Human Law* (Lincoln and London, 1877).

22 For an account of Sanderson which sees him as adopting an abnormally purist moral stance even by the standards of other casuists of the day, see Mosse, *op. cit.* (note 12), 141–4. Compare Slights, *op. cit.* (note 12), 422–3. It seems to me typical of Boyle that it was this approach that he chose to sponsor.

writings, or from extrapolation from the views of colleagues with which Boyle may or may not have agreed; in addition, unwarranted presumptions are frequently made as to the targets at which such writings were aimed. Thus the views of the so-called 'Boyle circle' around 1660 are extrapolated from the rather secularist opinions of Sir Peter Pett which there is no reason to think that Boyle wholeheartedly seconded: Jacob's principal evidence that he did so is a reading of the second section of Part II of *The Usefulness of Experimental Natural Philosophy* which depends on taking passages out of context and ignoring the main thrust of that work.[23] Similarly, Jacob's conviction that the figure whom Boyle had in mind in sponsoring Sanderson's lectures was Henry Stubbe seems to me fanciful in the extreme, while, though it is true that in the mid 1660s Boyle *did* become publicly embroiled with Stubbe through the Greatrakes affair, Jacob's portrayal of the latter almost entirely overlooks the convoluted nature of Boyle's attitudes, in many ways the most striking feature of the episode.[24]

The interpretation of Steven Shapin and Simon Schaffer is sometimes vulnerable to similar criticism. An instance is provided by their account of Boyle's refusal to take holy orders when it was suggested to him that he might do so after the Restoration. By way of explanation for this, they cite Boyle's view that, by speaking as a layman, he might claim an impartiality as an orthodox apologist which clergymen lacked. On the other hand, they completely ignore what he claimed to be 'his main reason' for the decision, namely his feeling that he had no vocation for the ministry, and that, if he accepted ordination without this, he 'should have lied to the holy Ghost'.[25] As such examples illustrate, very frequently the 'social functionalist' view is based on a much more selective reading of sources than the reader is ever told.

Boyle and oaths

It is for this reason that I think that it is so important to stress casuistry. For this is not just a closet concern of Boyle's, significant solely for the increase in historical verisimilitude that it gives. Rather, it is important because it actually affected and inhibited his attitudes and behaviour in crucial ways which have not hitherto been properly understood. Perhaps the most striking instance of this – if not the easiest to interpret – is Boyle's refusal to become President of the Royal Society in 1680. Boyle

[23] See Jacob, *Robert Boyle*, *op. cit.* (note 2), 133–44, esp. pp. 141–3; Boyle, *Works*, ed. Thomas Birch (2nd edn; 6 vols., London, 1772), **2**: 392ff.; [Sir Peter Pett], *A Discourse Concerning Liberty of Conscience* (London, 1661), a work displaying the rather secularist view of religious commitment also seen in Pett's *The Happy Future State of England* (London, 1688). On Pett, see especially Mark Goldie, 'Sir Peter Pett, sceptical Toryism and the science of toleration in the 1680s', *Stud. Church Hist.*, 1984, **21**: 247–73.

[24] Jacob, *Robert Boyle*, *op. cit.* (note 2), 164–76, esp. 172. For the alternative interpretation see particularly Eamon Duffy, 'Valentine Greatrakes, the Irish stroker: miracle, science and orthodoxy in Restoration England', *Stud. Church Hist.*, 1981, **17**: 251–73, esp. p. 269; see also N. H. Steneck and Barbara Kaplan, 'Greatrakes the stroker', *ISIS*, 1982, **73**: 161–85. For further criticism of Jacob's views, see my *Establishing the New Science: The Experience of the Early Royal Society* (Woodbridge, 1989), ch. 2, and 'Science and heterodoxy: an early modern problem reconsidered', in D. C. Lindberg and R. S. Westman (eds.), *Reappraisals of the Scientific Revolution* (Cambridge, 1990), 437–60.

[25] Shapin and Schaffer, *op. cit.* (note 3), 314 and ch. 7; Boyle, *op. cit.* (note 23), **1**: lx.

was elected to the Council and nominated for the presidency on St Andrew's Day that year, but, after a typical period of prevarication during which he attended at least one meeting, on 18 December he wrote to Robert Hooke, then Secretary, declining the office.[26] Boyle gave as the reason for his refusal of the honour his 'great (and perhaps peculiar) tenderness in point of oaths', language strongly reminiscent of the casuistical material already surveyed.

His exact motives are unclear, though a brief sketch of the context may go some way towards clarifying them. Oaths were controversial in seventeenth-century England, for two reasons: one because of those who took them too seriously, the other due to those who did not take them seriously enough. To deal with the latter first, there was much concern at the time that oaths were being trivialized by their casual use: indeed, Boyle himself had written a treatise on this subject in the late 1640s which was published after his death, his *Free Discourse against Customary Swearing*. In it, he gave a series of arguments against such use of oaths in fashionable circles, echoing the views of many at the time that it was because oaths were so serious a matter – religious acts in which God was invoked as a witness – that their abuse was so offensive: 'the Lord will not hold him guiltless that taketh his name in vain', in the words of the Ten Commandments (*Exodus*, 20.7).[27]

On the other side there were some who thought that oaths established such a sacred obligation that they were reluctant to use them at all: this was a point of view for which there was some scriptural support, including a passage from the Sermon on the Mount which could be seen as outlawing oaths altogether (*Matthew*, 5.34), though it was more commonly interpreted at the time as referring only to vain swearing.[28] The most notorious of such refusers of oaths were the Quakers, who took the view that men should be of such integrity that their affirmation should be dependable without the need for additional sanctions of this kind. Such a position was widely regarded at the time as seditious: for instance, in a book published in 1662, Bishop John Gauden attacked those people 'who refuse all *legal Oathes*, upon *scruples* of *Conscience*, and so threaten either to subvert our *Laws*, or to obstruct all *judicial proceedings*'. Though Gauden applauded the Quakers for their abhorrence of profane and trivial swearing, he considered them superstitious in their rejection of the practice altogether.[29]

Boyle's views seem to have had something in common with those of the Quakers. Indeed, it is conceivably significant that Gauden dedicated to him the very attack on Quaker attitudes that I have just quoted, since Boyle was not a particularly common dedicatee of non-scientific books.[30] Unlike the Quakers, Boyle did not wholly reject

[26] Thomas Birch, *The History of the Royal Society* (4 vols., London, 1756–7), **4**: 58, 60; Boyle, *op. cit.* (note 23), **1**: cxix.
[27] Boyle, *op. cit.* (note 23), **6**: 1ff, p. 4. Compare, for example, Robert Sanderson, *De Juramento* (London, 1647; Eng. trans., London, 1655), esp. pp. 259–63. It is probably significant that it was Sanderson's *De Juramento* that first attracted Boyle's attention to him (Walton, *op. cit.* (note 20), 399, 423).
[28] See, for instance, the sermons recorded in E. S. de Beer (ed.), *The Diary of John Evelyn* (6 vols., Oxford, 1955), **4**: 385, v. 423–4.
[29] John Gauden, *A Discourse concerning Publick Oaths* (London, 1662), sig. a3v, 17–18 and *passim*. On Quaker attitudes, see Richard Bauman, *Let your Words be Few* (Cambridge, 1983), ch. 7.
[30] For a list of works dedicated to Boyle, see J. F. Fulton, *A Bibliography of the Honourable Robert Boyle* (2nd edn, Oxford, 1961), 155–70.

oath-taking: indeed, in his *Discourse*, he had specifically distanced himself from those 'that indiscriminately condemn all oaths as absolutely and indispensably prohibited and abolished by the gospel'.[31] But he does seem to have felt abnormal scruples about taking oaths which were not strictly necessary, or which he had any reason to think that he might not be able to fulfil. One of the major issues raised in his casuistical interviews with Stillingfleet and Burnet was the extent to which he was bound by promises that he had made which might be to his disadvantage, while, in the autobiographical notes that he dictated to Burnet, Boyle recorded how 'He made no Vowes not knowing how his Circumstances might change but usually gave the 20th' (i.e., presumably, a twentieth part of his income).[32]

It is arguably in this context that one should read his attitude in 1680. It had been laid down in the Society's charter that, prior to taking office, the President should swear 'a corporal oath well and faithfully to execute his office' in front of the Lord Chancellor of the realm. This took the following form:

> I . . . do promise to deal faithfully and honestly in all things belonging to the trust committed to me as President of this Royal Society, during my employment in that capacity. So help me God![33]

Possibly, Boyle may have worried whether it was wise to guarantee to fulfil the obligations of this quite onerous office in so solemn a way – though it should be noted that this had apparently not prevented him accepting a comparable oath as a member of the Society's Council seven years previously, in 1673.[34]

More important is the fact that by 1680 the President of the Royal Society had to take the oaths prescribed by the Test Act of 1673 (at least, if it had not hitherto been clear that this applied to the President of the Royal Society, this was something that Boyle clarified by consulting no fewer than three lawyers, all of whom assured him that the act *did* affect him, as he explained in his letter to Hooke of 18 December). The Test Act insisted 'That all and every person or persons, as well Peeres as Commoners that shall beare any Office or Offices Civill or Military or shall receive any Pay, Salary, Fee or Wages by reason of any Patent of Grant from his Majestie or shall have Command or Place of Trust from, or under his Majestie' should take the oaths of supremacy and allegiance and receive the sacrament according to the usage of the Church of England.[35] Since Boyle specifically refers to oaths as the problem, it cannot have been taking the sacrament that caused him anxiety, while there is no reason to think that he disagreed with the content of the oaths that the act prescribed: both recited standard loyalist, anti-Catholic sentiments. Rather, it seems clear that it was the very issue of whether he should take such oaths at all that concerned him.

Ancillary evidence on this point comes from a letter from Henry Miles to Thomas

[31] Boyle, *op. cit.* (note 23), **6**: 26. This is described as 'that plausible error of our modern *Anabaptists*'.

[32] British Library Add. MS 4229, fol. 60.

[33] *The Record of the Royal Society of London* (4th edn, London, 1940), 228, 253.

[34] Birch, *op. cit.* (note 26), **3**: 114. Compare *The Record, op. cit.* (note 33), 229, 254.

[35] 25 Charles II, c. 2; cf. also 30 Charles II. stat. 1, c. 1: *Statutes of the Realm* (London, 1819), **5**: 782–5, 894–6. For the oaths of supremacy and allegiance, see *ibid.*, **4**: 352, 1074. For a manuscript version of part of these oaths in the hand of Sir Robert Southwell, referring to William and Mary, see Royal Society Domestic Manuscripts 5, fol. 30.

Birch, enclosing 'a Copy of a Q[uery] drawn up as if intended to be proposd to a Council for his Opinion concerning the necessity of taking the Test to qualifie him to Governor of the Corpor[ation] for propag[ating] of [the] Gospel in America', which he had found among Boyle's papers (though it is no longer extant). This read as follows:

> Q. whether the governour of a Company for management of a Charity (being a protestant) and acting by the rules of their Majesties Church, having nothing from their said majesties nor any salary & fee, or reward from them, or any other, and chosen by thirteen members of the said Company from time to time, be liable to take the tests, enjoyned by the late act for preventing dangers which may happen from popish recusants.[36]

It was presumably a similar question which Boyle asked his legal advisors in connection with the Royal Society presidency, and, being advised in the affirmative, decided against holding the office.

An interesting analogue to Boyle's scruples is provided by a case reported by Boyle's friend, the famous Presbyterian divine, Richard Baxter. This involved a prominent West Country politician, Thomas Bampfield, Speaker in Richard Cromwell's parliament, whom Baxter considered a 'prudent holy Man'. In Baxter's words:

> He is a Man of most exemplary Sincerity and Conscientiousness: He never took the Covenant, nor any other Oath in his Life, till he was a Member of the Parliament that brought in the King, and then he was put upon taking the Oath of Supremacy, which I had much ado (being my dear and much valued Friend) to perswade him to, so fearful was he of Oaths, or any thing that was doubtful and like to sin.

Subsequently, in connection with the Indemnity Bill of 1660, Bampfield opposed the imposition of the oaths of allegiance and supremacy on those who wanted to benefit from it, 'for then you press persons for the saving of their lives and estates to damn their souls'.[37] In Bampfield's anxieties, we see a close parallel to the likely feelings of Boyle.

Though Boyle's motives over the presidency have to be left a little indefinite, what is crucial is that in this instance Boyle's conscience made him act in a manner which it is very difficult to see as socially functional. One of the charges against the Quakers was that, however good their intentions, their rejection of oath-taking was subversive because of the indispensable role of oaths in underwriting social relations. John Tillotson, for instance, argued that the way in which the Quakers 'called in question the lawfullness of all Oaths' was 'to the great mischief and disturbance of humane Society'.[38] Perhaps no-one would have dared to make the same accusation against as

[36] Miles to Birch, 10 August 1743, British Library Add MS 4314, fols. 88v–9. A word is deleted before 'any salary'. Further evidence concerning Boyle's attitude is available from the 'oath' of secrecy that he imposed on his laboratory assistants (see Hunter, *op. cit.* (note 6), 406) which took the form: 'this I promise in the faith of a Christian, witnes my hand' (Royal Society MS 189, fol. 13).

[37] Richard Baxter, *Reliquiae Baxterianae*, ed. Matthew Sylvester (London, 1696), 1: 432 (I am indebted to John Spurr for this reference); B. D. Henning, *The House of Commons 1660–90* (3 vols., London, 1983), 1: 586.

[38] John Tillotson, *The Lawfulness, and Obligation of Oaths* (London, 1681), 8 and *passim*. See also Christopher Hill, *Society and Puritanism in Pre-Revolutionary England* (London, 1964), ch. 11, though Hill's overall interpretation is questionable; he also gives both Boyle's name and the date of publication of his *Customary Swearing* incorrectly (p. 412).

illustrious a person as Boyle (unless this was the point that Gauden was tentatively making by dedicating his book to him). But Boyle's conscience here *did* make him implicitly seditious by contemporary standards: that is to say, if his primary criterion had been social stability, he would not have gone down a path which was potentially disruptive according to the perceptions of the time. Such evidence is wholly at odds with a simplistically functionalist reading of Boyle.

Casuistry and natural philosophy

Boyle's scrupulosity also had implications for his intellectual activities. One example of this is Boyle's attitude to magic, which I have considered in detail elsewhere. This is illustrated by the autobiographical notes that he dictated to Gilbert Burnet, which may be supplemented by fragmentary survivals among his unpublished papers.[39] In recent years, we have become increasingly aware of the important contribution of ideas and information deriving from natural magic and alchemy to the thought of natural philosophers in this period. In Boyle's case, however, his moralistic attitudes made him much more alert than many contemporaries to the spiritual danger of the direct intercourse with the supernatural realm that he thought that magical practices entailed. The result was that he experienced scruples about activities of this kind, and the result may have been to limit his alchemical pursuits primarily to an attempt to verify empirically the process of transmutation. In contrast to Newton, whose profuse investigation of the mysteries of alchemy from the 1670s onwards has left so large a residue among his papers, Boyle's papers show little evidence of curiosity about such arcana. As a result, Boyle may have been deprived of fruitful insights into the operations of the natural world of a kind which recent scholarship has demonstrated that Newton derived from such sources.[40]

Moving to other facets of Boyle's natural philosophy, a 'scrupulosity' which bears a close relationship to the attitudes shown in his casuistry is much in evidence. Thus the lawyer and writer, Roger North, used this very word to describe Boyle's medical concerns, writing how 'Once in the company of old Doctor Denton [i.e., William Denton, royal physician] I was fleering [i.e., jeering] at the infinite scrupulosity Mr Boyle used about preserving his health'. He thus echoed a letter from William Petty of 16 April 1653 in which he criticized Boyle for 'your apprehension of many diseases, and a continual fear, that you are always inclining or falling into one or the other', and for 'practising upon yourself with medicaments (though specificks) not sufficiently tried by those, that administer or advise them'.[41] I am currently making a study of the issues surrounding Boyle's medical writings, but this valetudinarianism is a significant part of their background.

[39] See Hunter, *op. cit.* (note 6).
[40] Compare, for example, R. S. Westfall, 'Newton and alchemy', in Brian Vickers (ed.), *Occult and Scientific Mentalities in the Renaissance* (Cambridge, 1984), 315–35.
[41] Roger North, *Lives of the Norths* ed. A. Jessopp (3 vols., London, 1890), 3: 146; Maddison, *op. cit.* (note 5), 80–1.

Perhaps more surprising is the connection which may be demonstrated between Boyle's 'scrupulosity' and his scientific work. The unremitting pursuit of exactitude that he displayed in the laboratory parallels the pains that he took in salving his conscience. Indeed, he used the same terminology in describing both. Thus he told Burnet, looking back over his scientific career, that 'He made Conscience of great exactnes in Experiments'.[42] He also spoke of 'scruples' and 'scrupulosity' in an experimental context. This may be illustrated here by a single example, the description of his experiments concerning the apparent transmutation of water into earth in his *Origin of Forms and Qualities* (1666).[43] In this, these words occur again and again: thus he observed of the experiment 'that perhaps none but such a scrupulous person as I would think the prosecution of it other than superfluous'; subsequently, being forced to suspend operations 'before I had made half the trials I judged requisite in so nice a case, I have not yet laid aside all my scruples'; while within two paragraphs he is noting: 'Which scruple and some of the former I might have prevented, if I had had convenient metalline vessels wherein to make the distillations instead of glass ones'. Such talk of 'scruples' and 'scrupulosity' recurs throughout Boyle's published and unpublished writings on natural philosophy, and the significance of this has hitherto been overlooked. We tend to take Boyle's assiduity as an experimenter for granted, showing insufficient curiosity as to where these almost obsessive practices derived from. I would like to suggest that only one as assiduous in his spiritual exercises as Boyle would have thought it appropriate to employ such standards in the laboratory. Arguably it was his moralism which gave him the insistence on satisfying himself, if necessary going back over an issue time and time again, which is so typical of his science.

It could also be argued that Boyle's treatises owe something to the mental attitudes encouraged by casuistry. Even a casual survey of the literature of casuistry will illustrate how the casuistic approach was characterized by its constant effort to anticipate and deal with possible objections to an argument: an example is provided by Sanderson's lectures, the publication of which Boyle sponsored. There is an element of this in Boyle's scientific and philosophical writings: here too, there is an attempt to counter opposing views in anticipation in a manner that sometimes seems slightly artificial. Moreover if one studies the surviving drafts of Boyle's works, one can sometimes observe this taking shape, as, for example, in draft material for Boyle's *Free Enquiry into the Vulgarly Receiv'd Notion of Nature* (1686), which I am currently studying with Edward B. Davis. Here, a text which initially started in a relatively simple form became increasingly elaborate as more data was included to refute counter-arguments that occurred to Boyle during the course of composition. John Harwood has observed comparable characteristics in the early writings of Boyle which he has recently edited.[44]

Casuistry may also throw light on the origins of one of the crucial methodological

[42] British Library Add. MS 4229, fol. 60. [43] Boyle, *op. cit.* (note 23), 3: 102–9, pp. 105, 107.
[44] J. T. Harwood (ed.), *The Early Essays and Ethics of Robert Boyle* (Edwardsville and Carbondale, 1991), xliii and *passim*.

themes in Boyle's thought, his probabilism. The espousal of probabilist attitudes by Boyle and other late-seventeenth-century English scientists is commonplace, but the source of such attitudes is less clear. While some have postulated the influence of Gassendi and other continental philosophers, an alternative suggestion is that the roots of such a position is to be found in the approach of common lawyers.[45] It is worth pointing out here that the issue of whether certainty was either attainable or necessary was regularly discussed in the context of casuistry, and Boyle is as likely to have been familiar with such notions from his reading of casuistical literature as from the other sources that have been canvassed.[46] Moreover, that this approach had associations of this kind is suggested by a comment by Boyle's contact, the Oxford divine and natural philosopher, Robert Sharrock, who wrote in a letter to Boyle of 13 December 1661: 'And as to the Curablenesse of Cancers I have heard a story from a person of creditt, & I long to heare it seconded, that so it may bee to mee (according to the Jesuiticall canon of judging) a doctrine of probability'.[47]

This is speculative, but it suggests that, as an intellectual tradition, casuistry may prove more significant for Boyle's thought than has hitherto been acknowledged. More important is the moralism which made Boyle so passionately interested in cases of conscience in the first place, and which inculcated a similar intensity in his activity in other fields, not least his experimentation. A study of Boyle from this point of view provides a vivid, individual illustration of the role of the heightened religiosity of the Reformation era in forming the attitudes and approach of modern science.[48] Yet, if we appear to have here a clue to the source of Boyle's scientific originality, it is important to stress that these same attitudes were also responsible for a degree of conscientiousness which even Boyle's confessors apparently found tiresome, or his prevarication over the Royal Society presidency, thus illustrating the need to see the man as a whole in order to understand him. 'Dysfunctionalism' is a concept which may jar with some readers of this essay: but its intention is simply to illustrate the limitations of any interpretation of Boyle which picks and chooses among different facets of his life and thought, as is arguably the case with both the internalist and the social functionalist approaches to him. Only by seeing Boyle as a whole, taking the functional and the apparently 'dysfunctional' together, will we do justice to him; only thus will we do justice to the early modern culture from which he emanated.

[45] Sargent, *op. cit.* (note 1). See also Henry van Leeuwen, *The Problem of Certainty in English Thought, 1630–90* (The Hague, 1963) and Barbara Shapiro, *Probability and Certainty in Seventeenth-century England* (Princeton, 1983), esp. 37f.

[46] Margaret Sampson, 'Laxity and liberty in seventeenth-century English political thought', in Leites *op. cit.* (note 12), 72–118, esp. 78f; Shapiro, *op. cit.* (note 45), 37–8, 39, 69, 105–6. See also McAdoo, *op. cit.* (note 12), 88f.; Wood, *op. cit.* (note 12), 75f. For further discussion of this issue, see Ian Hacking, *The Emergence of Probability* (Cambridge, 1975), esp. 22f; D. L. Patey, *Probability and Literary Form* (Cambridge, 1984), esp. 50f; and Sargent, *op. cit.* (note 1).

[47] Royal Society Boyle Letters, v. 51.

[48] I ended the version of this paper that I delivered at the Oxford conference with some semi-humorous remarks about the celebrated thesis expounded by Robert K. Merton in his 'Science, technology and society in seventeenth-century England', *Osiris*, 1938, 4: 360–632 (reprinted in book-form, New York, 1970). Suffice it here to note that it seems to me that the issue of the relationship of Puritanism to the rise of science merits fresh scrutiny along the lines sketched here for Boyle.

10

Clandestine Stoic concepts in mechanical philosophy: the problem of electrical attraction

GAD FREUDENTHAL

Two chameleons secured themselves a place in intellectual history. The second is Woody Allen's.[1] Some three and a half centuries earlier, Pierre Gassendi had invoked the same reptile, albeit in order to shed light on the mysteries of nature, rather than on those of the human soul:

> If we attentively observe a *Chamaeleon* catching Gnats and other small Flyes in the Aer, for his food, we shall see him dart out a long and slender tongue, with a small recurvation at the tip, and birdlimed with a certain tenacious and inviscating moisture, wherewith, in a trice, laying hold of a Fly, at some distance from his mouth, he conveys the same into it with such cleanly speed, as exceeds the Legerdemane of our cunningst Junglers . . . And when we see a piece of Amber, Jet, hard Wax, or other Electrique, after sufficient friction, to attract straws, shavings of wood, quils, and other festucous bodies of the same lightness, objected within the orbe of their Alliciency, and that with a cleanly and quick motion: Why should we not conceive, that this Electricity or Attraction may hold a very neer Analogy to that attraction of Gnats, by the exserted and nimbly retracted tongue of a Chamaeleon?[2]

Thus Gassendi, Englished by Walter Charleton in 1654, giving a 'simile' through which 'the means of Attractions Sympathetical' will be shown to conform to the *Physiologia Epicuro-Gassendo-Charltoniana*. Gassendi's chameleon had a forerunner. In his *Two Treatises* published in Paris in 1644, Sir Kenelm Digby suggested that electrical attraction can be modelled after 'the little tender hornes of snails, [which] use to shrinke backe if any thing touch them, till they settle in little lumpes upon their heades'.[3]

[1] Woody Allen, *Zelig* (1983). See Woody Allen, *Three Films of Woody Allen: Zelig, Broadway Penny Rose, The Purple Rose of Cairo* (London, 1990).

[2] Walter Charleton, *Physiologia Epicuro-Gassendo-Charltoniana: or A Fabrick of Science Natural, Upon the Hypothesis of Atoms* (London, 1654; reprinted with a new Introduction by R. H. Kargon, New York and London, 1966), 345; punctuation partly added.

[3] Kenelm Digby, *Two Treatises. In the One of Which, The Nature of Bodies, in the Other of Mans Soule; is Looked Into* (Paris, 1644), 172.

That inhabitants of the *animal* kingdom are called upon to establish the new *mechanical* philosophy should give us pause. Can the upholders of corpuscularism legitimately conceive electrical attraction after the model of a chameleon's tongue or the horns of a snail? Charleton and Gassendi themselves seem to have had their doubts:

> All the Disparity, that can be objected, seems to consist onely in the Manner of their Return, or Retraction; the Tongue of the Chamaeleon being both darted forth, and retracted by help of certain Muscles, wherewith Nature, by a peculiar providence, hath accommodated that otherwise Helpless Animal: but, Electriques are destitute of any such organs, either for the Exsertion, or Reduction of their Rayes.[4]

Nonetheless, there is no reason for despair: can one not just forget about the muscles and think of those remarkable tongues or horns simply as elastic threads? The explanation of electrical attraction then lies to hand:

> so may the Rayes, which stream from an Electrique . . . be conceived to Reduce and Retract themselves, after the manner of Sinews and Lutestrings violently extended.[5]

However, Gassendi–Charleton foresaw that, as innovating theoreticians, they would have to confront uncomfortable questions from pedants of all sorts. To those who would wish to inquire how a thread, which *ex hypothesi* consists of hard corpuscles, can be elastic, they reply reassuringly:

> But, however this may be controverted, and the Way of all Electrique Attractions variously explicated, according to the various Conceptions of men . . . yet still it is firme and indubitable, that though the Attraction of straws by Amber, be in some sort Admirable, yet it is not *Miraculous*, as is implied in that opinion, which would have it to be by some *Immaterial* (i.e. *Supernatural*) Virtue; and that it is effected by some *Corporeal*, though both impalpable and invisible Organs continued from the Attrahent to the Attracted.[6]

To which we cannot but add: *quod erat demonstrandum*.[7]

I now wish to suggest that the electrical chameleon (or snail) has more to it than meets the eye and that 'deconstructing' it, to use a fine trendy term, is revealing. Digby, Gassendi, and Charleton, as also Sir Thomas Browne and Robert Boyle, pursue a research programme whose goal is to 'make occult qualities manifest' and they assume that certain macro-effects are caused by subtle *effluvia* imperceptibly emitted by some bodies. This, of course, is perfectly sound. The need to bring in the chameleon, however, points to a 'blind spot' in the research programme, an area of phenomena in the face of which the programme is helpless. As Gassendi and Charleton themselves were quick to recognize, *effluvia* have no muscles; but this means that their retraction remains unaccounted for. The reference to the elasticity of

[4] Charleton, *op. cit.* (note 2), 345. [5] *Ibid.* [6] *Ibid.*, 346.

[7] Robert Boyle rehearses precisely the same reasoning in his 'Experiments and Notes about the Mechanical Production of Electricity' of 1675; see *The Works of the Honourable Robert Boyle* (2nd edn, 6 vols., London, 1772), **4**: 347.

sinews and their like only exacerbates the problem, and the proponents of mechanical philosophy must retreat to a declaration of faith that although electricity is an 'admirable' phenomenon, it is none the less not miraculous.

The problem is in fact inherent to the very premises of the mechanical philosophy in the period before the introduction of interparticulate forces. The problem is this: the idea that all phenomena must be reduced to actions by contact of hard corpuscles in motion naturally leads to the assumption that what appears to the senses as an action at a distance is in fact brought about by invisible particles. In matters electrical, this suggests the view that rubbing an electric body causes an outflow of particles from it. But how can particles *issuing from* a body conceivably bring about a motion *toward* it? The theoretical model would lead us to expect precisely the opposite! This is the fundamental question which the chameleon's tongue and the snail's horn highlight rather than conceal.

In the seventeenth and early eighteenth centuries hardly anyone explicitly said that the emperor had no clothes, that is that the mechanical philosophy was incapable of accounting for attraction, electrical or other. Only Stephen Gray, it seems, had his eyes open: in a letter written in January 1707/8 but not published at the time, he first conjectured that the *effluvia* from the electric body are reflected by nearby objects: this conjecture would have had the welcome corollary that there are two streams of *effluvia* of which one is indeed directed toward the electric body; unfortunately, however, experiments quickly refuted this theory. Gray then goes on to say:

> I have therefore thought on an other Hypothesis . . . [namely,] that as all bodies Emitt, soe they Receive part of the Effluvia of all other bodies that Inviron them and that the attraction is made according to the current of these Effluvia; but then [honest Gray goes on to ask] how rubing the Glass, though it may cause a more copious and swift Eruption of the Effluvia, yet that it shold in like manner affect other distant bodies is hard to conceive.[8]

Indeed it is! Gray therefore preferred to leave the solution of the problem 'to the consideration of the Learned', who, however, wisely chose to let sleeping dogs lie.

The proponents of the mechanical philosophy in the mid seventeenth century in fact tried to give substance to the metaphorical models of electrical attraction by maintaining that the *effluvia* issuing from a rubbed electric body are not hard particles acting by impact, but rather *elastic* threads. Thus Gassendi and Charleton argue:

> That such tenacious Rayes [analogous to the tongue of a chameleon] are abduced from Amber and other Electriques, is easily convincible (. . .) from hence, that all Electriques are *Unctuous* and *Pinguous* Concretions, and that in no mean degree: and manifest it is, that a viscid and unctuous Bodie is no sooner Warmed by rubbing, but there rise out of it certain small *Lines* or *Threads*, which adhaere to a mans finger that toucheth it, and such as may, by gentle abduction of the finger, be prolonged to a considerable distance.[9]

[8] R. A. Chipman, 'An unpublished letter of Stephen Gray on electrical experiments, 1707–1708', *ISIS*, 1954, **45**: 33–40, on p. 36; punctuation partly added.
[9] Charleton, *op. cit.* (note 2), 346.

Sir Kenelm Digby's version of the same explanation is a little more elaborate:

> Let us then suppose, that there is a solide hard body, of an unctuous nature; whose partes are so subtile and fiery, that with a little agitation they are much rarifyed, and do breath out in steames (though they be too subtile for our eyes to discerne), like unto the steame that issueth from sweating men or horses . . .; but that these steames, as soone as they come into the cold ayre, are by that cold soddainely condensed againe; and by being condensed, do shorten themselves, and by little and little do retire, till they settle themselves upon the body from whence they sprung . . . If I say these strings of bituminous vapour should in their way outwardes meete with any light and spungie body, they would pierce into it, and settle in it; and if it were of a competent biggenesse for them to wield, they would carry it with them which way soever they goe; so that if they shrinke backe againe to the fountaine from whence they came, they must needes carry backe with them the light spungy body they have fixed their dartes in.[10]

The electrical threads, in other terms, are *elastic*, and they are elastic because they are *unctuous*.

This, indeed, was the most widely accepted theory of electricity in the second half of the seventeenth and early eighteenth century, particularly in Britain.[11] The chameleon simile made it possible to allow that elastic threads intervened where corpuscularism was found wanting. But we should now ask why the proponents of mechanical philosophy were so fond of the postulated unctuous, and therefore *ex hypothesi* elastic, threads. After all, there are also other elastic substances: why this almost unanimous consensus in choosing unctuous ones? Empirical considerations alone cannot provide the explanation: René Descartes, for one, rejected the theory of the unctuous threads with the argument that glass is electric but not unctuous.[12]

As I have already argued briefly elsewhere,[13] the account of electricity in terms of unctuous, and therefore elastic, threads was not a mere casual metaphor. Rather, it drew on an ancient and well-established chemical notion, on which, *faute de mieux*, our corpuscularians drew.

The history of the notion of unctuous moisture belongs to the history of chemistry within Aristotelian natural philosophy, specifically to the history of the explanation of cohesion. From the vantage point of Aristotle's theory of matter, the cohesion of

[10] Digby, *op. cit.* (note 3), 172.

[11] John L. Heilbron, *Electricity in the Seventeenth and Eighteenth Centuries* (Berkeley, 1979), 193–5, 202–5.

[12] René Descartes, *Principes de la philosophie*, (Paris, 1668) 4: *§§* 184–185; Descartes, let it be noted, was acquainted with the notion of unctuous moisture and sought to account for it mechanically: cf. e.g. *ibid.*, *§§* 62–63, 76. Robert Boyle sought to refute Descartes' argument in *op. cit.* (note 7), 345f.

[13] Gad Freudenthal, 'Die elektrische Anziehung im 17. Jahrhundert zwischen korpuskularer und alchemischer Deutung', in Christoph Meinel (ed.), *Die Alchemie in der europäischen Kultur- und Wissenschaftsgeschichte* (Wolfenbütteler Forschungen, Band 32) (Wiesbaden, 1986), 315–26; Freudenthal, 'The problem of cohesion between alchemy and natural philosophy: from unctuous moisture to phlogiston', in Z. R. W. M. von Martels (ed.), *Alchemy Revisited. Proceedings of an International Congress at the University of Groningen, 17–19 April 1989* (Leiden, 1991) 107–16. In what follows I draw on these two papers.

composite substances poses a serious difficulty: Why do not the four elements which make it up fly off up and down to their respective natural places? Why do not the four contrary qualities annihilate one another? What, in short, holds the body together, endowing it with transtemporal stability?[14] One account Aristotle gives of cohesion is the following: 'Earth has no power of cohesion without the moist. On the contrary, the moist is what holds it together; for it would fall to pieces if the moist were eliminated from it completely'.[15] But does not moisture evaporate? It certainly does – unless it is *unctuous* moisture. We in fact find that Aristotle himself at times draws on a distinction, going back to the ps.-Hippocratic treatise *On Fleshes* (end of the fifth century BC),[16] between 'ordinary', i.e. aqueous, moisture and unctuous (or fat, *liparos*) moisture, a moisture which does not easily evaporate, and which therefore allows the substance in which it inheres to resist corruption. For instance, plants live longer than animals 'because they have an oiliness and a viscosity which makes them retain their moisture in a form not easily dried up'.[17] And in general: 'fat things are not liable to decay'; indeed, 'a fat substance is incorruptible'.[18] Aristotle thus generalized the empirical fact that, as he himself observed, olive oil is not dried up by either heat or cold,[19] into a rudimentary theory of cohesion in terms of a non-evaporable unctuous moisture.

The notion of an incorruptible fat moisture, which resists evaporation and thus confers cohesion, came to the fore only in the Arab synthesis of Aristotelian natural philosophy and Greek and Syriac alchemy. This resulted from the spread of distillation and its application to a wide range of substances, mineral, vegetable, and animal.[20] The typical result of the fractionated distillation of organic matter was the following. After an aeriform, volatile and inflammable substance (which we know today to be sal ammoniac[21]), a soft fire would next give rise to a vapour which, upon condensation, became a clear liquid: this liquid was referred to as 'water'. Then, further and stronger heating would give rise to a coloured and unctuous liquid, referred to as 'oil' (*duhn*). At the end of the process, a dry powder was left behind at the bottom of the alembic.[22] Distillation was thus regarded as a process through which the two distinct kinds of moisture could be *isolated*: the aqueous, which evaporates easily and which condenses to water; and the unctuous, which evaporates only with difficulty and

14 I borrow this phrase from Montgomery Furth, 'Transtemporal stability in Aristotelian substances', *J. Phil.*, 1978, 75: 624–46. Compare also Mary Louise Gill, *Aristotle on Substance. The Paradox of Unity* (Princeton, 1989), 166ff, 213, 234.

15 Aristotle, *De gen. et corr.* 2.8, 335a1–3. All translations of Aristotle come from *The Complete Works of Aristotle. The Revised Oxford Translation*, ed. by Jonathan Barnes (2 vols., Princeton, 1984).

16 For text and translations compare K. Deichgräber, *Hippokrates Über die Entstehung und Aufbau des menschlichen Körpers (Peri Sarkon)* (Berlin and Leipzig, 1935); R. Joly, *Hippocrate, Des Chairs*, in *Oeuvres*, vol. XIII (Paris, 1978).

17 Aristotle, *De long. et brev. vit.* 6, 467a6ff.

18 Aristotle, *De long. et brev. vit.* 5, 466a23; *Historia animalium* 3.19, 521a1.

19 Aristotle, *Meteor.* 4.7, 383b34.

20 R. J. Forbes, *Short History of the Art of Distillation* (Leiden, 1948).

21 J. Ruska, 'Der Salmiak in der Geschichte der Alchemie', *Z.Angew. Chem.*, 1928, **41**: 1321–4.

22 Compare e.g. the 'Syriac and Arabic alchemical treatise', in P. E. M. Berthelot, *La Chimie au moyen âge* (3 vols., *Paris*, 1893; reprinted Osnabrück and Amsterdam, 1967), **2**: 184–5; J. Ruska, *Al-Rāzī's Buch Geheimnis der Geheimnisse* (Quellen und Studien zur Geschichte der Naturwissenschaften und der Medizin, **VII**) (Berlin, 1937), 78ff; 204ff; 207ff.

upon the disappearance of which the body disintegrates. Naturally, this unctuous fraction came to be considered as the perceptible cause of cohesion.

In the corpus of alchemical writings attributed to Jābir ibn Ḥayyān, for example, moisture is held to be isolated 'when oil is distilled until a very glutinous and elastic substance is obtained ... This substance never solidifies'.[23] Since oil is thus the principle of permanent liquidity, it is the cause of cohesion: it is in fact explicitly said to be the '*unctuous quality that brings about combination*'.[24]

Within Arab natural philosophy and alchemy, the notion of unctuous moisture thus became a theoretical concept in the full sense of the term, accounting for cohesion, elasticity, fluidity, and inflammability. Numerous explanations, in quite different contexts, drew on it. For instance, Ibn Sīnā uses it in the *Canon* to account for the formation of calculi, whose matter, he says 'is a thick unctuous moisture (*ruṭūbah lazijah ghalīzah*)', having its origin in thick nutriment.[25] Again, in the *Shifā'*, Ibn Sīnā advances the audacious view that mountains have not been created by God and that it is not His providence that maintains them against erosion: Ibn Sīnā's account of the continuous *natural* formation of stones and mountains is founded on the idea that they arise out of specifically unctuous clay – a matter, that is, which does not crumble upon desiccation.[26] The standard account of calcination is grounded in the same theory. Calcination, writes the famous physician, physicist, and alchemist al-Rāzī, is the 'destruction of bodies' (i.e. metals) through 'the burning of the sulphurs and oils they contain, [resulting in] their reduction to white lime whose parts cannot further be divided'.[27]

To come now to the Latin West, Albertus Magnus accepts Ibn Sīnā's petrology, and draws on the notion of unctuous moisture also in order to solve further difficulties arising within the Aristotelian theory of matter. Thus, starting from Aristotle's postulate that 'the material of all liquefiable things is one – that is, water', he faces the following problem:

> We know ... that watery moisture is easily converted into vapour ... But we see that metals retain their moisture even in hot fires. Therefore the moist materials of metals cannot be simple water ... [I]f we consider the [kinds of] moisture which are difficult to separate from things that naturally

[23] Paul Kraus, *Jābir ibn Ḥayyān: Contribution à l'histoire des idées scientifiques dans l'Islam*. Vol. II: *Jābir et la science grecque* (= *Mémoires présentés à l'Institut d'Egypte*, vol. 45) (Cairo, 1945; reprinted, Paris, 1986 and Hildesheim, 1989), 10.

[24] H. E. Stapleton, R. F. Azo and Hidāyat Husain, 'Chemistry in Irāq and Persia in the tenth century A.D.', *Mem. Asiatic Soc. Bengal*, 1927, 8: 317–418, at p. 338f.

[25] Ibn Sīnā, *Kitāb al-Qānūn fī-l-Tibb*, Book VI, Treatise 18, Article 2, Chapter 6 (Rome, 1593), p. 434.

[26] Compare E. J. Holmyard and D. C. Mandeville (eds. and trans.), *Avicennae De congelatione et conglutinatione lapidum, Being Sections of the Kitāb al-Shifā'* (Paris, 1927), 71ff. (Arabic text), 18ff. (English translation). I analyse Ibn Sīnā's petrology and geology in my paper '(Al)chemical foundations for cosmological ideas: Ibn Sīnā on the geology of an eternal world', in: Sabetai Unguru (ed.), *Physics, Cosmology and Astronomy, 1300–1700: Tension and Accommodation* (= Boston Studies in the Philosophy of Science, vol. 126) (Dordrecht, Boston and London, 1991), 47–73.

[27] Ruska, *op. cit.* (note 22), 126 (compare also p. 74); Stapleton *et al.*, *op. cit.* (note 24), 330. Similarly, the (Latin) Geber defines: 'Est ergo calcinatio rei per ignem pulverizatio ex privatione humiditatis partes cosolidantis.' ('The calcination of a thing by fire is, therefore, its conversion into powder, due to the removal of the humidity consolidating its particles.') Compare William R. Newman, *The Summa Perfectionis of Pseudo-Geber, A Critical Edition, Translation and Study* (Collection de travaux de l'Académie interntionale d'histoire des sciences, tome 35), (Leiden, 1991), 417, 704, respectively.

> contain them, we find that they are all unctuous and viscous. ... And therefore, since the moisture in metals is not torn out of them, even by strong heating, this, too, must be unctuous.[28]

To explain the cohesiveness of this 'viscous and sticky' moisture, Albertus draws on Aristotle's account of viscosity,[29] maintaining that 'its parts join with the earthy parts like the links of a chain'.[30]

Medieval Arabic and Latin alchemy and natural philosophy thus developed the ancient notion of unctuous moisture into a fully fledged theoretical concept accounting for inflammability, cohesion, and elasticity. In the Renaissance the notion was still alive and very well. Thus Conrad Gesner (1516–1565) describes distillation as follows:

> Of a plant or any other substance ordeined to be distilled, what parte of it is most meet to be extenuated and fyret (that is the purest parte, the lightest, the thinnest, the *moistest* and the most superficial parte . . .) being first of all fyret by the force of the heat, is lifted up; next suche other partes as in purenes cum nie to the first, and last such a *moisture* of the thinges as is more crosse *that held together the earthy partes*, a certain *fatness and oiliness*, by a stronger force of the fyre is separated, and taken up hooly; which once clean drawn forthe, the body remaineth dissolved and brought to ashes. [Italics added.][31]

Kenelm Digby gives much the same account.[32]

The seventeenth-century corpuscularians too were familiar with, and embraced, the notion of unctuous moisture and its explanatory import. Thus Walter Charleton writes:

> [T]here are *Two* sorts of Moisture, wherewith compact bodies are usually humectated; the one, *Aqueous* and *Lean*, the other, *Oleaginous* and *Fat*. The *First* is easily dissoluble and evaporable by heat, but not inflammable; the *other*, though it easily admit heat, and is as easily inflammable . . . is not easily exsoluble, nor attenuable into fume, in regard of the Tenacious cohaerence of its particles.[33]

For instance: 'Wood is sooner reduced to Ashes, than a stone: because that is compacted by much of Aqueous Humidity; this by much of Unctuous'.[34]

I thus conclude that the fashionable theory accounting for electricity in terms of elastic threads and for elasticity in terms of unctuous moisture has deep historical roots: far from being a mere occasional metaphor, it was grounded in received and well-established chemical theory. Indeed, the electrical theory owed its popularity precisely to its unanimously accepted theoretical underpinnings. It is also noteworthy that the invisible theoretical grounding of this electrical theory was to have highly visible effects during the eighteenth century. At the very beginning of the century, the

[28] Albertus Magnus, *De mineralibus*, Bk. III, Tract. i, Chap. 2; quoted after *Book of Minerals*, trans. Dorothy Wyckoff (Oxford, 1967), 156.
[29] Aristotle, *Meteorologica* 4: 9, 387a14ff.
[30] Albertus Magnus, *De mineralibus*, *op. cit.* (note 28), Bk. I, Tract. i, Chap. 2; quoted after *Book of Minerals*, 13.
[31] Conrad Gesner, *The Treasure of Evonymus* (translated by Peter Morwyng) (London, 1559), 2.
[32] Compare Betty Jo Teeter Dobbs, 'Studies in the natural philosophy of Sir Kenelm Digby. [Part I]' *Ambix*, 1971, **18**: 1–25, on p. 21.
[33] Charleton, *op. cit.* (note 2), 322. [34] *Ibid.*, 323.

research programmes of at least three investigators led them from an interest in 'phosphors' (luminosity produced by cold substances) to an interest in electricity: this *prima facie* unlikely transition becomes easily understandable if one considers that both luminosity and electricity received theoretical accounts in terms of their 'oiliness'.[35] Again, it has been noted that electricity and phlogiston were occasionally identified:[36] this, too, becomes comprehensible if (and only if) we consider that Becher's *terra pingua* and in its wake Stahl's phlogiston are but successors to the medieval notion of unctuous moisture.

Our story has still a further, as yet untold component: this is the Stoic connection announced in my title. I must however forewarn the reader that here my discussion will remain somewhat conjectural.

We start from a question: assuming unctuous moisture does not evaporate, how is it supposed to 'hold together' the components of a substance? The problem arises because Aristotelian natural philosophy strictly rules out the idea of the *interpenetration* of matter, the possibility of two bodies occupying the same place at the same time.[37] But if we suppose the unctuous moisture to be merely juxtaposed to the other components of a substance, how can it hold them together? Albertus Magnus, for one, was keenly aware of the problem, as is manifest in his endeavour to provide an answer:

> [W]e say that the cause of coherence and mixing is moisture, which is so subtle that it makes every part of the [element] earth flow into every other part; and this is the cause of the thorough mixing of the parts of the material. And in that case, if this moisture were not soaked all through the earthy parts, holding them fast, but evaporated when the stone solidified, then there would be left only loose, earthy dust.[38]

Again, stones are hard if their matter 'is very well mixed': 'First the moisture affected it, causing every part of the dry [material] to flow into every other part . . . For what is subtle and moist is capable of being well mixed, since it is active in penetrating the parts and even the smallest particles'.[39] The two components, the moisture and the dryness, then, 'each acts upon, and is acted upon, by the other', so that neither can be separated even by fire.[40] But how, we wish to know, can this account be reconciled with the Aristotelian postulate that interpenetrability is impossible?

A preliminary clue to an answer can be obtained from a careful look at the theories of Jābir ibn Ḥayyān. Jābir (to use this convenient name to refer to the authors of the corpus) established a one to one correspondence between the four fractions obtained in distillation and the four Aristotelian elements. The first distilled, volatile and inflammable, fraction was naturally taken to derive from fire; the next, watery, fraction

[35] Compare Gad Freudenthal, 'Early electricity between chemistry and physics: the simultaneous itineraries of Francis Hauksbee, Samuel Wall, and Pierre Polinière', *Hist. Stud. Phys. Sci.*, 1981, **11**: 203–29.

[36] W. H. Sudduth, 'Eighteenth-century identifications of electricity with phlogiston,' *Ambix*, 1978, **25**: 131–47.

[37] Compare e.g. Richard Sorabji, *Matter, Space, and Motion* (Ithaca, 1988), notably Chapter 5.

[38] Albertus Magnus, *De mineralibus*, *op. cit.* (note 28) Bk. I, Tract. i, Chap. 2; quoted after *Book of Minerals*, 12–13.

[39] *Ibid.* Bk. I, Tract. ii, Chap. 1; 37.

[40] *Ibid.*, Bk. III, Tract. ii, Chap. 1; 186–7.

was held to be the element water; the third, oily, fraction was associated with the element air; lastly, the dry residue which remained behind was identified with the element earth.[41] The *Ikhwān al-Ṣafā*, the Faithful Brethren, too, identified oil with the element air.[42]

There is something puzzling about this scheme, namely the identification of oil – the third and last substance to rise in distillation – with *air*, the second lightest element. Now it is possible that the reason for this identification is simply that the other possibility – identifying the second, watery, fraction with the element air, and the third, oily, fraction with the element water – was not very attractive either. It should also be remembered that Aristotle himself occasionally states that oil contains air.[43] I believe, however, that there is also another, by far more serious and significant explanation for this identification. I suggest the possibility that the association of air with unctuosity, the cause of cohesion, has to do with the theoretical import of the notion of air not in Aristotelian natural philosophy, but rather in Stoicism. The Stoics maintained the idea of interpenetration of bodies, the possibility of two bodies being totally blended – this in fact is one of the most original points of their physics. Specifically, the Stoics construed their divine *pneuma*, constituted of the elements fire and air, as a body that interpenetrates all other bodies,[44] endowing them with cohesion.[45] This subtle *pneuma*, the universal cause of elasticity and cohesion, was often confused with the element air, which at any rate was one of its two components. I thus suggest, first, the possibility that it is no mere coincidence that the Stoic *pneuma* shares with the unctuous moisture the interpenetrability and the capacity to bring about cohesion, and, further, that there may be a historical connection between the two concepts.

The influence of Stoicism on Medieval Arabic natural philosophy has not yet been studied systematically.[46] As a first small step in this direction consider the following,

41 Compare Kraus, *op. cit.* (note 23), 5f., 41ff.; compare also F. Rex, *Zur Theorie der Naturprozesse in der früharabischen Wissenschaft. Das 'Kitāb al-Ikhrāğ', übersetzt und erklärt. Ein Beitrag zum alchemistischen Weltbild der Gābir-Schriften* [= Collection des travaux de l'Académie internationale d'histoire des sciences, tome 22] (Wiesbaden, 1975), 42.

42 The *Ikhwān* refer to air itself as unctuous, and in fact think of unctuous moisture as consisting of unctuous particles of air, which they take to result from the evaporation of suitable substances; compare *Ikhwān al-Ṣafā, Rasā'il* (Beirut, 1957), 2: 106; German translation: F. Dieterici, *Die Naturanschauung und Naturphilosophie der Araber im zehnten Jahrhundert. Aus den Schriften der lautern Brüder* (Berlin, 1861), 113ff. These particles they take to combine with the particles of other elements. The *Ikhwān* indeed followed Jābirian doctrines on many matters: compare Yves Marquet, 'Quelles furent les relations entre "Jābir ibn Ḥayyān" et les Ihwān aṣ-Ṣafā?', *Stud. Islamica*, 1986, **64**: 39–51; also Marquet, *La Philosophie des alchimistes et l'alchimie des philosophes. Jābir ibn Ḥayyān et les 'Frères de la Pureté'* (Islam d'hier et d'aujourd'hui, **31**), (Paris, 1988).

43 Compare e.g. *Meteorologica* IV, 7, 383b22f; *Generation of Animals* II, 2, 735b29f; elsewhere Aristotle says that oil contains *pneuma* (*ibid.*, 735b24f) and that it is a 'combination of air and fire' (*Parts of Animals* II, 5, 651a24f).

44 Sorabji, *op. cit.* (note 37), Chap. 6.

45 Samuel Sambursky, *The Physics of the Stoics* (London, 1959), 2ff; Michael Lapidge, 'Stoic cosmology', in John M. Rist (ed.), *The Stoics* (Berkeley, 1978), 161–85.

46 Paul Kraus has already pointed out that in the corpus of writings attributed to Jābir ibn Ḥayyān the construal of the dry, the moist, the hot, and the cold as *substances* and not as qualities testifies to Stoic influences; see P. Kraus, *op. cit.* (note 23), notably pp. 168–73. For a masterly account of the influence of Stoicism on the medieval Latin West see Michael Lapidge, 'The Stoic inheritance', in P. Dronke (ed.), *A History of Twelfth-Century Western Philosophy* (Cambridge, 1988), 81–112.

from our vantage point particularly suggestive, passage from the Arabic version of Galen's *On Cohesive Causes*:

> The first philosophers of my acquaintance to speak of a cohesive cause (*sabab māsik*) were the Stoics. . . . Of the [four] elements . . . some they call material and some active and dynamic. They maintain that the material elements are held together (*mutamāsika*) by those that are dynamic, fire and air (*hawā'*) being dynamic and active in their view, while earth and water are material. They say that, when the elements are intermingled, the dynamic wholly penetrate the material, that is to say, air and fire penetrate water and earth. . . . The two active elements have fine parts and the other two thick parts. Every substance with fine parts the Stoics call spirit (*ruḥ*, i.e. *pneuma*) and they think that the function of this spirit is to produce cohesion in natural and in animal bodies.[47]

If we now assume that this Stoic view of *pneuma* as a subtle active substance, interpenetrating bodies and endowing them with cohesion, went into Jābir's alchemy, we will easily understand why Jābir and his followers identified oil, the cause of cohesion, with air.[48] We will also see that this Stoic view of *pneuma* may well lurk behind the following statement by Jābir: 'Every oil catches hold of, and enters into, and unites with, the "Bodies" (i.e. metals), so that the property of admixture is found in . . . Oils [only]'.[49]

Originally, it would thus seem, the notion of unctuous moisture derived its considerable explanatory import from Stoicism. This is no mere historical accident. Aristotelian natural philosophy was in fact incapable of accounting for cohesion and thus needed something to fill a theoretical gap. Stoicism was presumably the only physical doctrine having such an account to offer. Yet the Peripatetics could not borrow the Stoic explanation as such, because it explicitly presupposed interpenetrability. They could however assimilate the notion of unctuous moisture, because, on the face of it, it appeared Aristotelian, although in fact it was grounded in Stoic physics and also assumed interpenetrability. And so it happened that the notion of unctuous moisture became a vehicle, or rather a Trojan horse, by which distinctively Stoic ideas – the notion of *pneuma* as interpenetrating other substances and making them cohere – entered, albeit incognito, into medieval and later alchemy, chemistry, and natural philosophy.

The notion of unctuous moisture, we realize, could offer a valid explanation of cohesion only because it implicitly ran against a fundamental postulate of the Aristotelian theory of matter. This means that, the proclamations of its proponents

[47] Galen, *On Cohesive Causes*, ed. and trans. Malcolm Lyons (Corpus medicorum graecorum supplementum orientale II), (Berlin, 1969), 52–3. The importance of Galen's Arabic translations in transmitting Stoicism to the Arabs has been pointed out by P. Kraus, *op. cit.* (note 23), 326–30.

[48] Some points remain obscure, though. For instance, according to Galen's account, the Stoics associated the air's capacity to 'consolidate and thicken a substance' with its purported cold, whereas Jābir follows Aristotle in taking air to have the quality warm. See Galen, *op. cit* (note 47), 52, lines 9–10 for the text (53, lines 12–13 for the translation); Kraus, *op. cit.* (note 23), 5, 164, 173. Another problem is the different meanings of 'spirit' (*ruḥ*) in Stoicism and in alchemy.

[49] Stapleton *et al.*, *op. cit.* (note 24), 395.

notwithstanding, this notion could not consistently be accommodated within Peripatetic natural philosophy, and thus remained grafted upon it as an alien member. This was well observed by the seventeenth-century critic of Aristotle, the physician Edward Jorden:

> So for fat and unctuous substances, as Sulphur, Bitumen, Oyle, Grease &c., unto what Element shall we ascribe them? Not unto fire, because this is extreme hot and dry, that is temperate in heat, and very moyste. Moreover, fire would rather consume it, then generate it. . . . Ayre, if [it] have any ingenerate quality . . . it is cold and moyst . . . and therefore, as it cannot agree with fire, nor be a fewell to it, so it cannot be any materiall cause of fat, or oylie substance. . . . [B]eing of a watry nature, it cannot agree with oyle or fatnesse, or bee the matter of it. The like wee may judge of water. . . . As for earth, being cold and dry, and solid, it cannot be the matter of this which is temperate, and moyst, and liquid. Neither can all the Elements together make this substance. . . . So I cannot see how this oylie substance, which is very common in all naturall things, and wherein the chiefe faculties of every thing doth reside, as their *humidum radicale*, should be from the Elements.[50]

The corpuscularians too, as we saw, had no ready solution to the problems of elasticity and attraction. And so they, too, sought to incorporate the chemical notion of unctuous moisture within their theoretical framework. Consider Gassendi–Charleton:

> And thus much we learn in the School of Sense, that such bodies as are humectate with the Aqueous and Lean moisture, are easily capable of Exsiccation: but such as are humectate with the Unctuous and Fat, very hardly: Why? because the Atoms, of which the Aqueous doth consist, are more laevigated or smooth in their superfice, and so having no hooks, or clawes, whereby to cohaere among themselves, or adhaere to the concretion, are soon disgregated; but those, which compose the Oleaginous, being entangled as well among themselves, as with the particles of the body, to which they are admixt, by their Hamous angles, are not to be expeded and disengaged, without great and long agitation; and after many unsuccessful attempts of evolution.[51]

The cohesiveness of unctuous substances also explains why as a rule they are inflammable, i.e. why only they contain atoms of fire:

> [T]he Atoms of Fire cannot, in regard of their extreme Exility, sphaerical Figure, and velocity of motion, be in any but an *Unctuous* and viscous matter, such whose other Atoms are more hamous, and reciprocally cohaerent, than to be dissociated easily by the intestine motions of the Calorifick Atoms; so that some greater force is required to the dissolution of that unctuousness and tenacity, whereby they mutually cohaere.[52]

The corpuscularians, it would thus appear, reduced the received notion of unctuous moisture to the workings of atoms, thereby integrating it into their framework. But have they really? Only partly: the hooks and claws of the atoms may well account for

[50] E. Jorden, *A Discourse of Naturall Bathes, And Minerall Waters* . . . (London, 2nd enlarged edition, 1632), 76–7; see on him: Allen G. Debus, *The English Paracelsians* (New York, 1966), 162–4.
[51] Charleton, *op. cit.* (note 2), 322–3. [52] *Ibid.*, 297.

cohesion, but they do not account for *elasticity*. Indeed, elasticity, and with it electrical attraction, cannot be brought about by hard corpuscles acting only through impact or by means of branches and hooks, unless, of course, these branches and hooks are themselves construed as elastic. But why not, indeed: just think of the tongue of a chameleon . . .

In conclusion it may be said that the popular theory of electrical attraction adduced by most upholders of the mechanical philosophy was incompatible with the basic premises of atomism: this theory is thus an instance of what John Henry has fittingly described as a 'tradition of active principles in [pre-Newtonian] English matter theory'.[53] In this case, it would seem that the 'active principle' rested on ideas whose long forgotten origin was in fact distinctively Stoic. Of the thinkers we have considered, only Kenelm Digby, whose system of natural philosophy was eclectic,[54] escapes the charge of inconsistency, but not such professedly staunch corpuscularians as Gassendi, Charleton, and Boyle. In fact, scientists at times see themselves compelled to tinker with their theories by using concepts which are at hand, even when these concepts partly conflict with their overall scheme. This is how Stoic ideas acquired a clandestine afterlife in seventeenth-century mechanical philosophy.

[53] John Henry, 'Occult qualities and the experimental philosophy: active principles in pre-Newtonian matter theory', *Hist. Sci.*, 1986, **24**: 355–81.

[54] B. J. T. Dobbs, '[Part I]', *op. cit.* (note 32); and 'Part II', *ibid.* 1973, **20**: 143–63; 'Part III', *ibid.* 1974, **21**: 1–28.

11

Alchemy in the Newtonian circle: personal acquaintances and the problem of the late phase of Isaac Newton's alchemy

KARIN FIGALA AND ULRICH PETZOLD

When Isaac Newton moved to London in 1696, in order to dedicate himself to his tasks as newly appointed Warden of the Mint, this new career apparently put an end to the 'alchemist' Newton – or so the current interpretation would have it. This seems to be largely true, as far as his development as an 'experimental alchemist' is concerned. But the source materials regarding the continuation of Newton's purely theoretical alchemical studies from the time he arrived in London and through the years after 1700 prove to be problematic.[1]

The reconstruction of Newton's library by the late John Harrison reveals only four alchemical titles, all of which were published after 1700: a work by William Salmon; an expanded re-edition of George Starkey's *Marrow of Alchemy*; and two tracts by the pseudonymous 'Cleidophorus Mystagogus'.[2] The authors of this paper have shown

Translated by Gregory K. Dreicer.

[1] This view was established by Richard S. Westfall, especially in his *Never at Rest: A Biography of Isaac Newton* (Cambridge, 1980), 530–1; see also his 'Newton and alchemy', in *Occult and Scientific Mentalities in the Renaissance*, ed. B. Vickers (Cambridge, 1984), 315–35, on p. 332 seq., and 'Alchemy in Newton's library', *Ambix*, 1984, **31**: 97–101. Betty Jo Teeter Dobbs, too, covers the period from the 1660s to the early 1690s; see her *Foundations of Newton's Alchemy, or 'The Hunting of the Greene Lyon'* (Cambridge, 1975), also 'Conceptual problems in Newton's early chemistry: a preliminary study', in *Religion, Science and Worldview: Essays in honor of R. S. Westfall*, eds. M. J. Osler & P. L. Farber (Cambridge, 1985), 3–32, and, concerning the 1690s, 'Newton's alchemy and his theory of matter', *ISIS*, 1982, **73**: 511–38, and especially 'Newton's "Commentary" on the "Emerald Tablet" of Hermes Trismegistus: its scientific and theological significance', in *Hermeticism and the Renaissance: Intellectual History and the Occult in Early Modern Europe*, eds. I. Merkel & A. G. Debus (Washington, 1988), 182–91. Also Karin Figala, in her 'Newton as alchemist', *Hist. Sci.* 1977, **25**: 102–37, and '"Die exakte Alchemie von Isaac Newton". Seine "gesetzmäßige" Interpretation der Alchemie – dargestellt am Beispiel einiger ihn beeinflussender Autoren', *Verhandl. Naturforsch. Ges. Basel*, 1984, **94**: 157–228, mainly touched upon the problem of Newton's study of alchemy during his late years. For the most recent summaries of the common view see Derek Gjertsen, *The Newton Handbook* (London, 1986), art. 'Alchemical papers', and Jan Golinski, 'The secret life of an alchemist', in *Let Newton Be!*, eds. J. Fauvel, R. Flood, M. Shortland and R. Wilson (Oxford, 1988), 147–67.

[2] John Harrison, *The Library of Isaac Newton* (Cambridge, 1978); for further refs. to this work we use *HL*, followed by Harrison's item number, thus indicating individual books from Newton's library. For Salmon see *Medicina practica* . . . (London, 1707): *HL 1439*. With respect to its appendix, including translations of some basic texts of alchemy, we classify this book 'alchemical'; see John Ferguson, *Bibliotheca chemica*, 2 vols. (Glasgow, 1906, repr. London, 1954),

elsewhere, however, that, during his London years, Newton purchased a series of contemporary alchemical works, as well as several published many years before. The assessment of his library therefore becomes somewhat more complex.[3]

The chronological ordering of Newton's unpublished alchemical writings must also be considered questionable, as far as the notes from his later period are concerned. Yet the hope remains that through the still unpublished research of A. E. Shapiro, the watermarks in Newton's manuscripts will provide considerable assistance in the resolution of dating questions. Preliminary investigations suggest that it can scarcely be doubted that Newton penned a series of alchemical notes around 1700 and during a number of years afterward.[4] In the end, the correspondence of Newton's later years – collected and published thanks to the invaluable efforts of A. Rupert Hall and others – yields little that can help in our inquiry. Only one undated letter, which according to internal evidence was written *circa* 1702, contains clear alchemical references. It was sent to Newton by a certain 'W. Y.'

This paper is an attempt to prove the continued existence of the 'alchemist' Newton during his London years. We shall employ two pieces of evidence – i.e. two manuscript records, now both preserved in the Bodleian Library, Oxford – which without question originated *circa* 1702. First, we base our argument on a document which attests to the delivery of various French *alchemica* to Newton. We believe this document to be a late consequence of the friendship and collaboration of Newton with Nicolas Fatio de Duillier (1664–1753), a young mathematician from Geneva. The second piece of evidence includes the above-mentioned letter, as well as a series of notes with textual variants of a 'Processus mysterii magni philosophicus' – partly in autograph and partly written by another hand – which are preserved among Newton's unpublished papers. This letter, as well as the 'Processus', originated from the Dutchman William Yworth or Yarworth (*c.* 1650/60 – *c.* 1710), a pharmacist, chemist, and alchemist who also appears to be responsible for the pseudonymously published tracts of 'Cleidophorus Mystagogus' (i.e. the 'Key-Bearing Teacher of the Mysteries').[5]

Newton and Fatio de Duiller: the significance of French alchemical texts

When Harrison published his detailed work on Isaac Newton's library, he drew attention to an itemized bill for the shipment of books to Newton by an unidentified

2: 318. For Starkey see *A true light of alchymy ...* (London, 1709): *HL 1644*. For the pseudonymous 'Cleidophorus Mystagogus': *HL 1138* (publ. 1702), *HL 1302 & 1303* (publ. *n.d.*); see discussion below.

3 Karin Figala, John Harrison and Ulrich Petzold, '*De Scriptoribus Chemicis*: sources for the establishment of Isaac Newton's (al)chemical library', in Peter Harman and Alan Shapiro (eds.), *The Investigation of Difficult Things: Essays on Newton and the History of the Exact Sciences* (Cambridge, 1992), 135–79.

4 Private communications by Alan E. Shapiro to K. Figala and B. J. T. Dobbs (1987/88); we are deeply obliged to Prof. Shapiro for his kind co-operation. Further discussion based on his research will be possible as soon as his results are published.

5 For a previous statement of the identity of W. Y(ar)worth and 'Cleidophorus Mystagogus' see Karin Figala, 'Zwei Londoner Alchemisten um 1700: Sir Isaac Newton und Cleidophorus Mystagogus', *Physis*, 1976, **18**: 245–73. On the other hand, that account has, to some extent, to be replaced by our present discussion.

Table 11.1. Transcription of *MS New College 361/II* (Ekins Papers), ff. 78r–78v (Bodleian Library, Oxford)

		Books for Mr Newton		
[1]	1	Paracelsus opera omnia	fol	1–5–0
[2]	1	Traité des Monoyes		0–5–0
[3]	1	Experience sur l'Esprit Mineral		0–5–0
[4]	1	Medecine Metalique		0–5–0
[5]	1	O ____ Balsamique par du Chêne		0–3–6
[6]	1	Rhenani chimiatrica		0–4–
[7]	1	Filet dariane		0–3–
[8]	1	Pilote de l'onde vive		0–3–
[9]	1	Tombeau de Semira [*mis ?*]		–1–6
[10]	1	Philosophie Naturele de Trevisan		–2–6
[11]	1	Texte dalchimie		–2–6
[12]	1	Parnasse assiegé		–2–6
				3–2–6
[13]		Decouverte de la lumiere		–1–6
[14]		Phisique des anciens		–3–6
[15]		Traité des Pierres	[*?*] n°. fig	–3–6
[16]		Traité de Perspective	[*?*] 8vo	6–0

bookseller. Under the heading, 'Books for Mr Newton', sixteen works were listed with shortened titles and corresponding prices (see Table 11.1).[6]

The most conspicuous characteristic of this literary selection is the preference for French titles, as well as the fact that eleven of the sixteen works are alchemical or concern a related field; only the second and last four titles do not fall within this province. The period when the bill was made up can be limited to the years between 1701 (the latest publication date of the listed works) and 1705 (the year 'Mr' Newton was knighted). R. S. Westfall sets the date at between 1701 and 1702 – an assessment with which we concur, while distancing ourselves from Westfall's further interpretation that this list expresses Newton's general interest in the French language rather than in alchemy in particular.[7]

[6] Bodleian Library, Oxford: *MS New College 360/II* (Ekins Papers), f. 78r–78v; see Harrison, *op. cit.* (note 2), 9. The short-titles listed on the bill, with the exception of items [14] and [16], correspond to the following books in Newton's library: [1] *HL 1242* (publ. 1658); [2] *HL 237* (1692); [3] *HL 1372* (1668); [4] *HL 539* (1648); [5] *HL 540* (1626); [6] *HL 1397* (1668); [7] *HL 619* (1695); [8] *HL 1316* (1689); [9] *HL 511* (1689); [10] *HL 531* (1672); [11] *HL 1607* (1695); [12] *HL 1263* (1697); [13] *HL 635* (1700); [15] *HL 1675* (1701). The remaining two works are [14]: *La Physique des anciens* . . . (by D[ominique]. R[évérend].) (Paris, 1701, not *HL*), and [16]: *Traité de perspecitve* . . . (by B. Lamy) (Paris, 1701, not *HL*).

[7] See Westfall, 'Newton's library', *op. cit.* (note 1); Westfall classifies not more than nine titles 'alchemical' (without specifying his judgment); in our count, at least items [8] and [12] are included. In fact, [8] is an alchemical work, although its title suggests a treatise on tides; its second edn. (1689) had been issued, with a common title page, along with [9]. The 'medical' works, too, may be taken into account, since their authors ([1], [4] to [6]) were known to have dealt with alchemy. In addition to our count, one may consider the work by Révérend [14]; although this, first of all, is an anti-Cartesian pamphlet, it may have been of special interest to Newton as an apology for Hermetic philosophy.

When Newton accepted delivery of these books, which included nine French alchemical volumes, his library already contained at least as many related works in the same language. This number is verified in an inventory ('Lib. Chem.') of Newton's (al)chemical library, *circa* 1696–97.[8] Moreover, there were four additional books – among them one duplicate – which were not entered into the above listing, that is, they were either kept separately or came into his possession during the London years. Also left out of account are three works which may be classified as pure chemistry, including an essay on antimony by Nicolas Lemery, which did not appear until 1707.[9]

Newton's turning to French language (al)chemical literature was by no means an unique event; rather, it was a plan systematically followed. This is especially clear from the fact that the majority of these books have a post-1670 publication date and that a preponderance of these volumes appeared between the years 1686 and 1697. It seems therefore to be a matter of deliberate acquisitions by Newton, and not of chance finds which were of minor importance in his library. Furthermore, most of these volumes exhibit the typical sign of Newton's use – dog-eared pages – although this feature is apparent above all in the works acquired before 1696–97. Among the alchemical books delivered *circa* 1702, only two display their owner's 'ear-marks'. It appears rash, however, to conclude that Newton was uninterested in the contents of works which show no signs of use. These late-documented new entries amongst his alchemical books – apart from two by the Paracelsian Du Chesne (or Quercetanus) – were without exception French first editions of contemporary works, so that the purchaser could not judge their value beforehand. That at least one 'find' was among them, is shown by the example of the anonymous *Le Texte d'alchymie, et Le songe verde* (Paris, 1695), showing signs of intensive study by the owner.[10]

Newton's interest in French *alchemica* can be traced relatively far back. In the context of the Newton manuscript 'De Scriptoribus Chemicis' a series of bibliographical *desiderata*-lists can be reconstructed, which are based essentially (but not only) upon Pierre Borel's *Bibliotheca chimica* of 1654 and were re-drafted by Newton several times: *circa* 1670, in the late 1680s and again around 1690. French titles can already be found on all these lists, with some titles repeated in various lists, so that carelessness or coincidence in the choice of these works can be virtually ruled out.[11] This demonstrates that Newton's literature search from the beginning included alchemical texts in a foreign language which he read only with difficulty. His interest also

[8] Babson College, Mass.: *Babson [418]* (see note 22 below); see Harrison, *op. cit.* (note 2), 8–9. The nine French books in this inventory are: *HL 221* (publ. 1672–78), *HL 1309* (1682), *HL 127/169/130* (1659, 3 pts. in 1 vol.), *HL 437* (1678), *HL 1003* (1687), *HL 1311* (1651), *HL 950* (1686), *HL 1642* (1689), *HL 445* (1691).

[9] See *HL 1310* (2nd copy of *HL 1309*, *cit.* [8]), *HL 570* (1689), *HL 897* (1620), *HL 901* (1636); additional books on chemistry are *HL 670* (1676), *HL 908* (1682), *HL 939* (1707).

[10] *HL 1607* (item [11] in 'Books for Mr Newton'); signs of dog-earing can be found on more than 20 pages. Furthermore, the symbolic woodcut (on p. Aiiv) of Newton's copy is hand-coloured; whether this was done by Newton himself or not, is questionable. The second book with signs of use is *HL 531* (item [10]).

[11] For detailed discussion see Figala *et al. op. cit.* (note 3); Newton's bibliographical drafts, including notes on French books, can be found in Stanford University Libraries: *MSS Newton Collection (M132), container 2, folder 4* (*c.* 1670) and *folder 3* (late 1680s), and King's College, Cambridge: *Keynes MS 13* (middle section, *c.* 1690).

included manuscript texts, as is shown by his transcription of 'Le Procede Universelle [*sic*]', which is based on a source of unknown provenance. In fact, Newton's copy of this text unexpectedly breaks off; nevertheless this treatise fragment from the early 1670s appeared important enough to the copyist to cause him to preserve it in a bundle together with other Latin and English notes on alchemy from the same period.[12]

The point in time after which Newton was able to understand enough French to make use of the texts for his alchemical studies can only be determined indirectly. His rather limited knowledge of the language is well known. It is also known, however, that – at the latest – from the beginning of his friendship with the young Swiss Fatio de Duillier (around 1689), Newton had access to a collaborator who not only had command of the French language, but was also familiar with alchemical subject matter. Regardless of whether Fatio already had an inclination towards alchemy and hermetic philosophy when he came to England and promptly fell under Newton's influence, or whether Newton was his 'alchemical father',[13] it must be emphasized that the period of their collaboration became one of the most intensive phases of Newton's alchemical studies. In spite of the later – and still not really comprehensible – break between the two mathematicians, this partnership found a final echo in an 'Ecloga', written by Fatio on the occasion of the first anniversary of the death of his one-time mentor.[14]

The influence of Fatio can in any case be established, as it was he who provided Newton with a copy of a large collection of French alchemical texts. Early in the year 1692/3 at about the high point of their correspondence, Newton offered to buy, for an apparently generous sum, the two-volume *Bibliothèque des philosophes (chimiques)* (Paris, 1672–78) from Fatio, who was then staying in London. Newton explicitly noted that Fatio would hardly have a further need for the books, and this tone suggests that he was very eager to acquire the work.[15] In fact, there is proof of Newton's study of the *Bibliothèque*, as documented by some of his own alchemical manuscripts of the early 1690s (Keynes MSS 28 and 45): at the end of several drafts he added cross-references to the *Bibliothèque*.[16] It seems significant, however, that the acquisition of these volumes – together with a quantity of an (al)chemical remedy – presents us with the first appearance of the discussion of this theme in the correspondence between

[12] Jewish National and University Library, Jerusalem: *MS Var. 259*. 'Le Procede Universelle [. . .]' (*MS. Var. 259–1*) was attributed to a certain 'Jodocus a Rehe' (see lot 35 in Sotheby, *Catalogue of the Newton Papers sold by order of the Viscount Lymington* . . . (London: 1936), and Appendix A in Dobbs, 'Foundations', *op. cit.* (note 1); this statement needs further support. For another French fragment concerning alchemy, in this case not in Newton's hand, see *Babson [416] (A:2)* (see note 22 below).

[13] Westfall, *Never at Rest*, *op. cit.* (note 1), 529; Westfall, too, considers an active role of Fatio in the study of French *alchemica* during the early 1690s. On Fatio as a translator see, summarily, Gjertsen, *op. cit.* (note 1), art. 'Languages'.

[14] See Karin Figala and Ulrich Petzold, 'Physics and poetry: Fatio de Duillier's "Ecloga" on Newton's "Principia" ', *Arch. Int. Hist. Sci.*, 1987, **37**: 316–49.

[15] Newton to Fatio, 14 Feburary 1692/3: '[. . .] have also two Chymical books of your's wch I beleive wch be of no use to you'; see *The Correspondence of Isaac Newton*, eds. H. W. Turnbull *et al.*, 7 vols. (Cambridge, 1959–77), 3: 245–6 (no. 404), and subsequent letters by Fatio (nos. 408–411).

[16] *Keynes MS 28A* ('Tabula Smaragdina. Hermetis Trismegisti [. . .]'; *c.* 1690/91), f. 2v: 'See ye ffrench Bibliotheque [. . .]'; *Keynes MS 45* ('Practica Mariae Prophetissae [. . .]'; early 1690s), f. 2v: 'Extat hoc opus [. . .] et in Bibliotheca gallica'.

Newton and Fatio. Until this time, alchemy had scarcely been mentioned; now a lively exchange began. It included especially detailed statements concerning new routes of experimentation, knowledge of which Fatio acquired from an unnamed 'friend'. This apparently was important for Newton, because he incorporated information provided by Fatio into his extensive 'Praxis' draft.[17] The ultimate intensity of their correspondence on alchemical matters cannot now be judged with absolute certainty, because virtually all of the significant letters which have been preserved originated during the four-month period from February 1692/3 until May 1693; it is unlikely that this represents the full extent of their written communication.[18]

The evolution of the surviving correspondence suggests to us, however, that Newton and Fatio undertook joint alchemical researches during the latter's stay at Cambridge. At the very least, it is possible they carried out a joint literature study, which would have included scrutiny of French texts; this supposition is supported by the fact that Fatio had brought his copy of the *Bibliothèque des philosophes (chimiques)* to Cambridge and left it there upon his departure. At approximately this time Newton must have acquired most of the French books, which were documented in his (al)chemical library during 1696 or 1697, at the latest; the publication dates are in the main from the 1680s to the early 1690s. But it can be shown that there were complete or partial Latin editions of at least four of these works in his possession.[19] On the other hand, this gave Newton the further possibility of gaining access to the French language, and thereby to the texts' alchemical contents, which was in any case an exceptionally difficult venture. A sure indication – insofar as scattered notes can be considered as such – of Newton's increasing ease with the French language is offered by his above-mentioned bibliographical *desiderata*-lists. While the examples from the early 1670s (similarly to the approximately contemporary 'Procede Universelle' manuscript) are sprinkled with mistakes and corrections, as well as deficient translation attempts, the notes of the later years appear in a flowing hand and display a surer grasp of French orthography.

Finally, the significance of the French texts in Newton's alchemical studies is borne out by handwritten drafts whose dates and sources can be satisfactorily determined. B. J. T. Dobbs has already called attention to important 'Hermes' texts, which Newton translated at the beginning of the 1690s based on the versions presented in the *Bibliothèque*.[20] Our knowledge of the book's provenance permits us to postulate that Fatio shares responsibility for Newton's versions of these texts.

17 Newton, *Correspondence*, *op. cit.* (note 15), 3: 265–70 (nos. 414–415); to the first of Fatio's letters (4 May 1693) Newton made reference in *Babson [420]* (see note 22 below) ('Praxis'), f. 13, which presumably was composed by him as an original essay.

18 Some references, mostly on personal matters, by Fatio suggest 'missing links' within the surviving correspondence; see outline in Figala and Petzold, *op. cit.* (note 14), on 318–21, and the instructive summary by Gjertsen, *op. cit.* (note 1), art. 'Fatio de Duillier, Nicolas (1664–1753)'.

19 Cf. refs. above (note 9); *HL 437* (Luigi de' Conti, *Discourse philosophique . . .*): for Latin edn see *HL 436*; *HL 1311* (Jean d'Espagnet, *La philosophie naturelle restablie . . .*): Latin edn included in *HL 220* (*Bibliotheca chemica contracta*); *HL 127/169/130* (*Azoth . . .*): *see e.g. HL 1130* or *1131* (*Musæum hermeticum*, including the Basil Valentine tracts) and *HL 1608* (*Theatrum chemicum*, including the Latin version of the 'Azoth' tract); for Latin edn of *HL 445* (Sendivogius's works) see e.g. *HL 1192*. Of course, we may eventually find further overlaps in Newton's library.

20 *Keynes MS 28*; cf. note 16; see Dobbs, 'Newton's Commentary', *op. cit.* (note 1).

A more extensive example of such translation activity is provided by Newton's version of the anonymous 'Lumen de tenebris'. This work, which consists of a poem in three 'chantes' or 'canzoni' with detailed commentary, was indeed obtainable in an Italian/Latin version – *Lux obnubilata suapte natura refulgens* (Venice, 1666) – but was available to Newton only in a French translation entitled *La Lumière sortant par soi-même des tenebres* (Paris, 1687).[21] Newton prepared English excerpts based on this edition; his wording of the Latin title indicates that the original publication was not accessible, which is to say, was not known to him.

Newton's *La Lumière* notes are preserved in two different versions. The first bears the heading 'Ex Lumine de tenebris' and consists of roughly translated excerpts which Newton then proceeded to work out more fully on the same manuscript sheet. The second version, which unfortunately was separated during one of the numerous reorderings of Newton's papers, is considerably more extensive and contains no remarkable correction and emendation. Its title, 'Out of La Lumiere sortant des Tenebres', as well as the subheading of its main section, 'Out of the Commentator on La Lumiere sortant de [*sic*] tenebres', clearly allude to the French edition as the basic source. In dating the second version, the later rather than the earlier 1690s come into consideration;[22] thus these translations would have been produced after Newton's *Bibliothèque*-based 'Hermes' notes, that is, only after his documented contact with Fatio.

That the study of French *alchemica* was more to Newton than a kind of foreign language lesson is further shown by his classification of *La Lumiere*, which appears within 'Authores optimi' in one of Newton's drafts[23] on the development of chemical literature. The book delivery of *circa* 1702 thus appears consistent with the path followed since the beginning of the 1690s, which enabled Newton to include at least one contemporary foreign language in his alchemical research.

Newton and William Yworth: the last attempt at alchemy?

At about the same time that the large parcel of 'Books for Mr Newton' was delivered (*circa* 1701 or 1702), several documents appeared which preserve the last traces of an intense association of Newton with a person of unquestionable alchemical orientation. The star witness is an undated letter,[24] signed 'W. Y.', which displays exactly the

[21] *HL 1003*; the rare Latin edn was re-issued with *Ginæceum chimicum* . . . (Lyons, 1679), but nevertheless is not to be found in Newton's library.

[22] The first (rough) version is now *Babson [414]*, f. 1r–1v; the second (reworked) version begins on two folios, now in the Jewish National and University Library, Jerusalem: *Yahuda MS Var. 1/Newton MS 30*, and continues on the remaining folios of *Babson [414]*. Because of this separation, the compilers of *A Descriptive Catalogue of the Grace K. Babson Collection of the Works of Sir Isaac Newton* . . . (New York, 1950), on p. 191, were not able to realize the actual nature of their item [414].

[23] *Keynes MS 13* (late section; after 1700), f. 4r.

[24] *MS New College 361/II*, f. 89v; published in Newton, *Correspondence*, *op. cit.* (note 15). The letter was printed in 7: 441 (ed. A. Rupert Hall and Laura Tilling) and dated by the editors '?1705'.

same handwriting as two versions of an alchemical tract entitled 'Processus mysterii magni philosophicus', which were found among Newton's own alchemical manuscripts. The location of a third variation, also written in someone else's (Yworth's?) hand, has not yet been established.[25] But we have been able to identify an additional fragment as a part of the 'Processus' *corpus* which can be attributed to the same writer.[26] In these tract variants the author does not hide behind his initials, but reveals himself to be a certain William Yworth, also spelled as 'William Yarworth'; in two cases he dated his work: 1701 and 1702, respectively.[27] The latter year is also the earliest possible date for the above-mentioned letter, according to the editors of the Newton correspondence. It must be emphasized at the outset that neither the letter nor the 'Processus' manuscripts bear the name of the addressee; that these materials were found to be in Newton's possession is the only direct proof of his link with their author. His receipt and reworking of the 'Processus' (see below) as well as details regarding a second work by the same author, are clear indications that Newton was in fact the correspondent of that little known alchemist William Yworth alias 'Cleidophorus Mystagogus'.

Who was this person, author of a remarkably extensive series of published works – both alchemical and non-alchemical – which appeared during a period of fifteen years? The attempt at biographical reconstruction unfortunately must rest on scanty details which Yworth himself added to his books, from the very first, poorly printed, *Bacchean Magazine* (1690) to a two-part *Compleat distiller/Pharmacopœa spagyrica nova* (1705); more specifically, on the datings of introductions ('To the Friendly Reader') and/or 'Advertisements' which concerned proposed publications or the proprietary medicine trade.[28]

According to these, Yworth was Dutch by birth[29] and for some time resident in Rotterdam, where he gave his house the name 'Collegium Chymicum'.[30] Between the autumn of 1690 and the summer of 1691, Yworth apparently moved to England, where he previously may have visited and made acquaintances.[31] (Furthermore, there are very general references to 'travels in Europe' in his works.) In June 1691, he stated

[25] See Sotheby, *op. cit.* (note 12), lots 116 to 118. Lot 117 is now at Yale University Library, New Haven: *Mellon MS 80*; see *Alchemy and the Occult. A Catalogue . . . of the Collection of Paul and Mary Mellon . . .*, eds. I. McPhail *et al.*, (4 vols, New Haven, 1968–77), **4**: 484–6, including a facsimile (incorrectly attributing the MS to Sotheby lot 116). Lot 118 is now *Keynes MS 65*. The location of lot 116 (bought in 1936 by Maxwell) is unknown; thus all subsequent refs. to this version are based on the description in Sotheby's sale catalogue.

[26] *Keynes MS 91*; formerly last ('not autograph') section of Sotheby, *op. cit.* (note 12), lot 28.

[27] *Keynes MS 65*: signed 'William Yworth' (ff. 1r, 2v); *Mellon MS 80*: signed 'Wiliam Yarworth V.D.' (ff. 7r, 9v). The latter is dated '1702' (f. 6r); for the '1701' version see Sotheby lot 116.

[28] For Yworth's writings see Ferguson, *op. cit.* (note 2), **2**: 558–9, and D. G. Wing, *Catalogue of Books printed . . .1641–1700*, no. 213–220.

[29] W. Y., *A new treatise of artificial wines, or A Bacchean Magazine, in three parts . . .* (London, 1690), p. [B9v], preface signed: 'W. Y. Worth, Geboortigh tot Shipham / September 8. 1690'.

[30] *Ibid.*, closing preface: 'Written in the English Tongue, by the Author, from the Original, as it was delivered at his House at the Sign of the Collegium Chymicum, Rotterdam.'

[31] Cf. *ibid.*, pp. 61 seq., advertisement of remedies, 'Written from the Sign of the Collegium Chymicum from Rotterdam', including 'Greetings to these esteemed Friends, viz. J. P. [or R?] Hoddges, J. M. Baker, J. Leversidge, J. Van Ravatt [. . .]'.

that he was 'now resident at London'[32] and in the following year precisely noted the location 'at the Blue Ball and Star in S[t]. Paul's Shadwel [Lond].'[33] At a later, undetermined, point in time, Yworth took a new and final London residence in the 'Blue Ball and Star at the Corner of King-street in upper Morefields, London.'[34] Subsequently, that is, only a few years after he turned up in Isaac Newton's circle, Yworth obviously left the capital; in 1705 his son, Theophrastus Yworth, mentioned in the introduction to a new edition of one of his father's works, which he was preparing, that the elder Yworth was still living, but could be reached only via his son.[35] In fact, William Yworth resurfaced in Woodbridge, Suffolk, no later than 1709, the year in which he received – on the seventh of July – an episcopal licence for the practice of surgery.[36] And that is the last date in the life of William Yworth for which we have evidence. Shortly after 1710, the younger Yworth, Theophrastus, published a short pamphlet in which he unambiguously referred to his father in the past tense.[37] The actual life span of William Yworth can thus be confined to the period between 1650 or 1660, and approximately 1710.

Brief consideration of the seemingly eccentric forms of the name 'Yworth' is here appropriate; not least, since that spelling led to some confusion in cataloguing Newton's papers: though Thomas Pellet and Thomas Pilkington in 1727 used the proper 'Yworth', William Mann Godschall and Samuel Horsley in 1777 left a blank space following 'William'.[38] In his pamphlets, Yworth chose – in addition to the initials 'W. Y.' – the forms 'Y-worth' or 'Y.worth'; in manuscript he preferred

[32] W. Y., *A new art of making wines, brandy and other spirits compliant to the late Act of Parliament* . . . (London, 1691), p. [B8v], closing preface: 'Written and abundantly enlarged by the Author, so that the Original Copy that was deliver'd at his House, at the Collegium Chimicum at Rotterdam, is not comparable to it, &c.', signed: 'W. Y-worth, Geboortigh Van Shipham, & Van Rotterdam, Borger. Now Resident at London, June 6. 1691, at the Academia Spagirica Nova, [. . .]'; on p. 121 Yworth refers to '[. . .] Physitians, that I have met with in my European Travels.' The remainder of this book was issued as 'Second Edition', including a new 'Dr. Worth's [*sic*] Letter [. . .] to W. R. Gent.': W. Y. M.D. [*sic*], *The Britannian Magazine: or, A new art* . . . (London, 1694); a third edn (London, not seen) is dated to *c*. 1700.

[33] W. Y.worth, *Chymicus rationalis: or, The fundamental grounds of the chymical art* . . . (London, 1692), p. [A7v], closing preface: 'Written in S. Paul's Shadwel, from the Academia Spagyrica Nova [. . .]'. Further advertisements of remedies in 1692 all name the 'Blew Ball and Star in S[t]. paul's Shadwel'; see W. Y.worth, *Cerevisarii comes: or, The new and true art of brewing* . . . (London, 1692), p. 122; W. Y-worth, *Introitus apertus ad artem distillationis; or The whole art of distillation practically stated* . . . (London, 1692), p. [A8v].

[34] W. Y-worth, *The compleat distiller* . . . (London, 1705), p. a2, undated preface signed: 'From my House, the Blew Ball and Star at the corner of King-street in upper Morefields, London.'

[35] *Ibid.*, pp. [a7v]–[a8v]; in an 'Advertisement' Theophrastus Y = Worth, then living in 'King-Street', confirmed that all letters 'shall be faithfully Communicated to my Father'. This *Compleat distiller* is a revised 'Second Edition' of the *Introitus apertus* (1692), *op. cit.* (note 33), enlarged by a 'Second Part': *Pharmacopœa spagyrica nova: or, An Helmontian course* . . . (London, 1705), and edited by Yworth jr.

[36] See P. J. & R. V. Wallis (*et al.*), *Eighteenth Century Medics (subscriptions, licenses, apprenticeships)*, 2nd edn (Newcastle, 1988), 677; we have, so far, not examined further records, such as burial records etc.

[37] Theophrastus Yworth, *A Brief . . . Account of the Vertue, . . . of Certain . . . Medicines, Faithfully prepared as in my Father's Days* (Copy in British Library). This pamphlet must have been printed after 1710, since Yworth jr. refers to a remedy which had been published (*i.e.* in *New treatise* (1690), *op. cit.* (note 29)) 'more than twenty Years' before. Furthermore, he directs attention to some more acquaintances of his father's: Charles Marshall (1637–96), a medical practitioner and quaker (see *DNB*) with whom Yworth sr. seems to have had some quarrels concerning the invention of remedies; also ten physicians who had signed an undated (before 1705) 'Testimony' on Yworth and his medicines.

[38] See *Keynes MS 127A*, entries of 'Processus mysterii magni philosophicus': lists a/b (1727), no. 73; lists c/d (1777), item [bureau] B. [drawer] 6.

'Yworth', and only once applied the spelling 'Yarworth'.[39] His son Theophrastus, too, made use of the form 'Y-worth/Yworth'. If the family's origin in the Netherlands is taken into account, one may think of variations such as 'Ijword', Ijwaard(e)', or 'Ijvaert'. Although direct proof is still lacking, what – at first sight – appears to be a code name may actually be an Anglicization of a genuine Dutch name.[40]

The life and times of Yworth may be better understood through a survey of the contents of his writings. As an author of technical literature Yworth apparently followed two paths: firstly, the description of distilling processes and beverage manufacture; secondly, the preparation – and sale – of chemical remedies. In modern terms these interests may be characterized, respectively, as chemical technology and pharmacy.[41] Both had a distinctly financial aspect, for as a result of the 1689 import ban on French products and related regulations enacted by Parliament concerning the production of domestic distillate, a strong need had clearly arisen for distillation handbooks.[42]

Yworth's alternative field of action was an extensive market, which historians of medicine seem to have largely neglected until recent times: the manufacture and sale of proprietary medicines. Judging from the lists (headed 'Catalogus medicinarum') and advertisements which he included in all of his books, this may have been his chief source of income.[43] The basis of his business was his own laboratory, which he managed under the sign of 'Academia Spagirica Nova',[44] as he earlier had the 'Collegium Chymicum' in Rotterdam. In fact, it seems that Yworth was not the sole head of his London firm, because, at the least, one Thomas Newton was listed as 'our Operator' and agent. Finally, the younger Yworth, Theophrastus, was apparently brought into the enterprise; no later than 1705, his address and shop sign were identical to those of his father's former place of business.[45] Several copperplate

[39] See *Mellon MS 80*, *op. cit.* (note 25).

[40] The problem of Yworth's genuine name and his Dutch years is still unsettled; further examination may follow J. van Lieburg's survey, 'Die medizinische Versorgung einer Stadtbevölkerung im 17. Jahrhundert: Die Quellen- und Forschungssituation für Rotterdam', in *Heilberufe und Kranke im 17. und 18. Jahrhundert: Die Quellen- und Forschungssituation*, eds. W. Eckert and J. Geyer-Kordesch. (Münstersche Beiträge z. Geschichte u. Theorie d. Medizin, **18**) (Münster, 1982), 29–48.

[41] In Yworth's own understanding, the classification was 'chymical' and 'spagyrical', i.e. to separate and to combine substances; see the first chapter in *Chymicus rationalis* (1692), *op. cit.* (note 33).

[42] 'An Act for the Encouraging the Distillation of Brandy and Spirits [. . .]' (1690), printed in *The Statutes of the Realm. . . .*, vol. 6 (1891; reprinted London, 1963), 236–8; cf. ref. on the title page of *New art* (1691), *op. cit.* (note 32).

[43] See especially *New treatise* (1690), *op. cit.* (note 29), pp. 61 seq., indicating the opening of the London business, and *Cerevisarii comes* (1692), *op. cit.* (note 33), p. [122], giving prices and doses; 'The Friendly Cordonium Lenitivum', 'Spiritus Phrophylactic [*sic*] Imperially', etc. (1690) or 'Pilula Anodyna Specificata [. . .] or the Friendly Balsamick Pill', 'Species Nostra Mineralis, or the Spagyrical Triumphant Powder', 'Cordonum Regale Lenitum, or the Royal Purging Cordial' and 'Spiritus Odontugiasus or Mouthwash' (1692), etc. may serve as some few examples of curious, imaginative 'trade names'; for background information on the sale of special remedies see Roy Porter, *Health for Sale. Quackery in England 1660–1850* (Manchester, 1989), and the studies by Harold J. Cook, e.g. his 'The Rose case reconsidered: physicians, apothecaries, and the law in Augustan England', *J. Hist. Med.*, 1990, **45**: 527–55.

[44] The specification 'at the Academia Spagirica [or: Spagyrica] Nova' is used throughout in the publications of 1691 and 1692; it was dropped in 1705.

[45] See 'Advertisement' by Theophrastus Yworth (1705), *op. cit.* (note 35): '[. . .] faithfully prepared by Theophrastus Y=Worth, [. . .] at the Blew-Ball and Star [. . .]'; Thomas Newton, 'Chimicomedicus & Discipulus Spagirica Philosophia [*sic*]', 'our Operator', is first named in *New art* (1691), *op. cit.* (note 32); other agents or associates are: John Baker, 'Periwig-maker, at the Wool-pack in the Strand, near to the Savoy-Gate, London' (*ibid.*), who may be identical with one J. M. Baker mentioned in 1690 (*op. cit.* (note 31); see also Wallis, *et al.*, *op. cit.* (note 36), 394); John Spire,

engravings included in two of William Yworth's most mature works, *Introitus apertus ad artem distillationis* and *Chymicus rationalis* (both published in 1692), are extremely illuminating in this context (*see* Figs. 11.1 and 11.2).[46] We find a vivid rendering of his two principal interests as distiller and medical practitioner, as well as an idealized representation of his laboratory. Furthermore, each plate is adorned with his coat of arms; whether this last feature was genuine, or was added only in order to lend an air of respectability, remains an open question. But apparently Yworth used the very same coat of arms as a certificate of authenticity for his proprietary medicines to be sold by himself or his agents.[47]

Just as open is the extent of Yworth's education. He preferred to describe himself as 'Professor and Teacher of the said [i.e. spagyric] art' or 'Professor Medicinae'; no evidence of such training, such as a provable academic degree or good knowledge of Latin, has been found.[48] Further self-characterization as 'Philosopher by Fire' or 'per Ignem Philosophus',[49] clearly reveals his sources. With this appellation, Yworth has positioned himself squarely within the tradition of George Starkey and, indirectly, of Joan Baptista van Helmont. In fact, the preface to his earliest technological work already shows a decided alchemical background imbued with religious apologetics of the art. This side of Yworth is made more obvious through a *corpus* of projected writings which he advertised as early as 1690. All of them are full of Paracelsian and Helmontian code names and *termini technici* – that much can be gathered merely from the announced titles.[50]

But what is more important is that, in 1691, in one part of a projected six-volume set of 'Magician's Magazine', Yworth, now styling himself 'Hermetical Disciple', stated his authorship of two tracts which can be identified as works later published pseudonymously by 'Cleidophorus Mystagogus'. These are: 'Trifertes Soladinis, or a Declaration of the Fiery Spirits [. . .]' (issued in 1705 as *Trifertes sagani*) and 'Mercury's Caduce Rod, or Aquila Hermetica [. . .]' (issued in 1702 and 1704 as

'Chymical Physitian at Horsly-down, Southwark' (e.g. *Chymicus rationalis* [1692], *op. cit.* (note 33); see also Wallis *et al.*, *op. cit.* (note 36), 563); and also the printers/booksellers Andrew Sowle (in 1690) and John Taylor (1692).

46 Frontispiece in *Introitus apertus/Compleat distiller* (1692/1705), *op. cit.* (notes 33–35), engraved by Michiel van der Gucht (1660–1725), and folding table in *Chymicus rationalis* (1692), *op. cit.* (note 33), (following p. 6); *Introitus apertus/ Compleat distiller* includes four additional engravings, showing chemical apparatus. At least one of the vessels shown within the 'Academia spagirica nova' depiction (upper right) is obviously copied from the work of Johann Rudolph Glauber; see plate in *The Works* . . . (London, 1689), Pt. I (following p. 1), transl./ed. by Christopher Packe.

47 We have not been able to trace any source for this design; for the use as a certificate of authenticity see the advertisement in *Introitus apertus* (1692), *op. cit.* (note 33), of bottles, 'sealed with our coat of arms, which is [. . .] also on the bill of direction: That you may not be mistaken, and buy slops instead thereof [. . .]'.

48 See e.g. signature of preface in *New art* (1691), *op. cit.* (note 32): 'Professor and Teacher [. . .]'; title-pages of *Cerevisarii comes* (1692), *op. cit.* (note 33): 'Medicin. Professor, Ingenuarum Artium Studens [. . .]'; *Introitus apertus/ Compleat distiller* (1692/1705), *op. cit.* (notes 33–35): 'Medicinæ Professor in Doctrinis Spagyricis [. . .]'; signature of 'Dr. Worth's Letter [. . .]' in *Britannian Magazine* (1694), *op. cit.* (note 32): 'Spagirick Professor and Teacher [. . .]'; the 'M.D.' used on the title-page (*ibid.*) seems to be fictitious.

49 See e.g. title-pages of *New treatise* (1690), *New art* (1691), *Chymicus rationalis* (1692), *op. cit.* (notes 29, 32, 33): 'Spagyrical Physitian and Philosopher by Fire'; *Cerevisarii comes* (1692), *Introitus apertus* (1692), *op. cit.* (note 33): '[. . .] & per Ignem Philosophus'.

50 See *New treatise* (1690), *op. cit.* (note 29), p. 58 (3 items); see also Ferguson, *op. cit.* (note 2); *New art/Britannian Magazine* (1691/94), *op. cit.* (note 32), 'Advertisement', pp. [154–64] (4 items, most comprehensive quotation of titles); *Introitus apertus/Compleat distiller* (1692/1705), *op. cit.* (notes 33–35), pp. 187–9/274–6, letter to an unnamed friend (10 items); the 1705 edn advertises (p. 276) another title.

Figure 11.1. Representation of the twofold profession of William Yworth: distillation products in pharmacy (above) and chemical technology (below). Frontispiece, Yworth, *Introitus apertus*, (London, 1692/1705), engraving by M. van der Gucht. (Reproduced by courtesy of the British Library.)

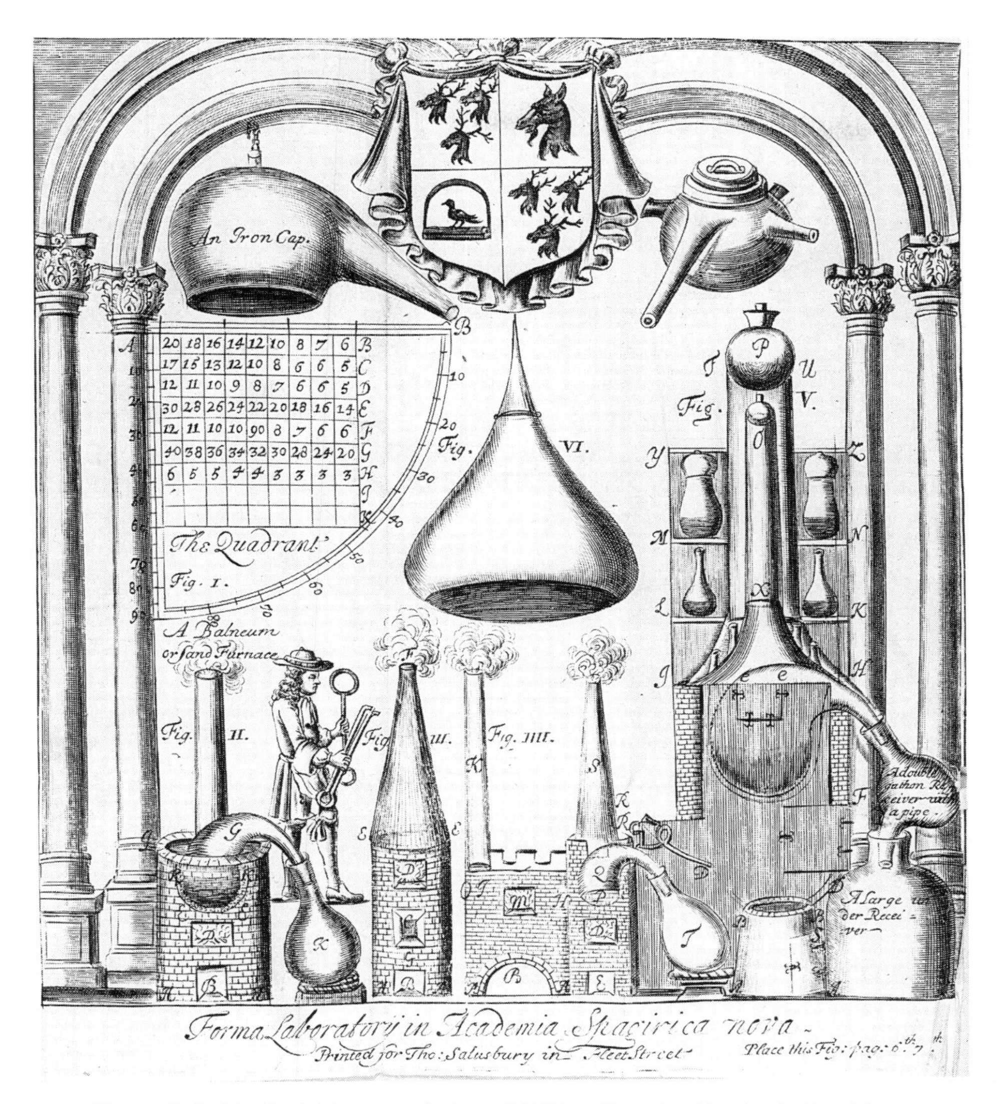

Figure 11.2. Idealized laboratory design of William Yworth's 'Academia Spagirica Nova', with his coat of arms. Folding plate, *Chymicus rationalis* (London, 1692), unsigned engraving. (Reproduced by courtesy of the British Library.)

Mercury's caducean rod).[51] Yworth nowhere gave any hint why he had decided to use a pseudonym for the publication of his purely alchemical writings, though the

[51] See 'Advertisement' in *New art/Britannian Magazine* (1691/94) *op. cit.* (note 50), item 4, p.1 to 6, esp. p.1/4; and 'Cleidophorus Mystagogus', *Mercury's caducean rod: or, The great and wonderful office of the universal mercury, or God's Viceregent* . . . (London, 1702); remainder reissued as 'Second Edition' (London, 1704); 'Cleidophorus Mystagogus', *Trifertes sagani, or Immortal dissolvent. Being a brief . . . discourse of . . . preparing the Liquor Alkahest* . . . (London, 1705); see also Denis I. Duveen, *Bibliotheca chemica et alchemica* (2nd edn, London, 1965), 629–30. And, vice versa, in *Mercury's caducean rod* (1702), following p. 76, the author advertises 'to make ready for the Press' another two titles,

identification of the true author should have been possible even to his contemporaries. With these tracts, however, Yworth completed the transformation from author of books on practical chemistry to Paracelsian–Helmontian occult physician, and finally to follower and theorist of transmutatory alchemy; and in this last capacity he contacted Newton.

In his early London years Yworth apparently sought to find a powerful supporter in the person of Robert Boyle. This, at any rate, is how we interpret his dedication in *Chymicus rationalis* (1692), which was – at least in print – not received by the man so honoured before his death in 1691.[52] Yworth's letter to Newton,[53] however, shows that he had in this case met with success, because the tone and certain turns of phrase indicate that correspondence or meetings had already taken place and, moreover, that Newton had supported Yworth in a criticial financial situation.[54] The letter, in addition, contains clear indications that it was definitely written in 1702: a book mentioned by Yworth, and sent with the letter, could only have been *Mercury's caducean rod*. His reference to a 'General Epistle' can now be understood, for in this book 'A Philosophical Epistle' in fact was included with separate pagination. Newton, moreover, must have received a preprint (or manuscript version) of the 'Epistle', and had offered a critical assessment, if Yworth's response is to be understood: 'thou wast Pleased to tell me, that this Age was not worthy of my General Epistle'.[55] Only thereafter – how long thereafter is unclear – did Yworth/Cleidophorus begin the 1702 printing of the entire work.[56]

There is a second letter which most likely was addressed to Newton, namely, the dedication of a handwritten 'Processus mysterii magni philosophicus' from the pen of Yworth.[57] The tone of this – undated – text is very similar to the letter of 1702, but must have been drafted somewhat later, because in the course of the subsequent 'Processus' manuscript, the author explicitly referred to 'my Mercury's Caduce Rod, and General Epistle extant'[58] – this conclusively settles the identification of Yworth

one of which ('Historia Nova de Thesauro Britanniæ') had been announced by Yworth in *Introitus apertus* (1692), *op. cit.* (note 50).

[52] See John F. Fulton, *A Bibliography of the Honourable Robert Boyle*, (2nd edn Oxford, 1961), 169, no. 294, including quotation from the dedicatory epistle, signed 'W. Y.'

[53] 'W. Y.' to [Newton], [*c* . 1702], *op. cit.* (note 24); all quotations from Newton, *Correspondence* and the original MS.

[54] Thus we read Yworth's phrase 'I have Presum'd to send to thee for the wanted Allowance [. . .]'.

[55] Thus the original MS; the Newton, *Correspondence* omitted the word 'Epistle'; see also transcript in Figala, *op. cit.* (note 5), 260.

[56] 'A Philosophical Epistle, Discovering the Unrevealed Mystery of the three Fires of the Sophi', signed (p. 32*) 'Cloidophorus [*sic*] Mystagogus', and bound with *Mercury's caducean rod* (1702/04), *op. cit.* (note 51); for Newton's copies see *HL 1130/1302* (1702 edn), *1303* ('Epistle' offprint). There is, unfortunately, an introduction to the main part, headed 'A General Epistle to the Reader [. . .]' (unpag., running title 'To the Reader'), nevertheless, it seems questionable whether this should have been pre-printed. 'W. Y.'s' remarks on the 'Epistle', together with some references to subject matters ('from Acetum to Elixer [. . .] ye Production of Azoth'), are clear indications that the book in question must have been the alchemical work ('this being ye first Book [*sic*] wch hath pass'd out of my hand, wch I present as my Mite into Minerva's Treasure') rather than a copy of the technological *Compleat distiller* (1705 edn); see Newton, *Correspondence*, *cit.* (note 24), n. 2.

[57] 'W. Yworth' to [Newton], [*c.* 1702]; cf. *Keynes MS 65* (*op. cit.* [note 25]), ff. 2r–2v; compare transcript Figala, *op. cit.* (note 5), 258 n. 46; the title-page, *in extenso*, reads: 'Processus Mysterii Magni Philosophicus or An open Entrance to ye great Mysteries of ye Ancient Philosophers – Delivered in plain and linear words in my labours for ye Benefit and instruction of my Honour'd Friend – By William Yworth'.

[58] *Keynes MS* 65, f. 11r; same phrase in *Mellon MS 80* (dated 1702), f. 22v.

with 'Cleidophorus Mystagogus'. Here again several lines of the dedication point to the direct participation of Newton in the preparation of the 'Processus', if one believes the appreciative words of the undersigned 'Affectionate Friend William Yworth': 'at your Request and in Gratitude to the favorable Aspect which you have Carryed towards me in my Indefatigable Search.' The presence of further 'Processus' variants in Newton's *corpus* of alchemical papers shows that such words must be regarded as more than merely the obsequious homage of a charlatan.

Apart from the aforementioned version with the dedicatory text dating from approximately 1702, Newton came into the possession of another – dated – 1702 variant, which, however, is prefaced by a dedication to the children of the author, 'Wiliam Yarworth';[59] in addition, one variant from 1701 existed.[60] All extant versions appear to be fragments; even the most detailed draft from about 1702 abruptly breaks off in the tenth chapter, whereas the dated variant from 1702 comprises six complete chapters. In any case, subsequent pages have not been found.[61] Far-reaching agreement between these two versions can be established; nevertheless, the chapter sequence was altered and additional paragraphs were added, while others were omitted. The earliest variant (1701) is quite different from all others, in that it was apparently subdivided according to 'Experiments' rather than split into 'Chapters'. Although we have been unable to obtain access to the entire version, we have identified what appears to be a part of this – or a related – variant; it contains the last page of a section as well as the complete following section, entitled 'Experim^{t}. 4th.' This fragment partially restores passages which in fact are present in the dated 1702 version, but not in the more extensive undated one.[62] Finally, one last variant in Newton's own handwriting survived; this differs considerably from all known autograph versions by William Yworth in both content and style.[63] In order to convey an idea of the complexity of the entire *corpus*, we present a tentative concordance of the preserved variants according to the arrangement of the chapters or paragraphs (see Table 11.2).

Newton's own version of the 'Processus' shows clear signs that, rather than merely copying existing texts, he reworked the material: he improved numerous passages, made additions, and underlined complete paragraphs. The most conspicuous signs are

59 *Mellon MS 80* (dated 1702; *op. cit.* (note 25)), ff. 8r–9v: 'A short Epistle to my Children', signed 'Wiliam Yarworth V.D.'; title-page (f. 7r): 'Processus Mysterii Magni Philosophicus or An open Entrance to the Mysteries of the Ancient Philosophers. / Delivered in plaine and linear words in my Labours to my Children for their Benefit and Instruction, &c. / Written for fear of Death before I finish ye same By me / Wiliam Yarworth. V.D.'; fly leaves (ff. 2r; 6r): 'This is the true and Principal Book'; 'Processus Mysterii Magni. / A. Dom. 1702'.

60 Formerly Sotheby, *op. cit.* (note 12), lot 116 (location unknown; *op. cit.* (note 25)).

61 *Keynes MS 65*: breaking on f. 38v, with catchword ('The'), subsequent ff. lost.; *Mellon MS 80*: booklet, text ending on f. 52v (middle), following ff. (53–192) blank; Sotheby, *op. cit.* (note 12), lot 116 (1701) has been described as 'incomplete', being a booklet of 107 pp. of text, no ref. to blank pp.

62 *Keynes MS 91*, *op. cit.* (note 26), f. 1r: end of chapter (inc. 'The Pondus in Dissecting the Subject Matter [. . .]'), additional 3 lines in Newton's hand ('In these recconings a pound is taken for 16 ounces Troy. The spirit of wine of 10 destillations weighed 3 lb 2 oz 1 dwt or 50 $\frac{1}{20}$ ounces troy, besides the corrosive oyle'); ff. 1v–2v blank; ff. 3r–6v (middle): complete chapter (headed 'Experimt. 4th. Being the Corolary of all ye. former, Containing a true Process of the whole Worke'); ff. 7r–8v blank; textual variants (ff. 5r–6v) in *Keynes MS 65*, ff. 37v–38r, and *Mellon MS 80*, ff. 49v–52v.

63 *Keynes MS 66* (formerly Sotheby, *op. cit.* (note 12), lot 119); no title page, ff. 6r–6v, 8r–8v blank, text ('Chap. V') breaking on f. 7v (1 line, incomplete).

Table 11.2: William Yworth's 'Processus mysterii magni philosophicus' as *work in progress*: A tentative concordance of textual variants (all copies from the alchemical papers of Isaac Newton)

(a) *Mellon MS 80* (1702)	(b) *Keynes MS 65* (*c.* 1702)	(c) *Keynes MS 66* (Newton's draft)	(d) [Sotheby lot 116] [1701: Experiment 1–9] / (e) *Keynes MS 91* (undated fragment)
Chap. I	Chap. I	Chap. I [part]	(Fragment/end of chap.)
Chap. II	Chap. II		
Chap. III	Chap. III	Chap. II	
	Chap. IV	---	
	Chap. V		
---	(verses)	---	
Chap. IV	Chap. VI	Chap. III*	(additions/Newton)
Chap. V	Chap. VII	Chap. IV*	
---	Chap. VIII [inserted]	---	
Chap. IV [part]	Chap, IX	Chap. V*	Experiment 4th.
	Chap. X [part] [?]		

* correlation in short sections only

the indications of weight obtained by experiment, which appear in the concluding sentences of several chapters.[64] Yworth, in his own versions, had described his experiments in detail, but almost exclusively in qualitative terms; the exceptions, for the most part, occur in the margins of his undated 'Processus' variants.[65] On the contrary, Newton with his own hand added precise indications of weight to Yworth's short 'Experiment' fragment, and these are, in fact, almost identical with a passage in his own version.[66] Although the idea of the *pondus* – signifying proportions of alchemical principles in the composition of bodies and in composite bodies – indeed played a central role throughout in Yworth's texts, it remained completely vague and

[64] See *Keynes MS 66*, ff. 2v (end of 'Chap. I'), 4v (end of 'Chap. III'); for extensive interlining by Newton see f. 7r (second paragraph of 'Chap. V').
[65] See Yworth's marginal additions in *Keynes MS 65*, ff. 7r, 8r.
[66] See addition to *Keynes MS 91*, *op. cit.* (note 62), and corresponding (obviously added) lines in *Keynes MS 66*, f. 4v; last paragraphs on f. 2v are in close correspondence to *Keynes MS 91*, f. 1r (except Newton's additions).

purely qualitative; a shift towards the quantitative takes place only in Newton's version.

Obviously, the date of Newton's manuscript cannot be precisely determined in relation to Yworth's 1701/02 drafts. It may be conjectured, however, that Newton's influence on Yworth included encouragement to further – and quantitative – experimentation. Nothing points to Newton's having done experimental work of his own during this period, although there is some indication that Yworth did this work in his stead; there is no doubt that experiments were conducted in connection with the 'Processus'. Most probably Yworth communicated experimental data to Newton to supplement the expositions presented in the 'Processus' drafts. In exchange, Newton used these data to draft a reworked and concise variant; the basic ideas of the tract remained unaltered. Judging from the theoretical background of his work, Yworth appears as a quite serious 'Philosopher', who essentially based his work on contemporary English writings, above all on the ideas presented in the works of 'Philalethes'.[67] These were in fact so familiar to him that he could even cite an unpublished commentary on the work of Michael Sendivogius, ascribed to the famous 'Philalethes'; the very same (anonymous) commentary, moreover, most probably served as the basis for Newton's own important 'Sendivogius explained' manuscript.[68] Furthermore, the entire 'Processus' *corpus* reads as an attempt to continue the ideas which Yworth (as 'Cleidophorus Mystagogus') had published in *Mercury's caducean rod* at about the same time (1702). Newton perused this tract, although – with the exception of short notes concerning the titles of works quoted in the text – we do not have Newton's typical handwritten citations and reading notes.[69] His reception of 'Cleidophorus Mystagogus', however, is thoroughly documented by his own version of Yworth's 'Processus'.

Newton's 'Processus' manuscript breaks off abruptly in the middle of a sentence. And Yworth must have disappeared from Newton's circle just as abruptly, because Newton seems not to have followed his publications further. After *Mercury's caducean rod*, no later work of this author can be found in Newton's library: neither the 'Cleidophorus Mystagogus' tract *Trifertes sagani* (1705), nor Yworth's *Compleat distiller* (1705), the second part of which – *Pharmacopœa spagyrica nova* – is, in essence, equivalent to his alchemical writings.[70] Apparently, Newton's interest had

[67] For a list of authorities quoted in *Mercury's caducean rod* see Figala, *op. cit.* (note 5), 264 n. 58; the 'Processus' variants are based on almost the same sources. Their corresponding to Newton's own (al)chemical library stock is remarkable.

[68] 'Philosophical Epistle', *op. cit.* (note 56), p. 23*: 'Sandivogius saith, That Nature hath but one [*i.e.* vessel], but for brevities sake we use two, which Philalethes in his Comment on him, calls the one the Star of Mars and Venus, and the other Mercury of the Philosophers'; compare Newton's transcript of the anonymous 'Sendivogius explained' tract, *Keynes MS 55*, f. 10r: 'The vessel of nature is only one, but we use two for brevities sake, yt is ye [star] of [Mars] & [Venus] & ye [Mercury] of Phers'; see also Figala, 'Die exakte Alchemie', *op. cit.* (note 1), 183, 216 n. 87, and Westfall, *Never at Rest*, *op. cit.* (note 1), 526 n. 199, who drew attention to further extant manuscript versions of the commentary.

[69] See Harrison, *op. cit.* (note 2), descriptions of *HL 1138/1302*; no further references to Yworth/Cleidophorus in Newton's alchemical papers have been found.

[70] Only one of Yworth's (non-alchemical) works seems to have been in Newton's possession: a copy of *Introitus apertus* (1692), *op. cit.* (note 33); see *HL 1760*. The late *Pharmacopœa*, *op. cit.* (note 35), appears to be a construct of

died out. If it had not, then – despite the disappearance of Yworth from London in 1705 – some trace of his later publications should surely have been found in Newton's possession.

Conclusion

That remark brings us to the end of our present reflections on the late phase of Newton's alchemical studies. That Newton continued such studies during his London years is an established fact; that they were of lesser importance to him than during the most intense period, in the 1690s, is also beyond question. The search for definite signs of extended alchemical activity is by no means intended as an attempt to make Newton out to be a classical alchemist rather than an innovator in natural philosophy. On the contrary, it is aimed at discovering the purpose of Newton's 'Hunting of the Greene Lyon', that is, determining where and to what degree his (al)chemical studies found their way into his scientific thought.

R. S. Westfall is a dedicated and astute champion of the thesis that an irreversible break in the career of the 'alchemist' Newton took place in the 1690s. Westfall claims that a breakthrough was achieved when 'with his *quantified concept of force*, he [Newton] had extracted the essence of the art'.[71] Though we do not wish to reject this interpretation, we should like to direct attention to a second goal of the Newtonian integration of (al)chemy and scientific theory: his *quantified concept of matter*. From the work of A. Rupert Hall and Marie Boas Hall onwards, Newton's theory of matter has often irritated historians of science;[72] it appears to us that it is significant that the genesis of this component of the Newtonian system notably coincides with the decrease in Newtonian alchemy: as the alchemical ideas faded from Newton's thought, the scientific concepts began to take a more well-defined shape. In fact, documentary evidence exists for Newton's having had a quantitative theory of matter as early as 1705; it was presented to the public in a chemical context in the second edition of the *Opticks* (1717).[73] According to the judgement of historians of science, Newton had by that time clearly made the change from alchemy to chemistry.[74] Whether he himself was aware of the transformation in his thought remains an open question. The fact is, Newton based his studies less on chemical than on alchemical texts, and he availed

Paracelsian/Helmontian dispensatory and contemporary Hermetical theory; compare e.g. alchemical poems, quite similar in style, in *Pharmacopœa* (1705), 239–42, and in 'Processus', *Keynes MS 65*, ff. 11v–12r.

[71] Westfall, 'Newton and alchemy', *op. cit.* (note 1), 332.

[72] See *Unpublished Scientific Papers of Isaac Newton. A Selection from the Portsmouth Collection in the University Library, Cambridge*, eds. A. Rupert Hall and Marie Boas Hall (Cambridge, 1962), 183–228; for a summary of different approaches see H. H. Kubbinga, 'Newton's theory of matter', in *Newton's Scientific and Philosophical Legacy*, eds. P. B. Scheurer and G. Debrock. (Archives internationales d'histoire des idées, 123) (Dordrecht, 1988), 321–41.

[73] On an entry in David Gregory's diary (Dec. 12, 1705) and Newton's 'Queries', appended to his *Opticks*, see Figala, 'Die exakte Alchemie', *op. cit.* (note 1), esp. 162–73, and Arnold Thackray, *Atoms and Powers. An Essay on Newtonian Matter-Theory and the Development of Chemistry*. (Harvard Monographs in the History of Science) (Cambridge, Mass., 1970), chap 2.

[74] See e.g. Marie Boas, 'Newton's chemical papers', in *Isaac Newton's Papers & Letters on Natural Philosophy*, ed. I. Bernard Cohen (Cambridge, 1958), 241–8, on Newton's late 'De natura acidorum'; also Thackray, *op. cit.* (note 73).

himself of the assistance of a natural philosopher such as Fatio on the one hand, and a 'hermetical philosopher' such as Yworth, on the other. Thereby, Newton – in his personal acquaintances as well as in his theoretical designs – exemplarily forged the link between Renaissance and Enlightenment.

Note: After finishing our manuscript, it came to our knowledge that the above-mentioned 'lost' version of W. Yworth's *Processus mysterii magni philosophicus* (formerly Sotheby Lot 116) is now preserved at Hampshire Record Office: *NC 17*; cf. *Sir Isaac Newton: A Catalogue of Manuscripts and Papers collected and published on microfilm by Chadwyck-Healey*, ed. P. Jones (Cambridge, 1991). The manuscript is also available on microfilm (*Sir Isaac Newton: Manuscripts and Papers*, Chadwyck-Healey: Cambridge – reel 43), as have most of the Newton papers considered above (excluding material from the Babson Collection). For our discussion of Fatio de Duillier and French *alchemica* see also B. J. T. Dobbs, *The Janus Faces of Genius. The Role of Alchemy in Newton's thought* (Cambridge, 1992), esp. 170–85.

12

Newton's subtle matter: the *Opticks* queries and the mechanical philosophy

R. W. HOME

There is an immense literature dealing with Newton's attitude towards hypotheses and, in particular, towards the idea of action at a distance. It is hard to imagine that there is anything new to be said on the question. Nevertheless, given the importance of the issue – it amounts to asking how best to characterize the end result, intellectually speaking, of the Scientific Revolution – I believe it is worth pursuing. I wish in this paper to challenge some elements of the consensus that has emerged among historians of science about Newton's views concerning forces acting at a distance. I shall argue, in fact, that Newton was far more reluctant to accept such things than is generally believed. (So, too, for that matter, were most of his disciples.) My evidence derives from various late-in-life statements by Newton, many of them still not fully published but others extremely well known, concerning the likely existence and properties either of a subtle spirit associated with samples of ordinary matter, or of an all-encompassing subtle aether. There has been a tendency among historians not to take these statements seriously, to see them as a product of Newton's old age, an over-defensive reaction on his part to the criticisms levelled at him by his continental critics. I, however, see them – or at least some of them – as providing, instead, important insights into his natural philosophy.

A caveat, however, should be entered at once. The fact that Newton from time to time wrote of a subtle *spirit* as the cause of various natural phenomena has sometimes led to misunderstandings as to the materiality of the cause. These have even extended to such notable Newtonian scholars as Koyré, who once suggested that for Newton, the 'electric and elastic spirit' of the 1713 *Principia* was to be equated with the spirit of God,[1] and McGuire, who has represented it as 'an electrical *arche* connecting mind with matter'.[2]

More recently, those seeking to reconstruct Newton's image into that of a latter-day magus have tended to find comfort in his phraseology here. To do so, however, is a mistake, and the discussion that follows should not be seen as giving support to such

[1] A. Koyré, *From the Closed World to the Infinite Universe* (Baltimore, 1957), 234.

[2] J. E. McGuire, 'Force, active principles, and Newton's invisible realm', *Ambix*, 1968, **15**: 154–208, p. 176.

views. First, Newton himself makes it clear, in a draft intended for *Principia*, that, for him, the word 'spirit' can refer to material entities:

> Vapours and exhalations on account of their rarity lose almost all perceptible resistance, and in the common acceptance often lose even the name of bodies and are called spirits. And yet they can be called bodies in so far as they are the effluvia of bodies and have a resistance proportional to density.[3]

In addition, when Newton in various unpublished drafts discusses, in more detail, the mode of operation of this particular spirit, the terms he uses are all appropriate to a material rather than an immaterial agency. It 'expands' and subsequently 'condenses' and 'shrinks' back into the body whence it came. It can be 'agitated', whereupon it 'will rush out . . . with a soft crackling noise' and 'push against the finger'. On occasion Newton describes its 'turbulent motions', which he compares to 'a wind'. He also says it is 'susceptible of a vibrating motion like that of air'.[4] It is clear, therefore, that while Newton's electric spirit is (as he puts it) 'much more subtile than common Air or Vapour',[5] the difference is one of degree, not kind. The Newton who emerges from these passages is no magus but someone more akin to an ordinary old-fashioned mechanical philosopher taking his images from seventeenth-century precedessors such as Descartes. (Though not, it should be added, his views on the ultimate source of the action involved. Newton's subtle matter is elastic and Newton, as we shall see, unlike Boyle and Charleton and also Descartes, did not evade the problem of explaining this but was prepared to suppose a repulsive force acting everywhere between its particles.)

Stages in Newton's thinking

Let us, however, return to the main thread of the argument. The general consensus, parts of which will be challenged here, about the evolution of Newton's thinking, begins by holding that in his early years he was profoundly influenced by mechanical philosophers such as Descartes, Gassendi, Gassendi's English mouthpiece Walter Charleton, and Robert Boyle. In particular, he embraced the fundamental principle of the mechanical philosophy, the reducibility of all natural events to the motions and impacts of particles of matter.[6] At this period he accepted the existence of an all-encompassing material aether and also that of 'various other aethereall Spirits' that resided in the pores between the particles of ordinary bodies. Best known of the documents where he sets out such views are the 'Hypothesis explaining the Properties of Light' of 1675[7] and (in somewhat different terms) his letter to Boyle of 1679.[8]

[3] University Library Cambridge, Add. MS 3965.13, fol. 437v; quoted by McGuire, 'Body and void in Newton's *De mundi systemate*: some new sources', *Arch. Hist. Exact Sci.*, 1966, 3: 206–48, p. 219 (Latin original, p. 245).

[4] University Library Cambridge, Add. MS 3970, fols. 241v–241r, 293v, 295.

[5] *Ibid.*, 293v.

[6] Richard S. Westfall, 'The foundations of Newton's philosophy of nature', *Brit. J. Hist. Sci.*, 1962, 1: 171–82; also Westfall, *Never at Rest: A Biography of Isaac Newton* (Cambridge, 1980), 83–93; A. R. Hall, 'Sir Isaac Newton's notebook, 1661–65', *Camb. Hist. J.*, 1948, 9: 239–50; and J. E. McGuire and Martin Tamny, eds., *Certain Philosophical Questions: Newton's Trinity Notebook* (Cambridge, 1983).

[7] *The Correspondence of Isaac Newton* (Cambridge, 1959–77), 1: 362–86.

[8] *Ibid.*, 2: 288–95.

Yet by the early 1680s, it is agreed, Newton abandoned his belief in a universal dense aether along the lines of that described by Descartes, and began to speak of forces acting between separated particles of matter. It has been convincingly argued, by Westfall and Dobbs in particular,[9] that Newton's intense engagement with alchemy during the preceding ten years and more helped prepare the way for this, in so far as alchemy proclaimed the existence of active principles associated with matter and by doing so accustomed Newton to thinking in such terms. More recently, Dobbs has sought to confine the influence of alchemy in this regard to Newton's way of thinking about the interactions of matter at the microscopic level, arguing the importance of a version of ancient Stoic natural philosophy in Newton's also developing at this period his notion of a universal force of gravity acting between separated particles of matter.[10]

A further factor, hitherto seemingly unremarked, that is likely to have influenced Newton's thinking in this regard, is his growing understanding of the principles of mechanics. This would in itself have left him increasingly dissatisfied with the habit of mechanical philosophers such as Descartes of treating subtle matter as a source of new activity whenever their 'mechanical' explanations of particular phenomena required this. Consider, for example, Descartes's account of fire, which is altogether typical of his approach to such questions. According to this, all that is required to convert ordinary, inert matter into active fire is to surround its particles with particles of the 'first element' or subtle matter.[11] According, however, to the principles that Newton came to grasp ever more securely during these years, this kind of thing would simply not do. To introduce new motion into a system required the action of a force: the mere presence of additional subtle matter would not suffice.[12]

What is certain is that by the mid-1680s, Newton became convinced, as a result of his work on the motions of bodies in resisting media for Book II of *Principia*, that the observed motions of the planets were inconsistent with the existence of a dense, space-filling interplanetary aether of the kind invoked by Descartes and other mechanical philosophers. At the same time, in Book III of *Principia*, Newton announced his celebrated conclusion that throughout the Universe a gravitational force acted between separated particles of matter. Abandoning the aether left him seemingly without the possibility of explaining this force in mechanical terms. And yet, as he made clear in letters he wrote to Richard Bentley in the early 1690s, Newton remained totally convinced that gravity must have a cause of some kind:

> That gravity should be innate inherent & essential to matter so yt one body may act upon another at a distance through a vacuum wthout the mediation of any thing else by & through wch their action or force may be conveyed from one to another is to me so great an absurdity that I beleive no man who has in philosophical matters any competent faculty of thinking can ever fall

[9] Westfall, *Never at Rest*, *op. cit.* (note 6), 299–307, 389–90; B. J. T. Dobbs, 'Newton's alchemy and his theory of matter', *ISIS*, 1982, 73: 511–28.

[10] Dobbs, 'Newton's alchemy and his "active principle" of gravitation', pp. 55–80 in P. B. Scheurer and G. Debrock, eds., *Newton's Scientific and Philosophical Legacy* (Dordrecht, 1988).

[11] Descartes, *Principles of Philosophy*, trans. V. R. and R. P. Miller (Dordrecht, 1983), 219–28.

[12] See Gad Freudenthal's paper in this volume.

> into it. Gravity must be caused by an agent acting constantly according to certain laws . . .[13]

Though he feigned agnosticism as to what the cause might be – 'whether this agent be material or immaterial is a question I have left to ye consideration of my readers' – the answer to which Newton had been driven is manifest: the cause of gravity cannot be material.

Subsequently, as is well known – above all in the Queries added for the first time to the Latin edition of *Opticks* published in 1706 – Newton invoked many other unexplained forces, mostly acting at the microscopic level, to account for such things as cohesion and various optical and chemical phenomena. However, he qualified his doing so in a well-known passage:

> How these Attractions may be perform'd, I do not here consider. What I call Attraction may be perform'd by impulse, or by some other means unknown to me. I use that word here to signify only in general any Force by which Bodies attract one another, whatsoever be the Cause. For we must learn from the Phaenomena of Nature what Bodies attract one another, and what are the Laws and Properties of the Attraction, before we enquire the Cause by which the Attraction is perform'd.[14]

Yet most commentators have placed little weight on this and other like remarks of Newton's, seeing them as little more than symptoms of his habitual caution when it came to committing himself publicly to controversial ideas. As will emerge in what follows, I think this is a mistake.

Shortly after publishing the Queries in which this passage appears – that is, in about 1707 – Newton once more began actively to consider possible causes for many of the forces he had invoked. He was led to do so by some spectacular new electrical experiments devised by his protégé Francis Hauksbee, that reawakened his interest in his earlier ideas about subtle spirits associated with ordinary gross matter.[15] He included a paragraph about this in the General Scholium added as a conclusion to the second, 1713, edition of *Principia*. Soon afterwards, he added several new Queries to the 1717 edition of *Opticks* in which he speculated anew about an all-encompassing aether as the possible cause of gravity as well as of various optical and physiological phenomena.

Subtle matter *versus* aether

Historians who have emphasized Newton's earlier commitment to forces acting at a distance have been somewhat embarrassed by these late-in-life Newtonian specula-

[13] Newton, *op. cit.* (note 7), **3**: 240, 253–4.

[14] Newton, *Opticks: or a Treatise of the Reflections, Refractions, Inflections and Colours of Light* (New York, 1952; based on the 4th edition, London, 1730), 376.

[15] Henry Guerlac, 'Francis Hauksbee, expérimentateur au profit de Newton', *Arch. Int. Hist. Sci.*, 1963, **16**: 113–28; Guerlac, 'Sir Isaac and the ingenious Mr Hauksbee', pp. 228–53 in I. Bernard Cohen and R. Taton, eds., *Mélanges Alexandre Koyré: Tome I, L'Aventure de la science* (Paris, 1964); R. W. Home, 'Francis Hauksbee's theory of electricity', *Arch. Hist. Exact. Sci.*, 1967, **4**: 203–17; Home, 'Newton on electricity and the aether', pp. 191–213 in Z. Bechler, ed., *Contemporary Newtonian Research* (Dordrecht, 1982).

tions, and have tended, as indicated already, to dismiss them as an aberration.[16] Such a view may, in fact, be justified in the case of the 1717 'aether Queries'. To judge by the manuscript evidence, these were indeed a concession on Newton's part, added to the work very much at the last moment.[17] I believe it is a mistake, however, to see all of Newton's remarks about subtle matter from this period in the same light. The majority, far from being hastily thrown together, were the fruit of at least ten years of speculation and inquiry, much of it pre-dating the main Continental onslaught on the doctrine of forces, by a Newton still capable of prodigious feats of intellectual activity. The work from which they emerged, far from being dismissed, should thus be seen as yet another constructive phase – perhaps the last major constructive phase – in Newton's thinking about matter and its powers.

Unfortunately, despite the effort Newton devoted to this inquiry and the many drafts he prepared in which he expounded his views – drafts that survive among his papers in Cambridge – he never allowed more than hints of his ideas to appear in print. One may surely see in this an acknowledgement on Newton's part that his ideas as yet lacked sufficient empirical grounding to enable him to sustain them in print; and yet the consistency with which he maintained these particular opinions over many years suggests a firm belief that they were nevertheless correct.

When it comes to understanding Newton's ideas here, perhaps the most important point to recognize is that the aether of 1717 is not at all the same thing as the subtle matter of the 1713 *Principia*. This has often been overlooked, even though the point emerges from a careful reading of Newton's published statements and does not depend upon recourse to the manuscript record. Significantly, the relevant passage in *Principia* – the very last paragraph in the book – occurs immediately after the paragraph containing the famous phrase, 'hypotheses non fingo'. That phrase appears in a discussion of the force of gravity. When it comes to various other natural powers, Newton reveals himself quite happy to frame (or 'feign'[18]) the hypothesis that they are brought about by the action of a

> certain most subtle spirit which pervades and lies hid in all gross bodies; by the force and action of which spirit the particles of bodies attract one another at near distances, and cohere, if contiguous; and electric bodies operate to greater distances, as well repelling as attracting the neighboring corpuscles; and light is emitted, reflected, refracted, inflected, and heats bodies; and all sensation is excited, and the members of animal bodies move at the command of the will, namely, by the vibrations of this spirit, mutually propagated along the solid filaments of the nerves, from the outward organs of sense to the brain, and from the brain into the muscles.[19]

Consistent with what Newton had said in the preceding paragraph, gravity does not appear amongst the phenomena to be explained in terms of this subtle spirit: that is,

[16] See McGuire, *op. cit.* (note 2), 176, and Westfall, *Never at Rest*, *op. cit.* (note 6), 644, 793.

[17] Guerlac, 'Newton's optical aether: his draft of a proposed addition to his *Opticks*', *Notes Rec. Roy. Soc. Lond.*, 1967, **22**: 45–57.

[18] See A. Koyré, *Newtonian Studies* (London, 1965), 35–6.

[19] Newton, *Mathematical Principles of Natural Philosophy*, trans. Andrew Motte, rev. Florian Cajori (Berkeley and Los Angeles, 1934), 547.

the spirit being invoked here has nothing to do with the explanation of gravity. However, electrical attraction does appear among the phenomena listed – which is why, at the very end of the paragraph and hence of the book, Newton describes the spirit as 'electrical' (i.e. responsible for the phenomena of electricity) as well as 'elastic'. Importantly, the spirit is said to be confined to 'gross bodies' and their immediate surroundings: it does not extend through all of space.

By contrast, the aether that Newton proposes in 1717 is not confined in this way but extends (as it must, of course, if it is to account for gravity) through all of space: 'Doth it not readily pervade all Bodies?', asks Newton, 'And is it not (by its elastick Force) expanded through all the Heavens?'[20] Moreover, while this aether is said to be the cause not only of gravity but of the various optical and physiological effects attributed in 1713 to the subtle matter, it is specifically not proposed as the explanation of certain other phenomena listed in 1713, namely electricity, cohesion, and the emission of light from luminous bodies. Quite the reverse, indeed, at least so far as electricity is concerned, as becomes clear when Newton seeks in Query 22 to give at least a modicum of plausibility to his notion of an aether that is both extremely rare and extremely elastic. To do this, he cites instances already 'known' in which matter retains its activity when reduced to a very low density, and one of these is the subtle matter which, he assumes, gives rise to electrical motions:

> If any one would ask how a Medium can be so rare, let him . . . tell me, how an Electrick Body can by Friction emit an Exhalation so rare and subtile, and yet so potent, as by its Emission to cause no sensible diminution of the weight of the electrick Body, and to be expanded through a Sphere, whose Diameter is above two Feet, and yet to be able to agitate and carry up Leaf Copper, or Leaf Gold, at the distance of above a Foot from the electrick Body?[21]

There is no suggestion here that the aether and this electrical 'exhalation' are one and the same thing. On the contrary, the electrical exhalation is taken for granted as something different, the nature of which is well known and understood, the properties of which provide a useful analogy to buttress Newton's account of the aether.

From the passage just quoted, it is in fact clear that in 1717 Newton had in mind two different kinds of subtle matter: the all-encompassing aether on the one hand, the spirit responsible for electrical phenomena on the other. This is confirmed by the manuscript evidence, which shows that Newton intended to include additional Queries in the 1717 *Opticks*, over and above those describing the aether, Queries in which he would set out his ideas on the subtle electrical matter that he believed lay hidden in the pores of ordinary 'gross' matter. Only at the very last minute, it would appear, did he decide, after all, not to publish these, thus leaving the passage quoted above as the only clue from which an attentive reader might divine what he had in mind.

Armed with this knowledge, we are better able to appreciate Newton's remarks at the end of the 1713 *Principia*. It emerges that Newton was not introducing here a

[20] Newton, *op. cit.* (note 14), 349. [21] *Ibid.*, 353.

preliminary version of the 1717 aether, but rather a subtle matter in the separate existence of which he had long believed and in which he continued to believe, even after he introduced the new aether of 1717. The point was clear enough to J. T. Desaguliers, who, as the Royal Society's Demonstrator of Experiments at the time, was in a position to know: 'whether the Medium acting upon Light be the same as the Aether hinted at for the Cause of Gravity, is very cautiously insinuated by our incomparable Philosopher'.[22] Newton in 1717 merely tentatively transferred to the latter – the aether – a few of the explanatory responsibilities previously given to the subtle matter. His doing this meant, of course, that the passage at the end of *Principia* no longer represented his most recent thinking on the subject. Consistent with this, there is some evidence that Newton envisaged deleting the entire paragraph from the next edition of his great work.[23] In fact, however, when this appeared, in 1726, the passage remained intact. This suggests that, by then, Newton had probably changed his mind yet again about the aether, and reverted again to explaining the various optical and physiological effects in terms of the subtle spirit associated with gross matter. It is another reason for regarding the 1717 aether as an aberration and not really representative of Newton's thinking.

Explanations in terms of subtle matter

On the other hand, when one focuses on the idea of a subtle spirit (or spirits) associated with matter, and ignores the distractions of the aether and possible causes of gravity, important continuities in Newton's thought emerge. I have presented evidence in various papers published during the past few years that Newton's ideas about the causes of electricity and magnetism changed scarcely at all from the 1670s to the end of his life.[24] Magnetism had its own subtle spirit that circulated through and around magnets very much in the manner proposed by Descartes (though without some of the complexities of the Cartesian account). Electrical attraction (and later the repulsion as well which, following Hauksbee, Newton recognized as succeeding this) was due to a different subtle matter exhaled from a body as a result of friction and then collapsing back to it. In both cases, and at all stages of his life, the account is thoroughly mechanistic, with no hint of any thought of attributing the effects to new forces acting at a distance. In the case of electricity, Newton is very explicit about this, asserting baldly at one point that 'electrical bodies could not act at a distance without a spirit reaching to that distance'.[25] And, as remarked already, how that spirit then brought about the effect is also asserted very explicitly, in thoroughly mechanistic terms:

[22] Desaguliers to Sir Hans Sloane, 4 March 1730/31, quoted by Guerlac, *op. cit.* (note 17), 51.
[23] A. Koyré and I. B. Cohen, eds., *Isaac Newton's Philosophiae naturalis principia mathematica: The Third Edition (1726) with Variant Readings* (Cambridge, 1972), 2: 764.
[24] Home, 'Newton on electricity' *op. cit.* (note 15); Home, ' "Newtonianism" and the theory of the magnet', *Hist. Sci.*, 1977, **15**: 252–66; Home, 'Force, electricity, and the powers of living matter in Newton's mature philosophy of nature', pp. 95–117 in Margaret J. Osler and Paul Lawrence Farber, eds., *Religion, Science and Worldview: Essays in Honor of Richard S. Westfall* (Cambridge, 1985).
[25] University Library Cambridge, Add. MS 3970, fol. 241v.

> Is not electrical attraction and repuls^{n} performed by an exhalation wch is raised out of the electrick body by friction & expanded to great distances & variously agitated like a turbulent wind, & wch carrys light bodies along with it & agitates them in various manners according to its own motions, making them go sometimes towards the electric body, sometimes from it & sometimes move with various other motions? And when this spirit looses its turbulent motions & begins to be recondensed & by condensation to return into the electric body doth it not carry light bodies along with it towards the Electrick body & cause them to stick to it without further motion till they drop off?[26]

A similar conclusion emerges with respect to certain other natural powers that Newton discusses. When it comes to accounting for the generation of living forms, for example, Newton in his manuscripts adopts the mechanistic (or at least pseudo-mechanistic) 'preformation' theory widely accepted in his day.[27] And even in the 1704 *Opticks*, at the height of his supposed commitment to unexplained forces, Newton, though careful not actually to embrace such a mechanism, is still prepared at least to raise the possibility that the reflection and refraction of light might be due to a 'refracting or reflecting medium' occupying the spaces between the 'solid parts of the refracting or reflecting Body'.[28]

It would appear, therefore, that in considering Newton's attitude towards the various forces of nature, his disclaimers about their causes, such as the one quoted earlier: 'How these attractions may be performed I do not here consider' should be given more weight than they usually receive. Whatever his attitude to gravity, Newton did not at all suppose that other forces, the existence of which he invoked, were irredeemably non-mechanical.

Indeed, had he done so – had he, for example, attributed the 'amber effect', the drawing of little light objects to rubbed amber or glass, to an unexplained electrical attraction – he would have been saying no more than that electricity acted because it acted. And he would then have been open to precisely the same objection that seventeenth-century scientists and philosophers frequently levelled at the Aristotelian doctrine of substantial forms. Molière's parody, attributing the sleep-inducing effect of opium to the drug's 'dormitive power', is well known,[29] while Hobbes inveighed against the Aristotelian notions of heaviness and lightness on the ground that this was merely to say 'that bodies descend, or ascend, because they doe'.[30] Newton himself made the same point in a famous passage in Query 31 of *Opticks*: 'To tell us that every Species of Things is endow'd with an occult specifick Quality by which it acts and produces manifest Effects, is to tell us nothing'.[31]

[26] *ibid.*, 293v. [27] Home, 'Force, electricity, and the powers of living matter', *op. cit.* (note 24), 111–14.

[28] Newton, *Opticks*, *op. cit.* (note 14) (1704 edn), Book II, 80–81. The passage appears unaltered in later editions of the work (pp. 280–1 of the 1952 reprint edition).

[29] Molière, *Le Malade imaginaire* (1673), interlude at the end of the play, see Molière, *Oeuvres complètes*, ed. G. Couton (2 vols, Paris, 1971), **2**: 1173 and *The Plays of Molière*, ed. A. R. Waller (Edinburgh, 1926), **8**, 328; but see Keith Hutchison, 'Dormitive virtues, scholastic qualities, and the new philosophies', *Hist. Sci.*, 1991, **29**: 245–78.

[30] Hobbes, *Leviathan* (London, 1651; reprinted, Oxford, 1909), 529.

[31] Newton, *op. cit.* (note 14), 401. The question of specific *versus* universal causal principles is discussed, in relation to an eighteenth-century example, by Keith Hutchison, 'Idiosyncrasy, achromatic lenses, and early Romanticism', *Centaurus*, 1991, **34**: 125–71.

Unexplained forces

Newton was in fact much more unwilling than has often been supposed to invoke a specific kind of in-principle-inexplicable force to account for each different category of natural event. Unexplained forces remained full of mystery for him; indeed his conviction that the most famous of them, gravity, brought him close to the activity of God himself, is well known,[32] and there is no reason to suppose that others would have had less profound implications.

Prima facie, then, one might expect that Newton would have had recourse to entities of this kind only for the most universal of natural powers; and this does, indeed, seem to be the pattern that emerges. It is also precisely the pattern he proclaimed in Query 31 of *Opticks*: 'to derive two or three general Principles of Motion from Phaenomena, and afterwards to tell us how the Properties and Actions of all corporeal Things follow from these manifest Principles, would be a very great step in Philosophy'.[33] The attraction of a single substance, iron, to the peculiar kind of stone that we call a magnet is not an effect of sufficient universality. Neither is the 'amber effect'. However, as remarked earlier, a force of repulsion acting between the particles of a universal and all-pervading subtle matter, and ultimately responsible not just for macroscopic electrical attractions but for the wide range of natural phenomena listed in the 1713 General Scholium, might be general enough. Similarly, the various specific forces associated with living forms are unlikely to be the end point of the analysis, whereas there may well be some very general power that distinguishes living from non-living matter. Something presumably does; and whatever it be, it is likely to be closely linked to the activity of God as the creator of life and to be irreducibly non-mechanistic in character.

Meanwhile, Newton's methodology leads him to invoke forces of all kinds in studying natural processes, and encourages him to focus on discovering the mathematical laws according to which those forces act, as he had done so successfully for gravitation, rather than on explaining their action. So far as the magnetic force is concerned, Newton tells us that he himself had tried to determine experimentally the law according to which this varies with distance, finding that 'in receding from the magnet [the force] decreases not as the square but almost as the cube of the distance, as nearly as I could judge from some rude observations'.[34] He also incited various followers (especially Brook Taylor) to continue these efforts.[35] But in all these cases,

[32] Westfall, *Never at Rest*, *op. cit.* (note 6), 509, quotes a corollary that Newton drafted, but never published, to his proposition asserting the proportionality of gravity to mass (*Principia*, Proposition VI, Book III): 'Corol. 9. There exists an infinite and omnipresent spirit in which matter is moved according to mathematical laws'. Cf. also Dobbs, 'Newton's alchemy' *op. cit.* (note 10).

[33] Newton, *op. cit.* (note 14), 401–2.

[34] Newton, *op. cit.* (note 19), 414. This report did not appear in the first edition of Newton's book, but was added in the second (1713) edition; cf. Koyré and Cohen, *op. cit.* (note 23), **2**: 576.

[35] Francis Hauksbee, 'An Account of Experiments concerning the proportion of the Power of the Load-Stone at different Distances', *Phil. Trans.*, 1710–12, **27**: 506–11; Brook Taylor, 'An Account of an Experiment made by Dr. Brook Taylor assisted by Mr. Hauksbee, in order to discover the Law of Magnetical Attraction', *ibid.*, 1714–6, **29**: 294–5 and 'Some experiments relating to magnetism', *ibid.*, 1720–21, **31**: 204–8; also Taylor's report in Royal Society of London, *MS Letter Book original, XIV*, 394.

as indicated above, we ought to take more seriously than has usually been done Newton's oft-expressed caveat about the causes lying behind the forces of which he spoke. Some of those forces might, he suggested, be 'perform'd by impulse'. If the interpretation suggested here is correct, Newton would in fact have thought that an impulsive mechanism was involved in most instances. I have elsewhere presented evidence, indeed, that he did think so in certain leading cases. He would also not have been surprised, however, if, as these mechanisms were investigated, they were found ultimately to depend upon a small number of irreducibly non-mechanical powers of quasi-universal application. At that point, he would have said – and only at that point – we are indeed approaching a knowledge of the First Cause, which is God.

Finally, it should be remarked that most of Newton's eighteenth-century disciples and followers were similarly reluctant to grant ontological significance to their practice of invoking a variety of forces acting at a distance in their analyses of natural phenomena. In this sense, the mechanical philosophy that had lain at the heart of the Scientific Revolution continued to hold sway for much longer than has often been supposed. Certainly this was so in respect to electricity and magnetism, where prevailing ideas about the underlying causes of the observed effects changed very little from those held by Newton (and hence, I have argued, from those held by Descartes) until after mid-century – with a caveat, always, about the ultimate source of the activity of the subtle matter held to be involved.[36]

Eventually, of course, Aepinus and others did come to invoke actions at sensible as well as microscopic distances, in ways that made it almost impossible to conceive that there might still be some underlying material cause (though even in 1759 Aepinus felt obliged to issue disclaimers about the possibility that there might be one).[37] I would argue that such a step eventually became acceptable because by then not just a few small-scale effects but a very wide range of phenomena – some of them, like the powerful electrical shocks delivered by the Leyden jar, of a quite spectacular character – had come to be known, and it came to be seen that they could (apparently only) thereby be reduced to order. To use Newton's terminology, no longer was it a matter of invoking an 'occult specifick quality' to account for just one or two effects; rather, it was now a question of introducing 'general principles of motion' to reduce a whole field to order by explaining various 'properties and actions of all corporeal things'. Eventually, one investigator after another could be driven in this way to tolerate the previously almost unthinkable.

[36] Home, '"Newtonianism"', *op. cit.* (note 24); also Home, 'Introduction' to *Aepinus's Essay on the Theory of Electricity and Magnetism* (Princeton, 1979), chaps. 2, 4.
[37] *Ibid.*, 240.

13

Huygens's reaction to Newton's gravitational theory

ROBERTO DE A. MARTINS

Much has already been written about the reception of and resistance to Newton's *Principia* – especially in France – and about Huygens's criticism of Newton's gravitational theory.[1] Such studies have usually focused upon the differences between Newton's and Descartes's conceptual world views that affected these reactions. Newton did not provide a mechanical explanation of gravitation, while in Descartes's theory gravity was an effect due to the motion of a vortex of subtle matter around the Earth.[2] Different conceptual views were, of course, instrumental in producing strong resistance to Newton's theory. Some papers have further pointed out that there were other aspects of Newton's work that also contributed to its slow acceptance – such as its mathematical difficulty.[3]

The part played by such factors cannot be ignored. None the less, we may ask whether it is possible to assert that people who did not accept Newton's gravitational theory at once were misled either by prejudice or by the weakness of their mathematics? Clearly this is not so. Other factors were also at work.

The present chapter proposes to analyse Huygens's reasons for rejecting Newton's theory, with the aim of showing that, in a sense, there were good methodological reasons for his resistance to accepting it.

The author acknowledges the support received from the Brazilian *National Council for Scientific and Technological Development* (CNPq), and grants from VITAE and FAPESP Foundations that allowed him to present this paper at the conference on 'The Scientific Revolution'.

[1] The classic reference is J. Brunet, *L'Introduction des théories de Newton en France au 18ème siècle tome 1 – Avant 1738* (Paris, 1931). See also: René Dugas, *La Mécanique au XVIIème Siècle* (Paris, 1954); Alexandre Koyré, *Newtonian Studies* (Chicago, 1968); I. Bernard Cohen, *The Newtonian Revolution* (Cambridge, 1980); A. Rupert Hall, 'Newton in France: a new view', *Hist. Sci.*, 1975, **13**: 233–50; E. A. Fellmann, 'The *Principia* and continental mathematicians', *Notes Rec. Roy. Soc. Lond.*, 1988, **42**: 13–34.

[2] René Descartes's vortex theory was first fully described in his *Principia Philosophiae*, (Amsterdam, 1644). See Eric J. Aiton, *The Vortex Theory of Planetary Motion* (London, 1972).

[3] Greenberg has recently stressed this aspect of Newton's theory of the flattened Earth: J. L. Greenberg, 'Isaac Newton et la théorie de la figure de la Terre', *Rev. Hist. Sci.*, 1987, **40**: 357–66.

The choice of Huygens as a case study

Why study the reaction of Huygens, instead of someone else? Several circumstances determine our choice. Many of their contemporaries would have deemed Huygens a fit judge of Newton's theory.

Let us look at a letter from Fatio de Duillier to Huygens, written when the former was in England, on 24 June 1687:[4]

> I have been three times at the Royal Society where I have heard both very good proposals and some platitudes. Some of those gentlemen that compose it have an extremely favourable prejudice about a book from Mr. Newton that is now in press and will be issued in three weeks. . . . I have seen part of this treatise and it is certainly very fair and full of many valuable propositions, but I wish, Sir, that the Author had taken some advice from you about this principle of attraction that he assumes between celestial bodies.

So, according to Fatio de Duillier, Huygens might be a good judge of Newton's work. Were these no more than flattering words? I think not. Let us briefly recall Huygens's qualifications.

First: he was one of the few mathematicians of that time who could understand and master the *Principia*. Second: he was receptive to Newton's work, as his letters show. Third: he had proposed an earlier theory of gravity and discussed the flattening of the Earth independently of Newton. All these circumstances – which perhaps make Huygens unique in his time – contribute to our choosing him as a privileged expert whose opinions are worthy of attention.

Let us begin by clarifying the points just made about Huygens's reputation.

Most contemporaries would have been in agreement on the first point: Huygens was generally accepted as being a very good mathematician.[5]

Huygens is now best remembered for his wave theory of light. Indeed, his *Traité de la lumière* was a wonderful work. It did not merely put forward a suggestive theory of the nature of light. It was remarkable in expressing the theory in mathematical language and in being able to provide the extremely difficult proofs concerning double refraction in Iceland spar. Anyone who once tried to follow Huygens's demonstrations would certainly have been struck by his mastery of geometrical methods. From this point of view, his *Treatise on light* may be said to be far more advanced than Newton's *Opticks*. This is not the relevant point here, however, since Huygens's treatise appeared only after Newton's *Principia*. At that time, the basis for Huygens's eminence was his *Horologium oscillatorium*.

This work, published in Paris in 1673, represented an important step in the development of classical mechanics. Its geometrical style, stressed in its very

[4] Letter 2465, from Fatio de Duillier to Huygens, *OCCH*, **9**: 167.

[5] According to Whiteside, 'in Newton's own lifetime only a handful of talented men . . . had, each in his own way, achieved a working knowledge of the *Principia*'s technical content' – and these few were Huygens, Leibniz, Varignon, de Moivre and Cotes – in Fellmann, *op. cit.* (note 1), 13.

title,[6] presents clearly stated propositions of increasing difficulty that are proved in what was recognized at the time as a rigorous way. It is this same style that was later used by Newton in the *Principia*. Although parallels can be traced with Galileo's *Due nuove scienze* (Leiden, 1638), there is a wide gap between the two works. Any high-school pupil can now understand Galileo's book. The same cannot be said of the *Horologium*. Like the *Principia*, Huygens's work includes several very difficult demonstrations – a challenge to mathematicians of the day.

It is also important to notice that Huygens's mechanics is very close to Newton's. It does not include all of Newton's ideas, of course. However, most of Newton's mechanics was already there – and all of Huygens's propositions are compatible with the *Principia* (a statement that does not apply, for example, to Galileo's works).

To drive this point home, let us compare Huygens with Descartes. There can be no doubt that Descartes was a good mathematician, but his natural philosophy was qualitative (except his geometrical optics) and his mechanics is weak and incompatible with that of Huygens and Newton. It is well known that there is a wide gap between Descartes's *Principia* and Newton's. On the other hand, Huygens and Newton are much closer. They belong to the same world. Newton would certainly have been proud to have written the *Horologium oscillatorium*.

This is the natural philosopher and mathematician that we propose as an expert assessor of Newton's gravitational theory. Very well: he seems to have the necessary competence. Is he sufficiently impartial?

One might think that the opposition between Newton's and Huygens's theories of light might predispose the Dutch scientist against the theory of gravitation – resulting in an attitude something like Leibniz's systematic opposition to Newton. That was not the case, however. Huygens praised Newton's optical researches highly. Besides, his travelling to England, in 1689, merely to meet the author of the *Principia*, is clear evidence that there was no personal prejudice of that kind. One may quote, for instance, Huygens's letter of late December 1688 to his brother Constantijn, who was then in England:[7]

> I think that the Royal Society is now on vacation. Nevertheless, you may have some chance of seeing Mr. Boyle and some other members. I would like to be in Oxford, only to meet Mr. Newton, whose beautiful inventions found in the work that was sent to me, I extremely admire. Maybe I shall send you a letter and you will find an easy way to deliver it to him.

Huygens's theory of gravity

The kind of evidence presented in the previous section allows us to accept that Huygens was a competent and unprejudiced judge of Newton's work. Let us now

[6] The full title is: 'Horologium oscillatorium sive de motu pendulorum ad horologia aptato *demonstrationes geometricae*' (my emphasis), (Paris, 1673). English translation in J. Yoder, *Unrolling Time: Christiaan Huygens and the Mathematization of Nature* (Cambridge, 1984).

[7] Letter 2529, of 30 December 1688, from Christiaan to Constantijn Huygens, *OCCH*, **9**: 304.

briefly study his own work on gravity, to compare his method with Newton's, as applied to the same subject.

While he was living in Paris, in 1669, Huygens presented to the French Academy of Sciences a short work called 'On the cause of gravity'.[8] This small essay was intended to provide a mechanical explanation of the force that draws heavy bodies towards the Earth. The main ideas were explicitly Cartesian. About 18 years later, Huygens was led to think about the influence of the Earth's rotation upon gravity. The stimulus for this new work was information from Richer, published in 1686 by Mariotte.[9] In 1672 Richer had found that a pendulum had to be shorter at Cayenne (compared to Paris) to beat seconds. It was only natural that Huygens, the great pendulum authority, should try to explain this phenomenon. He attributed it to the centrifugal force caused by the rotation of the Earth. This also led him to suggest the Earth was flattened and to investigate the relation between latitude and gravity.

After the publication of Newton's *Principia*, Huygens wrote an additional final section to his essay. It was printed under the name *Discours de la cause de la pesanteur*, in 1690, together with his *Traité de la lumiere*.[10] At the end of this work, Huygens compares his own theory with Newton's. He also presents an analysis of the motion of bodies in resisting media – a subject also treated by Newton. In this instance, his results completely confirmed those of the *Principia*.

Like Descartes, Huygens states that there is no void in the Universe and there is a fluid invisible matter moving around the Earth. This fluid is described by Huygens as consisting of small particles circulating around the Earth, in all directions, with a very large velocity. On account of this motion the fluid matter endeavours to recede from the centre. It produces an opposite thrust, towards the centre, upon bodies that are at rest inside the fluid. Notice that Huygens does not accept Descartes's vortices turning about an axis.[11]

Huygens assumes that the particles of this fluid matter are able to pass undisturbed through solid matter. They are even able to move through and inside the Earth – otherwise, there could be no gravity inside a well.[12]

In this theory, there is no assumption about the force or mutual influence between two bodies at a distance from one another; there is no reference to the motion of planets around the Sun; and no assumption about a dependence of force on distance from the Earth. For Huygens, the Earth's gravity is constant, both inside and outside its body.

In the part of the *Discours* written before the publication of Newton's *Principia*, Huygens assumes at the outset that the Earth is spherical, and computes the changes

[8] This first essay and a discussion of its context can be found in *OCCH*, **19**: 617–45. The later development of Huygens's ideas on gravity is documented in vol. 21, p. 377–426.

[9] See *OCCH*, **9**: 130–1. See also: C. Wolf, 'Mémoires sur le pendule', in: Société Française de Physique, *Collection de Mémoires relatifs à la Physique* (5 vols., Paris, 1884–91), **4**: B-13.

[10] C. Huygens, *Discours de la cause de la pesanteur* (Leiden, 1690). This work was reprinted in *OCCH*, **21**: 427–88. All references in the present article are to the pages of the original edition.

[11] *Ibid.*, 131–7. [12] *Ibid.*, 139.

in the length of a seconds pendulum as a function of latitude. The only cause affecting the motion of the pendulum is assumed to be the centrifugal force due to the rotation of the Earth. From these hypotheses, he computes that the seconds pendulum should be shorter at the equator than at the pole by one part in 289. He also shows that, due to the rotation of the Earth, the surface of the seas cannot be spherical and there must be an increase in the radius of the Earth from the pole to the equator – but he does not compute the quantity of this polar flattening.

It has already been pointed out[13] that Huygens's theory is very weak compared with Newton's. The latter's treatment of these same problems is very sophisticated. From the hypothesis of forces varying inversely as the square of the distance, Newton first computes the gravitational properties of a homogeneous ellipsoid.[14] Under the assumption that this applies to the Earth, he studies the equilibrium of columns of liquid from the surface to the centre of the Earth.[15] By an ingenious process he computes the difference between the equatorial and polar radii as equal to 3/689ths of the polar radius. He also computes the variation of gravity with latitude[16] and obtains a polar gravity greater than the equatorial gravity by one part in 230. Measurements made in the eighteenth century led to results close to Newton's predictions.[17]

After reading Newton's *Principia*, Huygens wrote a final section to his *Discours*. He uses Newton's method of considering the equilibrium of water columns and computes the figure of the Earth. However, he computes no change in gravity caused by this new shape: he still assumes that gravity is constant. Indeed, he even states that gravity is not produced by the Earth but is in fact itself responsible for the shape of the Earth. He computes a small difference between polar and equatorial radii – just one part in 578. He compares his own work with Newton's:

> I have supposed that gravity is the same both inside the Earth and at its surface; . . . Mr. Newton . . . makes use of a completely different assumption – I will not examine it here, because I do not agree with a Principle that he assumes in this and in other computations. This is: that all small parts that one may imagine in two or more different bodies attract or tend to mutual approach. I could not admit this, since I clearly saw that the cause of such an attraction cannot be explained by any principle of Mechanics or by the rules of motion. I am also not convinced of the necessity of the mutual attraction of whole bodies; for I have shown that, even if the Earth did not exist, the bodies would not cease to tend to a centre by the so-called gravity.[18]

[13] F. Mignard, 'The theory of the figures of the Earth according to Newton and Huygens', *Vistas Astr.*, 1987, **30**: 291–311.

[14] I. Newton, *Philosophiae naturalis principia mathematica*, (London, 1687; 2nd edn, Cambridge, 1713; 3rd edn, London, 1726), Book I, Proposition 91.

[15] *Ibid.*, Book III, Prop. 19.

[16] *Ibid.*, Book III, Prop. 20.

[17] For the empirical confrontation between Newton's and Huygens's theories, see: I. Todhunter, *A History of the Mathematical Theories of Attraction and the Figure of the Earth – from the time of Newton to that of Laplace* (2 vols., London, 1873; New York, 1962); R. de A. Martins, 'Huygens e a gravitação Newtoniana', *Cad. Hist. Fil. Ci.*, series 2, 1989, **1**: 151–84.

[18] Huygens, *op. cit.* (note 10), 159.

The contrast between Huygens and Newton

Part of the difference between Huygens's and Newton's views about gravitation (or gravity) is due to their conflicting methodologies. In the preface to his *Discours*, Huygens states his basic commitment to Cartesian principles:

> When Nature leads the so-called heavy bodies to the Earth, she acts by such secret and imperceptible ways that the senses have found nothing there, even though much attention and industry were employed. This constrained philosophers of past centuries to look for the cause of this admirable effect but in the bodies themselves and to ascribe it to some inner inherent quality that makes them to move downwards and to the centre of the Earth, or to an appetite of the parts to unite to the whole. That is not to expose the causes, but to suppose obscure and unknown principles . . . .[19]

After discussing some attempts to explain gravity, Huygens states:

> Mr. Descartes has recognized, better than those that preceded him, that nothing will be ever understood in Physics except what can be made to depend on principles that do not exceed the reach of our spirit, such as those that depend on bodies, deprived of qualities, and their motions.[20]

Any explanation of gravity by means of attraction was unacceptable to Huygens. Before reading Newton's book, he writes a letter to Fatio de Duillier in which we find the statement: 'I want to see Newton's book. I am glad that he is not a Cartesian, as long as he does not present to us suppositions such as that of attraction'.[21] After reading the *Principia*, Huygens gave a clear account of his opinion of Newton's ideas. He accepted that the following features of the gravitational theory had a solid basis.[22]

(a) The existence of forces between the Sun and the planets and among the planets themselves. Huygens supposed that he could explain these forces by an extension of his model for the Earth's gravity.

(b) The decrease of such forces in accordance with the inverse square law – a feature necessary to explain elliptical orbits and the relation between the Moon's orbital acceleration and terrestrial gravity.

(c) The non-existence of Descartes's vortices: this accounted for the planets moving freely, and was necessary to explain why the eccentricities and inclinations of the orbits were constant. This premise was also required for understanding the motion of comets.

Notice that all these features imply a considerable departure from earlier Cartesian ideas. So Huygens was clearly not a blind follower of Descartes, as one might perhaps have expected.

It might seem that rejecting Cartesian vortices and accepting the inverse square law of gravitation entails accepting the whole of Newton's gravitational theory. This is not

[19] *Ibid.*, 125. [20] *Ibid.*, 126.
[21] Letter 2746, of 1692, in *OCCH*, **10**: 354. [22] Huygens, *op. cit.* (note 10), 159–60.

so. There were two main points on which Huygens was in disagreement with Newton. The first one was that Huygens did not accept attractive forces acting at a distance. He conceived, as Descartes did, that forces must be transmitted by something material – some kind of aether. It was, in fact, possible to build a mechanical model for Huygens's aether compatible with the inverse square law. Leibniz and Fatio de Duillier were later to do so.[23]

The second difficulty was that Huygens did not accept that Newton's law of gravitation could be applied to the smallest parts of bodies. This was a basic premise used by Newton to compute the gravity of the Earth (both in the spherical and in the oblate spheroidal case) and its dependence upon distance. Accordingly, Huygens states that the inverse square law does not apply near the Earth and that gravity is constant.

There was, moreover, a third point that Huygens did not accept: Newton builds a huge edifice of mathematical propositions on the foundation of his theory. In doing so, he goes far beyond the domain where there was a reasonable basis for the gravitational theory – that is, in the field of planetary motion. In several of his letters, Huygens shows his dissatisfaction with this feature of Newton's work. Perhaps he thought that Newton was building on sand.

One may easily perceive some general methodological questions behind each of the points of disagreement. Such of these general questions as are not specifically related to any research field, but might be applied to any subject, can be put into the following form:

(1) Is it acceptable to construct a theory that describes effects derived from a cause that cannot be explained?
(2) Is it valid to generalize some results, obtained in a restricted domain, to all bodies, to all distances and to all circumstances?
(3) Is it valid to devote a large amount of work to the derivation of consequences of a theory whose basis is open to question, instead of applying oneself to elucidating the actual basis of the theory?

It seems as though Huygens were warning Newton: you should not do this, you are not following the correct method.

Why should Huygens and Newton be in disagreement about methodology?[24] Was it because Huygens was a Cartesian while Newton disliked the work of Descartes? It is not so simple as that. As we have seen, Huygens was not a blind follower of Descartes. Their methods were not exactly alike. And, as we have already seen, Huygens was certainly willing to change his ideas and accept several of the non-Cartesian features of Newton's work. Huygens was, indeed, a Cartesian – in some respects. However, this is

[23] Leibniz's work was published in 1689: 'Tentamen de motuum coelestium causis', in: G. W. Leibniz, *Mathematische Schriften*, ed. by C. I. Gerhard (Hildesheim, 1971), **6**: 144. For Fatio's work, see: Bernard Gagnebin, 'De la cause de la pesanteur', *Notes Rec. Roy. Soc. Lond.*, 1949, **6**: 105–60. A history of mechanical explanations of gravitation is presented in W. B. Taylor, 'Kinetic theories of gravitation', *Ann. Rep. Smithsonian Inst.*, 1876: 205–82.

[24] Any general discussion about what is *valid* in science – or, in general terms, any discussion about intrinsic scientific values – is a methodological debate.

not the cause of his attitude – it might be considered a symptom or description of his view, but not its cause. The question that needs to be asked is: Why did Huygens remain committed to the Cartesian method, instead of changing his mind? This question brings us to the general problem of the meaning of a 'scientific revolution'.

The Scientific Revolution of the seventeenth century

The process usually called *the* Scientific Revolution, that had its culmination in the period from Galileo to Newton, was both a revolution of concepts and of methods. The methodological revolution, like the conceptual one, did not stop with Galileo. Huygens, Newton and others developed new ideas and new research methods. The acceptance of both proved to be difficult – and this was particularly true as new methodological types[25] were successively introduced. Even those who took an active part in earlier stages of the revolution found it hard to accept new methodological rules: it proved easier to accept that some idea or theory was wrong than to accept that one could do research in new ways. The causes of this special difficulty were twofold: first, the creators of the Scientific Revolution regarded themselves as the founders of the *true* scientific method (as against the Aristotelian method); second, the scientific method was regarded as *restrictive*, prescribing some ways of doing research as correct and ruling out other ways as illegitimate. This difficulty in regard to methodology seems to be the reason for Huygens's negative reaction to Newton's theory.

What does a methodological revolution mean? It is the introduction of new ways of doing research. Those new ways are not incompatible with older ones – except if the former method is regarded as restrictive. Take as an instance the introduction of mathematical methods in some particular field. Quantitative reasoning is not incompatible with qualitative reasoning – except if one states that some particular field of study is, in principle, not subject to quantitative laws. Otherwise, the quantitative method can be introduced without any problems of compatibility. The same can be said about the introduction of other methodological types – different ways of doing research – such as the introduction of observation into a previously purely speculative field, or the introduction of experimentation where before only observation was used, or the building of special instruments for measurement or observation, and so on.

Galileo's work provides a clear example of the introduction of new methodological types in the early seventeenth century. However, one should not lay such stress upon his contribution to the new scientific method as to imply that the methodological revolution was also completed by him. There may well be as large a distance between Galileo and Newton as that usually recognized as separating the Aristotelians from Galileo.

Let us recall one instance: Galileo tried to prove, in the *Dialogo* (1632), that the

[25] Any description of a scientific method can be decomposed into statements about the intrinsic scientific value of 'elementary' procedures. The 'elementary' valuable scientific procedures are here called 'methodological types'.

rotation of the Earth could have no detectable influence upon the fall of a cannon ball from a tower. He also stated that the rotation of the Earth could not cause motion among bodies that are on its surface, but he did not compute numerical values for the effects of rotation of the Earth. Newton was also to agree that such effects were small. However, instead of dismissing them as negligible, he computed their size, and afterwards tried to observe their consequences.[26] What was he doing? To describe his activity in methodological terms, he was accepting the existence of small influences, computing their effects, predicting new phenomena and looking for them. He did this systematically in his work: he considered the effects of the resistance of air on the motion of the pendulum; he looked for small corrections in planetary motions due to forces between planets and due to the small motion of the Sun. Who else paid so much attention to secondary, perturbing causes, before Newton? This is just one of the new methodological insights that can be ascribed to Newton. Was he aware that this was a new and important step? We do not know. He does not theorize about it. He simply does it.

As to the methodological points of disagreement between Huygens and Newton, it is easy to understand Huygens's cautious, conservative position. Newton's step was a bold one. The lack of an explanation of gravitation did not prevent him proceeding with his work; he made what could be seen at the time as wild generalizations from laws arrived at by induction; and he devoted the greatest mathematical care to the development of detailed consequences of those laws. Huygens had good reason to inquire: How does Newton dare to introduce such new attitudes into Natural Philosophy?

Newton himself perceived (but not soon enough) that he had to present a defence of his method. This seems to be the reason why, from the second edition of the *Principia* onwards, he stated his methodological rules separately and explicitly and introduced the third and fourth of his *Regulae philosophandi*.

Newton's methodological rules

In the second and third editions of the *Principia*, the third book ('The System of the World') begins with a section called 'Regulae philosophandi'. It is a very short section (three pages). It presents four methodological rules.[27] The first and second were already present in the first edition of the *Principia* but were placed among the physical hypotheses of the System of the World. Between the first and the second editions, Newton had decided it was necessary to draw attention to his methodological rules and to introduce the third one (the fourth only appears in the third edition). It seems

[26] See, for instance: Angus Armitage, 'The deviation of falling bodies', *Ann. Sci.*, 1947, 5: 342–51.

[27] See the various versions of the 'Regulae' in the *variorum* edition of Newton's *Principia*: A. Koyré, I. B. Cohen and A. Whitman, *Isaac Newton's Philosophiae Naturalis Principia Mathematica, third edition (1726) with variant readings* (2 vols., Cambridge, 1972), 2: 550–6.

likely that this change is a response to the reaction of Huygens and other natural philosophers to Newton's work. It is easy to show that the content of the new rules does indeed provide an answer to Huygens.

The first and second rules are very short and are followed by just three lines of commentary. They state:[28]

> *Rule 1*: We are to admit no more causes of natural things than such as are both true and sufficient to explain their appearances.
>
> *Rule 2*: Therefore to the same natural effects we must, as far as possible, assign the same causes.

These rules present a principle of simplicity in natural philosophy, and are methodological counterparts to the old metaphysical assumptions about the simplicity of Nature. They say nothing about induction. Now, the third rule, which first appears in the second edition of the *Principia*, and its latter complement – the fourth rule – can be described as principles for inductive argument:

> *Rule 3*: The qualities of bodies, which admit neither intensification nor remission of degrees, and which are found to belong to all bodies within the reach of our experiments, are to be esteemed the universal qualities of all bodies whatsoever.
>
> *Rule 4*: In experimental philosophy we are to look upon propositions inferred by general induction from phenomena as accurately or very nearly true, notwithstanding any contrary hypotheses that may be imagined, till such time as other phenomena occur, by which they may either be made more accurate, or liable to exceptions.

According to these rules, if experience has shown that gravitation applies to all bodies within our reach, we may accept that gravitation applies to all bodies in the Universe. And this assumption can be made notwithstanding any hypotheses about possible restrictions to such induction – only experiment can undermine experiment. If one accepts these rules, one cannot reject Newton's extrapolations and his tedious computations of so many consequences of the law of universal gravitation. If Huygens accepted that the planets, the Earth and the Sun do attract one another according to the inverse square law, why did he deny that all pieces of matter have the same property? Evidently, because he does not agree with the third and fourth rules.

These methodological rules, in the form in which they are stated and used by Newton, are indeed new. They establish that induction is independent of causal explanation, independent of theory and of necessary knowledge. They afford the natural philosopher a new freedom to explore the consequences of his assumptions, if these are based upon induction. Nowadays it seems natural that a physicist should be allowed to work for years and to publish many papers upon doubtful (or even wild) premises. It was not so in the seventeenth century. Natural philosophers were looking for the truth, and that truth had to be based on acceptable principles from its very beginning.

[28] We use here Motte's translation of the *Principia*.

Conclusion

One cannot conclude that Newton's proceeding was completely unjustified; but there were methodological problems in his work, and Huygens had good reason to criticize him: Newton was doing something new, deviating from the old standards and introducing 'dangerous' procedures.

We may note that Newton was not doing something incompatible with the old method. He would also have agreed that an explanation of gravitation was desirable.[29] However, he would not have agreed that all work should be suspended until this problem was solved. He would also have agreed that tests of laws obtained by induction are desirable, but not that research should stop and wait until no doubts remained concerning the results of induction. If earlier methodology is regarded as restrictive, Newton was indeed transgressing the rules; but if methodological rules are regarded as statements of non-prohibitive *desiderata*, Newton was not transgressing the rules: he was just adding new *desiderata* to the repertoire of methodological types.[30]

Since Huygens rejected Newton's theory, it seems that in his view the methodological rules were restrictive. This, after all, is the commonest view of scientific method, even now. According to this interpretation, Huygens criticized Newton because he assumed a restrictive interpretation of scientific method and because Newton's work represented a new step in the continuing methodological revolution of the seventeenth century.

[29] As is well known, Newton entertained several different explanations of gravitation at different times. At some time, he accepted Fatio de Duillier's theory. See the 'Draft addition to the Principia' (University Library Cambridge, Add. MS. 4005, fols. 28–9), in: A. Rupert Hall and Marie Boas Hall, *Unpublished Scientific Papers of Isaac Newton* (Cambridge, 1962), 313.

[30] For a systematic study of methodological *desiderata*, see: R. de A. Martins, *Sobre o Papel dos Desiderata na Ciência* (Campinas, 1987).

14

The reception of Newton's *Opticks* in Italy

PAOLO CASINI

Background and personal relations

The contributions of Della Porta to the theory of vision and of De Dominis to dioptrics were familiar to Newton. He owned a copy of the *Magia naturalis* in the Dutch edition of 1651, and the treatise *De radiis visus*, published in Venice in 1611.[1] He knew about Grimaldi's experiments on diffraction only through Honoré Fabri's critical discussion in his *Dialogi physici*, as Rupert Hall has remarked in a recent paper.[2] But Newton might also have read a chapter devoted to Grimaldi's optical–astronomical research in the *Almagestum novum* by Riccioli, Grimaldi's pupil.[3] Moreover, a *Miscellanea italica physico-mathematica* (Bologna, 1692), also listed by Harrison as among Newton's books, contains a dissertation by Giandomenico Cassini about celestial and atmospheric refraction, as well as several observations on comets by Montanari, Cellius and Ponthaeus.[4] This served as a source-book for the theory of comets in the second edition of the *Principia mathematica*. So, directly or indirectly, Newton was well aware of Italian research prior to his own early work in optics.

On the other hand, no mention of his optical papers of 1672 and 1675 has so far been found in the Italian milieu. The Roman *Giornale de' Letterati*, by Francesco Nazari, reviewed the *Philosophical Transactions* frequently; but a strange gap eclipses Newton's youthful controversies.[5] However, towards the end of the century the

[1] See J. Harrison, *The Library of Isaac Newton* (Cambridge 1987), nos. 1340, p. 220; 535, p. 135.

[2] A. R. Hall, 'Beyond the fringe: diffraction as seen by Grimaldi, Fabri, Hooke and Newton', *Notes Rec. Roy. Soc. Lond.*, 1990, **44**: 13–23.

[3] Giovanni Battista Riccioli's *Almagestum novum* (Bologna, 1651) is also listed by Harrison, *op. cit.* (note 1) no. 1400, p. 227.

[4] *Miscellanea Italica physico-mathematica*, colligit Gaudentius Robertus Carm. Cong., (Bologna, 1692) (see Harrison, *op. cit.* (note 1), no. 1408, p. 228). This collection contains essays on various physico-mechanical questions by Domenico Guglielmini and Evangelista Torricelli; my reference here concerns some optical and astronomical tracts: J. D. Cassini, 'De Solaribus Hypothesibus et Refractionibus Epistulae tres, Eiusdem Motus Cometae anni 1664' (commenting upon the writings of Riccioli and Grimaldi on related matters); G. Montanari, 'Cometes Bononiae Observatae anno 1664 et 1665'; G. Montanari, J. D. Cassini, M. A. Cellii, J. D. Ponthaei *et al.*, 'De Cometis ann. 1664, 1665. 1680, 1861, 1682 Epistolae'.

Newton utilized these observations of comets in *Principia mathematica*, Lib. III, Prop. XLI, Probl. XXI (3rd edn, London, 1726).

[5] J. M. Gardair, *Le 'Giornale de' Letterati' de Rome, 1668–1681* (Florence, 1984); 55 and *passim*.

well-established Italian tradition of experimental optics was in decay, perhaps as a result of the Jesuits' extremely conservative attitude and of their firm control on college education. The French Jesuit Honoré Fabri served on the board of the Inquisition in Rome from 1646 to 1688, and in 1669 he published his anti-Cartesian, backward-looking theory of light and colours.[6]

Newton, however, was assiduous in sending presentation copies of some of his books to eminent Italian mathematicians and natural philosophers, as we can deduce from his personal relationships with Guido Grandi, Angelo Bianchini, and Antonio Conti.[7] The English *Opticks* of 1704 reached Florence just after publication, as the author's gift to Grandi.[8] Two years later, a group of virtuosi in Rome enjoyed the Latin *Optice*. Their warm praises reached Newton through a letter from a diplomat, his namesake, reporting in particular the enthusiasm of a 'Gallianus', a remarkable man whom the editor of volume four of Newton's *Correspondence* found 'difficult to identify'.[9]

Recent research has uncovered evidence that these early Roman Newtonians successfully repeated the basic experiments of the *Optice*. Two unpublished papers by Celestino Galiani (Fig. 14.1), dated 1707, contain extensive commentaries on experiments 1 to 7 of Book I.[10] Some years later Galiani, in a letter to Grandi, said: 'I performed those experiments with Monsignor Bianchini and found them true'.[11] For his own part, Francesco Bianchini confirmed personally to Newton that his experimental results had been replicated exactly ('eius experimenta a me tentata respondisse ad amussim suis observationibus').[12] This is a short passage from Bianchini's *Iter in Britanniam*, a manuscript document that tells of three visits he paid to Newton in January 1713, his attendance at Hauksbee's experiments at the Royal Society, his acquaintance with John Keill and John Flamsteed, and the friendly welcome he received in Oxford. Monsignor Bianchini, a distinguished astronomer and scholar, was probably charged by the Roman court with some diplomatic mission. He exchanged various books with Newton; among them, one copy of the Latin *Optice* was expressly intended for the Vatican Library.[13]

6 'Fabri, Honoré', *DSB*.

7 See 'Newton in Italia, 1700–1740', in my *Newton e la coscienza europea* (Bologna, 1983), 173–227. V. Ferrone, *Scienza natura religione. Mondo newtoniano e cultura italiana nel primo settecento* (Naples, 1982), 27ff.

8 See Newton's letter to Guido Grandi of 26 May 1704, in H. W. Turnbull, J. F. Scott, A. R. Hall, L. Tilling (eds), *The Correspondence of Isaac Newton* (7 vols., Cambridge, 1959–1977), 7: 434.

9 See Henry Newton's letter to Isaac Newton of December 1707; *ibid.*, 4: 506.

10 The MSS papers, entitled 'Animadversiones nonnullae circa opticem Isaaci Newtoni', and 'Differenza tra le scoperte di Newton e l'ipotesi cartesiana', are in Naples, Società Napoletana di Storia Patria, XXX.D.5 and XXX.D.2; for some quotations see Ferrone, *op. cit.* (note 7), 31–35.

11 Galiani's letter to Grandi of 21 April 1714: 'A proposito del suddetto libro del signor Newton de' Refractionibus etc, che io lessi fin da cinque o sei anni sono, e feci con Monsigr Bianchini alcune di quelle sperienze che ritrovammo vere, vorrei che V.P. R.ma favorisse scrivermi se sa che altri le abbiano esaminate e che giudizio ne han fatto'. C. Galiani and G. Grandi, *Carteggio 1714–1729*, F. Palladino and L. Simonutti eds, (Florence, 1989), 31. This book contains up-to-date biobibliographical references concerning the two men.

12 F. Bianchini, 'Iter in Britanniam', MS in the Biblioteca Vallicelliana, Rome (T.46, p 2a, f. 21–22); the passage has been partially published, with a commentary, by G. Costa, 'Documenti per una storia dei rapporti anglo-romani', *Saggi sul Settecento*, (Naples, 1968), 437. See also 'Bianchini, Francesco', *DBI* and Ferrone *op. cit.* (note 7), 59ff.

13 The copy is still extant there (Racc. Gen. Scienze, IV. 29); on the left of the title page there is a manuscript annotation, probably by Bianchini: 'Librum Optices / ab Auctore Viro Clarissimo / sibi contraditum Londini, / die

Leibnizians and Newtonians

Through his relations with Bianchini – and two years later with the famous and less trustworthy Abbé Antonio Conti – Newton probably hoped to diminish Leibniz's prestige, which was at its height in Italy during the calculus dispute. In Padua and Venice, after 1710, Jakob Hermann and Nicolas Bernoulli were the promoters of a fierce campaign against the method of fluxions. Accordingly, even the well-tested experiments with prisms became a matter of controversy in the struggle between the two parties. After Bianchini's visit to England, the pro-Newtonian circle of Galiani was to repeat a series of optical experiments in Rome under the auspices of an ecclesiastical patron, Cardinal Gualtieri. His Accademia Fisico-Matematica was very active there about 1714, and decided to devote its first meetings to the experimental study of light and colours, thanks also to its 'prodigious collection of instruments and apparatus'.[14] Galiani reported what Newton thought about the French reactions to his *Opticks*,[15] and, as a convinced supporter of Newton, argued against Edme Mariotte:

> Mariotte's objections against Newton's experiments are untrue; he says that if, after separation through the prism, one of the coloured rays – for instance the red – is refracted through another prism, this too will divide into rays of different colours. I say this is not true. The experiment has been made several times: red always remains red, it never divides into other rays.[16]

Leibniz himself suggested that his followers exploit the case. In a short anonymous review published in the *Acta Eruditorum* of 1713, he challenged Newton 'to be so gracious as to express his opinion on the objection the most ingenious Mariotte raises against him in his *Treatise of Colours*'.[17] Henry Guerlac and Rupert Hall have shown in detail how in France, thanks to Pierre Coste's translation (1722), the success of the *Opticks* finally overcame the established views inspired by Mariotte.[18] But in the same year 'another champion arose where Mariotte fell'.[19] Giovanni Rizzetti, a Paduan nobleman, follower of Fabri and Mariotte, and certainly an *agent provocateur* of the Leibnizians, rashly submitted to the Royal Society a general critique of Newton's

VII id. Februarii / MDCCXIII / ut in Bibliothecam Vaticanam / preservetur / ibidem custodiendum transmittit / Franciscus Blanchinus / III Kal. Octobris ejusdem anni'. Two books by Bianchini belonging to Newton are listed by Harrison, *op. cit.* (note 1), no. 185, p. 101.

[14] 'Il soggetto delle prime conferenze sarà la luce, che cosa elle sia, e delle sue varie proprietà. E come il Sig.r cardinale sopra tutto vuole che si facciano delle sperienze, per lo quale effetto ha egli fatta una prodigiosa raccolta di strumenti e ordegni ed è pronto a spendere tutto quello che mai vi vorrà, io proposi che intorno al lume in primo luogo si facessero quelle riferite dal Sig.r Newton nel suo libro *de Reflexionibus Refractionibus, etc.*' .Galiani's letter to Grandi of 21 April 1714, in Galiani and Grandi, *op. cit.* (note 11), 30.

[15] 'A Monsigr Bianchini, mentre era in Londra, disse il Signor Newton, che quei dell'Accademia delle Scienze di Parigi dicevano di averle rifatte e di non averle ritrovate vere'; *ibid.*, 31.

[16] Galiani's letter to Grandi of 9 February 1715, *ibid.*, 51.

[17] *Acta Eruditorum*, Oct. 1713, pp. 444ff. This review, as well those of Jan. 1705 and Feb. 1706, are mentioned by H. Guerlac, *Newton on the Continent* (Ithaca and London, 1981), 103, 116; Guerlac completely ignores the Italian Newtonians and their attitude toward the *Opticks*.

[18] See, besides Guerlac, *op. cit.* (note 17), A. R. Hall, 'Newton in France: a new view', *Hist. Sci.*, 1975, **13**: 233–50.

[19] J. L. Heilbron, *Physics at the Royal Society during Newton's Presidency* (Los Angeles, 1983), 92.

Figure 14.1. Celestino Galiani. Caricature, dated 27 November 1729.

theory of light and colours.[20] This late attack aroused Newton's anger and provoked elaborate refutations simultaneously both in London and in Bologna, where Rizzetti tried to obtain some support.

Rizzetti boasted of having performed several of Newton's experiments and found them all faulty or not conclusive. He rejected the dispersion theory in its entirety, and tried to explain prismatic colours by a modification hypothesis, based on aberration effects in white light passing through the crystalline humour and the lenses. In particular, he discussed Experiment 2 from the first book of *Opticks*, describing the apparently different positions of two portions of a coloured card, separated by a black thread, as seen through a prism. The blue half seemed much nearer to the lens than the red one, owing to the greater refrangibility of blue rays.

Newton and Desaguliers refute Rizzetti

Newton took the task of refuting his critics seriously. He wrote three drafts of an answer, quoting verbatim Rizzetti's misunderstanding of his principles. He criticized Rizzetti's experimental tests and theoretical conclusions in some detail, and recalled Mariotte's defeat. Newton remarked that his opponent's failure was due to his incompetence: 'If it did not succeed with him it was because he did not know how to try it'. And he continues:

> He should have told the world that Newton founded his theory of light & colours upon the experiment wch for its demonstrative evidence he calls experimentum crucis. This experiment he could not deny because it hath been tried again & again with full success both in France and in Italy as well as in England: & therefore he passes it by in silence & denies the conclusion drawn from it, & contrary to the Rules of logic, disputes against the conclusion without taking note of the premisses.[21]

The experiment in question was performed again before the Royal Society by Desaguliers, who used a dark lantern to illuminate a two-coloured card with black threads wrapped around it. Focusing successively on the red and the blue portions of the card at slightly different distances, their images appeared perfectly clear on a screen; the separating thread of black silk was visible on the red half, but imperceptible on the blue.[22] Rizzetti refused to surrender, and formulated twelve more objections. Desaguliers' answer was firm:

> As Signor Rizzetti had put the issue of the dispute upon the success of an experiment, which after repeated trials had succeeded contrary to his

[20] See ref. 23 below. Giovanni Rizzetti (1675–1751) is a rather obscure amateurish figure, belonging to the circle of the mathematician Jacopo Riccati of Castelfranco Veneto. See on him: G. Moschini, *Della Letteratura Veneziana* (Venice, 1806), **2**: 185. For a broad sketch of the milieu see M. L. Soppelsa, *Leibniz e Newton in Italia. Il dibattito padovano, 1678–1750*, (Trieste, 1989).

[21] Cambridge University Library, Newton MS. Add. 3970, f. 609r. I warmly thank Professor Maurizio Mamiani for his generous help in procuring a copy of this unpublished MS.

[22] J. T. Desaguliers, 'An account of an Optical Experiment made before the Royal Society, on Thursday, Dec. 6th, and repeated on the 13th, 1722'; *Phil. Trans.*, 1722, **32**: 206–8.

> opinion, he ought to acknowledge his mistake; and then I should willingly repeat all the other experiments which he had called in question, and endeavour to remove his other difficulties.[23]

Desaguliers really was to do so a few years later, but in a less friendly way. For the Paduan nobleman now turned for support to the Leibnizian headquarters in Leipzig. His paper *De Systemate Opticae Newtonianae et de Aberratione in Humore Chrystallino* – the same paper he had sent to London – appeared in the *Acta Eruditorum* in 1724, but only to meet with the severe criticism of the German mathematician Friedrich Richter.[24] An even more challenging move of Rizzetti's was the publication in 1727 of his treatise *De Luminis Affectionibus* (Fig. 14.2) containing a huge mass of experiments and a stubborn diatribe against Newton.[25] Desaguliers took the opportunity to dissipate the sense of outrage and to purify the memory of the great man, by performing a series of variants of the card experiment in the presence of Hans Sloane, the new President, and of many members of the Royal Society.[26] Among them there were four Italian Abbés who subscribed to Desaguliers' account of his performance. This is considered by Priestley, Montucla and other eighteenth-century historians as the definitive test of the different refrangibility of rays of light.

Optics in Bologna

What happened in Italy during these same years is less well known. Open acceptance of the law of gravitation was hindered by the fact that the Catholic Church still banned belief in the motion of the Earth. The optical experiments had been performed successfully by Catholic scholars, but in small private circles. They helped to introduce what might be called a 'soft' approach to the Newtonian system of the world, while public refutation of the theory of colours might also signify a more or less implicit opposition to the officially condemned hypothesis *de terra mota*. For instance, the didactic poem *De iride*, written by the Jesuit teacher Carlo Noceti in 1729, is a sign of the changing, very cautiously pro-Newtonian, attitude of the Collegio Romano. But this was to appear in print only in 1747, with a full commentary by Father Boscovich.[27]

We find a similar attitude in Bologna, a town situated in the Papal States, and the centre of a Galilean tradition since the mid seventeenth century, where authors were

[23] From Desaguliers' letter of 1722, quoted by himself in the further paper of 1728 (note 26); 597.

[24] Friedrich Richter, 'De Systemate Opticae Newtonianae', *Acta Eruditorum*, suppl. VIII, 1724; 227–46, F. Richter's defence of Newton appears *ibid.*, 313–19, 384–8, 394–8. On this episode see J. E. Montucla, *Histoire des Mathématiques*, (2nd edn, 4 vols., Paris, 1799–1802), 3: 558ff.

[25] J. Rizzetti, *De Luminis Affectionibus Specimen Physico-Mathematicum* (Treviso and Venice, 1727).

[26] J. T. Desaguliers, 'Optical Experiments made in the Beginning of August 1728, before the President and several Members of the Royal Society, and other Members of several Nations, upon the Occasion of Signor Rizzetti's Opticks, with an Account of the said Book', *Phil. Trans.*, 1728, **35**: 596–630.

[27] C. Noceti, *De iride et aurora boreali carmina ... cum notis R. J. Boscovich* (Rome, 1747); for a commentary, see 'Ottica, Astronomia, Relatività. Boscovich a Roma, 1738–1748', in Casini, *op. cit.* (note 7), 143–71.

DE LUMINIS
AFFECTIONIBUS
SPECIMEN PHYSICO MATHEMATICUM
IOHANNIS RIZZETTI
IN DUOS LIBROS DIVISUM
AC
EMINENTISSIMO PRINCIPI
SANCTÆ ROMANÆ ECCLESIÆ
CARDINALI DE POLIGNAC &c.
DICATUM

MDCCXXVII.
TARVISII ex TYPIS EUSEBII BERGAMI
VENETIIS Apud ALOISIUM PAVINUM.
SUPERIORUM PERMISSU.

Figure 14.2. Title page of Rizzetti, *De luminis affectionibus* (Treviso and Venice, 1727).

accustomed to a self-imposed censorship. As early as 1720, the Secretary of the Istituto delle Scienze, Francesco Zanotti, taught the theory of attraction and promoted some optical experiments. His pupil, the *enfant prodige* Francesco Algarotti, concerned himself with prisms as early as 1726. The history of his failures and successes is closely linked to manoeuvres made by Rizzetti, who was a Corresponding Member of the Istituto. Algarotti's experiments are described in the *Commentaria*.[28] The dates are not clearly given, but some details are worthy of note. At first the experiment with coloured cards and the *experimentum crucis* produced no clear results. The single red ray coming from the second prism showed some additional shades of blue or white. This seemed to support Mariotte's thesis. After several trials, Algarotti realized that his prisms were defective. When he tried again with a couple of first-quality English prisms, he and his fellow-academicians – including Giambattista Morgagni – were astonished to see the *experimentum crucis* exactly replicated, 'not once, but always, and a hundred times thereafter' ('neque semel sed semper, cum idem plus centies postea sumptum sit').[29]

Like Desaguliers, in the same year 1728, Algarotti worked to discredit Rizzetti and his treatise. The latter's unpublished correspondence shows that his complaints and obscure threats embarrassed the members of the Istituto and provoked a delay in the publication of the experimental results obtained by Algarotti. This was also the starting point of Algarotti's appeal to public opinion. He first conceived his *Dialoghi sull'Ottica Newtoniana* as an answer to Rizzetti. His Fontenelle-like, gracious and frivolous literary style was intended to avoid a growing storm. For, as a result of the political situation during the War of the Austrian Succession, the Catholic Church was becoming more and more suspicious about the influence of English ideas and particularly of the increasing appeal of Freemasonry.

Inquisition and the advent of a liberal Pope

The Inquisition was extremely active. In 1734 Locke's *Essay concerning Human Understanding* was put on the Index. The great historian Ludovico Antonio Muratori, one of the leading Catholic reformers, humbly apologized for having absorbed Locke's 'subtle poison'.[30] Antonio Conti and Celestino Galiani were simply denounced; the Neapolitan historian Pietro Giannone was seized and imprisoned for the rest of his days.[31] The climax was reached in 1738 with the Bull of Clement XIII against Free-

[28] For an account, see *De Bononiensi Scientiarum et Artium Instituto atque Academiae Commentarii* (Bologna, 1731), 1: 197–203. A MS correspondence between Rizzetti and the Secretary of the Accademia, Francesco Zanotti, concerning the experiments with prisms (conserved in the Archivio of the Accademia) may be useful for obtaining a clearer understanding of their respective attitudes; some letters of Algarotti, *Opere* (17 vols., Venice, 1784), **11**: 8, 31, 37, give further details.

[29] *De Bononiensi Instituto . . . Commentarii op. cit.* (note 28), 203.

[30] L. A. Muratori, *Opere*, G. Falco and F. Forti eds. (2 vols., Milan and Naples, 1964) **2**: 819ff.

[31] See: N. Badaloni, *Antonio Conti. Un abate libero pensatore tra Newton e Voltaire* (Milan, 1968), 190; and Pietro Giannone, *Vita scritta da lui medesimo* (Milan, 1960), 324.

masonry, and the trial and death of a Freemason hero, the Florentine poet Tommaso Crudeli.[32]

Algarotti wrote his dialogues on Newtonian optics in Rome and at Cirey, in Voltaire's own house.[33] His personal relations with Desaguliers, the Grand Master of the London Lodge, and with a group of English Freemasons in Florence, were no mystery for the Holy Office. Notwithstanding their literary smoothness, the *Dialoghi sopra la luce i colori* (Fig. 14.3) were considered to be the result of a Masonic plot, and were therefore put on the Index in May 1738, with the formula *donec corrigatur* ('until it be corrected').[34] It is not easy to decide, at present, whether Rizzetti's manoeuvres influenced the decision taken by the Holy Office, a tribunal which still presented a serious threat.

The complicated story that I have tried to summarize had a happy ending. Benedict XIV, the new Pope who was elected in 1740, was the enlightened Prospero Lambertini, the former Archbishop of Bologna and a lifelong friend of Galiani, Bianchini, and other Newtonians.[35] When he was a young dignitary of the Roman Court, he had been present at their optical experiments. When he was elected Pope, a sudden flowering of Newtonian studies took place, in the 1740s, in three distinct ecclesiastical centres. The French *Pères Minimes* Thomas Le Seur and Francisque Jacquier prepared their edition of Newton's *Principia Mathematica* in the convent of Trinità dei Monti.[36] The Jesuits of the Collegio Romano freely taught the theory of the rainbow and of colours, while the *genius loci*, Roger Boscovich, published his pro-Newtonian *Dialogi sull'Aurora Boreale* and selected the physical problem *de lumine* as the starting point of a new dynamical theory of matter.[37] The University of Rome appointed a Newtonian professor of physics and astronomy.[38]

Meanwhile, in Geneva an Italian refugee from Tuscany, Giovanni Salvemini da Castiglion Fiorentino (better known as Johannes Castilloneus) published the most comprehensive continental collection of Newton's optical, mathematical and historical

[32] See my 'The Crudeli affair: inquisition and reason of state', *Eighteenth-century studies Presented to Arthur M. Wilson*, P. Gay ed. (Hanover, New Hampshire, 1975); 133–52. For the 1738 Bull, see *Bullarium Romanum*, 24 vols. (Augusta Taurinorum, 1857–1872), **23**: 366ff., and L. von Pastor, *Storia dei Papi*, Ital. edn (Rome, 1933), **15**: 722.

[33] Voltaire, *Correspondence, definitive*, ed. by T. Besterman (107 vols Geneva and Oxford 1968–1977); see the letters to Thieriot of 3 November 1735 (no. 935), 16 and 18 March 1736 (nos. 1035, 1037), **51**: 3, 241, 389, 393. For a general assessment: A. R. Hall, 'La matematica, Newton e la letteratura', *Scienza e Letteratura nella Cultura Italiana del Settecento*, R. Cremante and W. Tega eds (Bologna, 1984), 29–46.

[34] M. De Zan, 'La messa al Indice del "Newtonianismo per le dame" di Francesco Algarotti', in Cremante and Tega, *op. cit.* (note 33), 133–47. The first edition of Algarotti's dialogues appeared in Venice, 1737. The work was soon translated into French, German and English; the only modern edition is included in the collection *La Letteratura Italiana. Storia e Testi*, **46**: 2, (Milan and Naples, 1969); reprinted in a single volume as: F. Algarotti, *Dialoghi sopra la luce e i colori* (Turin, 1977). A critical edition is badly needed for a collation of the various changes made by the author in the several editions of the work.

[35] 'Benedetto XIV', *DBI*, see 'Algarotti, Francesco', *DBI*; see also Pastor *op. cit.* (note 32), **16**: 1, 240ff.

[36] The so-called 'Jesuit edition', which deserves a serious study: *Philosophiae Naturalis Principia Mathematica perpetuis commentariis illustrata communi studio* PP. T. Le Seur & F. Jacquier ex Gallicana Minimorum Familia, Matheseos Professorum (4 vols, Geneva, 1739–1742).

[37] R. J. Boscovich, 'Dialogi sull'aurora boreale', *Gior. Lett.*, 1748, 192–202, 264–75, 293–302, 329–36, 363–8; a recent reprint: *Scienziati del Settecento*, M. L. Altieri Biagi and B. Basile eds. (Milan and Naples, 1983), 703–54. Also by Boscovich, *De lumine dissertatio* (Rome 1748); see *Casini* op. cit. (note 27).

[38] F. M. Renazzi, *Storia dell'Università di Roma* (4 vols, Rome, 1803–1806; reprint, Bologna 1971), **4**: 270ff.

IL NEWTONIANISMO
PER LE DAME
OVVERO
DIALOGHI
SOPRA
LA LUCE E I COLORI.

——— *quæ legat ipſa Lycoris.*
Virg. Egl. X.

IN NAPOLI
MDCCXXXVII.

Figure 14.3. Title page of *Il Newtonianismo par le dame* (Naples, 1737).

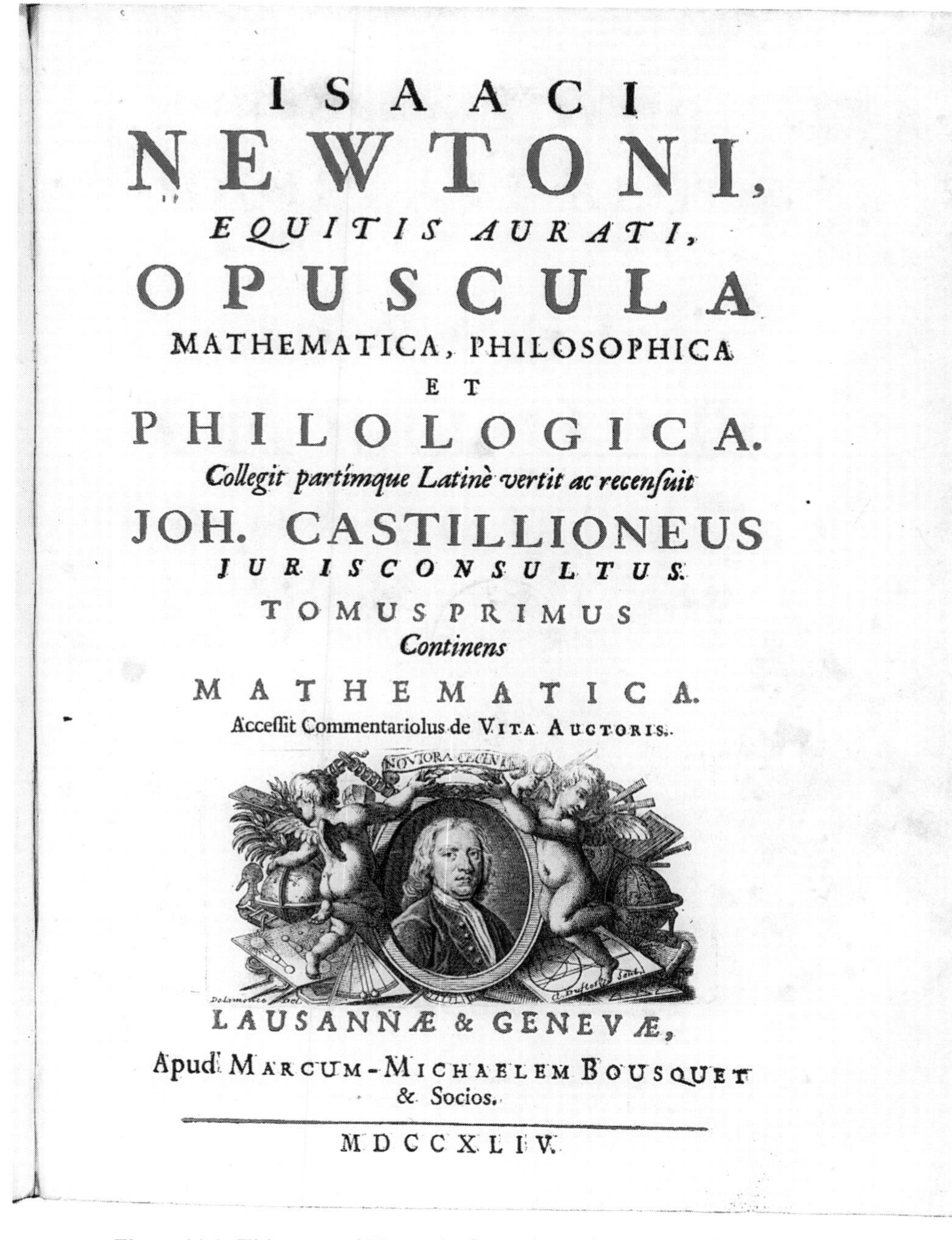

ISAACI
NEWTONI,
EQUITIS AURATI,
OPUSCULA
MATHEMATICA, PHILOSOPHICA
ET
PHILOLOGICA.
Collegit partímque Latinè vertit ac recensuit
JOH. CASTILLIONEUS
JURISCONSULTUS.
TOMUS PRIMUS
Continens
MATHEMATICA.
Accessit Commentariolus de VITA AUCTORIS.

LAUSANNÆ & GENEVÆ,
Apud MARCUM-MICHAELEM BOUSQUET
& Socios.

MDCCXLIV.

Figure 14.4. Title page of Newton's *Opuscula mathematica, Philosophica et Philologica* (Lausanne and Geneva, 1744).

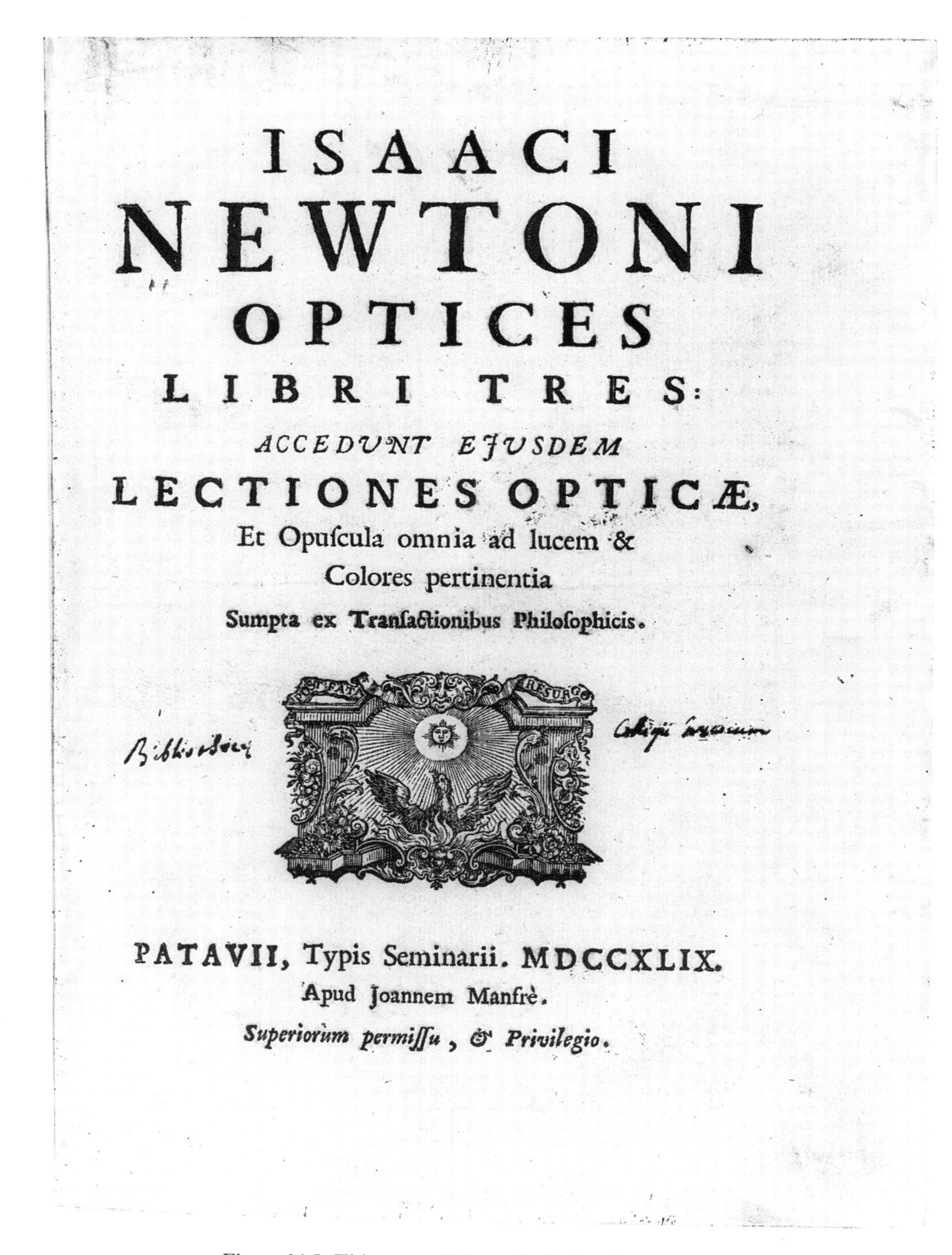

ISAACI
NEWTONI
OPTICES
LIBRI TRES:
ACCEDUNT EJUSDEM
LECTIONES OPTICÆ,
Et Opuſcula omnia ad lucem &
Colores pertinentia
Sumpta ex Tranſactionibus Philoſophicis.

PATAVII, Typis Seminarii. MDCCXLIX.
Apud Joannem Manfrè.
Superiorum permiſſu, & Privilegio.

Figure 14.5. Title page of Newton's *Optices* (Padua, 1749).

papers (Fig. 14.4).[39] Finally, a complete edition of all Newton's writings on optics (Fig. 14.5) appeared in Padua just before the middle of the century.[40] Thus Rizzetti, who refused to surrender, was defeated at home. His last attack – a pamphlet of 1741 – had no audience;[41] he turned at last to a hobby, the theory and practice of architecture. His name was almost forgotten, except by another opponent of Newton's optics, Goethe, who in his *Farbenlehre* (1808–1810), quoted with some respect the experiments described in Rizzetti's tract *De Luminis Affectionibus*.

[39] I. Newtoni *Opuscula Mathematica, Philosophica et Philologica. Collegit partimque Latine vertit ac recensuit Joh. Castilloneus* (3 vols., Lausanne and Geneva, 1744). For an account of the editor: G. Spini, 'Giovan Francesco Salvemini "de Castillon" tra illuminismo e protestantismo', *I Valdesi e l'Europa* (Società di Studi Valdesi, Torre Pellice, 1982), 319–49.

[40] *I. Newtoni Optices . . . accedunt . . . Lectiones Opticae et Opuscola Omnia ad Lucem et Colores pertinentia* (Padua, 1749); no name of editor is given.

[41] [Giovanni Rizzetti,] *Saggio dell'anti-newtonianismo sopra le leggi del moto e dei colori* (Venice, 1741).

15

Marsigli, Benedict XIV and the Bolognese Institute of Sciences

GIORGIO DRAGONI

The subject of this essay is an episode which illustrates the state of the study of natural philosophy in Bologna in the eighteenth century. The chief protagonist is Prospero Lambertini (1675–1758), who was Archbishop of Bologna from 1731.[1] On being elected Pope in 1740, he took the name of Benedict XIV. Although Lambertini's family connections and his contributions to political, diplomatic and religious activity have been studied in considerable detail,[2] he is as yet insufficiently well known for his achievements in promoting the study of natural philsophy. There will follow a brief account of his many acts of patronage, with particular emphasis on his decisions ('motu proprio'), which were of particular significance for experimental physics.[3]

Prospero Lambertini was a humane, cultured, learned and religious man. Surprisingly, he can be seen as, in some respects, deeply non-conformist, notably in having a clear and detailed programme for natural philosophy that can be described as Galileian and Newtonian. By studying Lambertini we shall be able to appreciate the nature of the intellectual history of Bologna. In this history Galileian ideas (or, more precisely, Baconian–Galileian–Newtonian ideas), though officially condemned by the Church (since the trial of Galileo in 1633), nevertheless were able to enjoy an extraordinary revival in the activities of some outstanding individuals. For instance, Bonaventura Cavalieri (1589–1647) a mathematician and follower of Galileo, who

I wish to express my gratitude to the Archivo di Stato Bologna (ASB), the Biblioteca Universitaria di Bologna (BUB) and the Archivio Segreto Apostolico Vaticano (ASAV) for allowing me to read, microfilm and compare manuscripts in their possession. I also thank the librarians and archivists for their courtesy and valuable help.

1 On Benedict XIV see *Diario Benedettino che contiene un'ampia serie di beneficenze fatte dalla Santità di N.S. Papa Benedetto Decimoquarto alla sua Patria*, Bologna, 1754; D. Majocchi, *Le vicende di alcuni antichi microscopi lasciati in dono al Museo di fisica della R. università di Bologna*, Bologna, 1928; L. von Pastor, *Storia dei Papi dalla fine del Medioevo, vol. XVI, Parte I, Benedetto XIV e Clemente XIII (1740–1769)*, Rome, 1933; E. Morelli, *Tre profili: Benedetto XIV, PS. Mancini, P. Roselli*, Rome, 1955; R. Haynes, *Philosopher King: the Humanist Pope Benedict XIV*, London, 1970; AA.VV., *Benedetto XIV (Prospero Lambertini)*, 3 vols, Ferrara, 1980; G. Dragoni, 'Vicende dimenticate del mecenatismo bolognese dell'ultimo '700: l'acquisto della collezione di strumentazione scientifiche di Lord Cowper, *Il Carrobbio*, 1985, 9: 67–85.

2 See particularly AA.VV. *op.cit.* (note 1).

3 *Lettere, brevi, chirografi, bolle ed apostoliche determinazioni prese dalla Santità di Nostro Signore Papa Benedetto XIV nel suo pontificato per la Città di Bologna sua Patria*, Bologna, 1749 and *Diario op. cit.* (note 1).

taught in Bologna from 1629 onwards; his pupil Geminiano Montanari (1653–1687) mathematician and astronomer; the astronomer Giovanni Domenico Cassini (1625–1712); the physician, anatomist and microscopist Marcello Malpighi (1628–1694) and, most importantly for this study, Luigi Ferdinando Marsigli (1658–1730). Marsigli – a pupil of the anatomist Antonio Borelli (1608–1679), of Malpighi and of Montanari – pursued a military career, rising to the rank of general in the army of the Emperor Leopold I. He nevertheless also found time to pursue his interests in natural philosophy, to which he made a number of important contributions, achieving an international reputation for his work on geography, hydrography and related subjects.[4] Unlikely as this may seem, to this list of distinguished Bolognese men of science we must add the name of a pope: Prospero Lambertini (Benedict XIV).

Benedict XIV: a biographical sketch

The early years of Prospero Lambertini's life were rather difficult. He was born in Bologna on 31 March 1675, into the old noble family of the Signori del Poggio, who used the surname Lambertini. On 17 March 1676, before Prospero was a year old, his father Marcello died, and on 19 June, after only three months of widowhood, his mother, Lucrezia Bulgarini, married the Marquis Luigi Bentivoglio. Historians have not yet discovered how Prospero and his elder brother were brought up, but it seems clear that he did not live with his mother. Very little is known about his childhood.

Marcello Malpighi, a native of Crevalcore, not far from the Lambertinis' ancestral seat of Poggio Renatico, may have played an important part as a symbolic figure in Prospero's early education, which took place in Bologna. The most formative part of his education was, however, at the Collegio Clementino (founded 1595), in Rome, where he was enrolled from 1689 to 1692.[5] During this period, the Collegio Clementino was the training ground of the Catholic intelligentsia. The pupils included future cardinals, archbishops, generals and men of letters. The main subjects studied were grammar, humanities, rhetoric and philosophy, but the curriculum also included Latin, Italian literature and history, along with fencing, riding, gymnastics, music, painting, drawing, mathematics and foreign languages.

After completing this extremely broad education, Lambertini embarked upon an ecclesiastical career, in which he slowly but surely rose to the highest office. He was appointed Archbishop of Ancona in 1727, and made a Cardinal by Pope Benedict XIII in 1728. In 1731 he became Archbishop of Bologna, a diocese second in importance

[4] Much biographical material on Marsigli can be found in A. McConnell, 'A profitable visit: Luigi Ferdinando Marsigli's studies, commerce and friendships in Holland, 1722–23' in C. S. Maffioli and L. C. Palm (eds.), *Italian Scientists in the Low Countries in the XVIIth and XVIIIth Centuries*, (Amsterdam and Atlanta, 1989) 189–206. Primary biographical sources are L. F. Marsigli, *Autobiografia di Luigi Ferdinando Marsigli*, ed. E. Lovarini (Bologna, 1930); L. D. C. H. D. Quincy, *Mémoires sur le vie de M. le Comte de Marsigli* (Zurich, 1741); G. Fantuzzi, *Memorie della vita del generale Co L. F. Marsigli* (Bologna, 1770).

[5] On the Collegio Clementino see L. Zambarelli, *Il nobile pontificio Collegio Clementino*, Rome, 1936; L. Montalto, *Il Clementino (1595–1875)*, Rome, 1938; G. L. Masetti Zannini, 'Prospero Lambertini e la sua educazione al Collegio Clementino (1689–1692)' in AA.VV. *op. cit.* (note 1), 141–60.

only to Rome. In 1740, at the age of 65, he was elected Pope, in succession to Clement XII. Nevertheless, for the following ten years he continued to act as Archbishop of Bologna. At this time Bologna was the second city of the Papal States.

Benedict XIV's acts of patronage

Even a simple list of Benedict XIV's chief actions in support of the study of natural philosophy suffices to demonstrate the degree to which his interest and attention were continually engaged in promoting and supporting scientific interests in his native city. We have ample evidence of the exceptional importance he attributed to the study of natural philosophy and to using appropriate means for fostering teaching, research and the dissemination of results to a wide public. The following summary shows something of the nature and range of his activity.

The year following his election as Pope, in 1741, Benedict XIV initiated a huge project for the reorganization of the library of the Istituto delle Scienze e delle Arti.[6] Under his orders, the following three years saw the construction of the 'Benedictine' wing of the magnificent reading room (Aula Magna) which is now an important part of the University Library.

In 1741, the Observatory (Specola) was equipped with instruments bought from the famous instrument maker Sisson of London.

In 1742, in accordance with the Pope's suggestion, the library and collections of the famous natural historian Ulisse Aldrovandi (1522–1605), bequeathed to the Senate of Bologna University in 1603, were donated to the Istituto delle Scienze e delle Arti.

Also in 1742, Ercole Lelli (1702–1766), who in 1732–4 had made the wooden sculptures for the anatomy theatre of the University (Archiginnasio), was commissioned to make wax anatomical statues for the Museum of Anatomy of the Academy of Sciences of the Institute of Bologna, by what is now the Istituto di Anatomia Umana. The series of figures, on which Lelli and his assistants worked from 1742 to 1751, shows how open-minded the Pope was, in comparison with other churchmen of the time, both in undertaking the project and in his willingness to make available to the public representations of the male and female human bodies and their internal organs.

In 1743, the Naturalia Museum, the Collection of natural objects gathered by Senator Ferdinando Cospi (1606–1686), was donated to the Istituto delle Scienze e delle Arti.

In 1744, the *Collected Works* of Galileo were published, in four volumes, in Padua. This edition, commissioned to commemorate the first centenary of Galileo's death, in 1742, included his *Dialogue concerning the two greatest world systems* (1632), which had not appeared in the 1655–6 Bologna edition of Galileo's works.

[6] On the Istituto see G. G. Bolletti, *Dall'origine e dei Progressi dell'Istituto delle Scienze di Bologna*, Bologna, 1751, reprinted, Bologna, 1977; AA.VV. *I Materiali dell'Istituto delle Scienze*, Bologna, 1979 (this has a large bibliography); M. Cavazza, 'Bologna and the Royal Society in the seventeenth century', *Notes Rec. Roy. Soc. Lond.*, 1980, **35**: 105–23; W. Tega (ed.), *Anatomie Accademiche*, 2 vols, Bologna, 1986–7.

Also in 1744, important instruments for teaching and experiments connected with the work of Galileo and of Newton were commissioned from famous Dutch instrument makers, on the advice of Petrus van Musschenbroek (1691–1761) and Willem Jacob's Gravesande (1688–1742).

In 1745, the new rules of the Academy of Sciences of the Institute of Bologna set up the Gabinetto di Fisica, a Cabinet of Physics which was not only a museum but also contained a laboratory. Thus physics became officially recognized as a special subject.

Between 1745 and 1747, three volumes of Acts of the Academy of Sciences were published, entitled *Commentarii*, that is the Proceedings of the Accademia delle Scienze dell'Istituto di Bologna. The first volume had been published in 1731 and a further five volumes were published in the last half of the eighteenth century. The *Commentarii* document the activities of Bolognese savants during a large part of the eighteenth century. In 1747, the complete workshop of Giuseppe Campani (1635–1715) was donated to the Istituto delle Scienze e delle Arti for the Cabinet of Physics. Campani, who worked in Rome, had been a skilled astronomer and one of the most famous makers of optical instruments – telescopes and microscopes – of the seventeenth century.[7]

In 1749, Benedict XIV decided to take 10,000 escudos from the Treasury of Bologna to pay for the improvement of the laboratories and the needs of the Professorship of Chemistry, which was first held by Jacopo Bartolomeo Beccari, in 1737/8.

In 1754, on the Pope's suggestion, Cardinal Filippo Maria Monti presented the Istituto delle Scienze e delle Arti with his library, comprising about twelve thousand volumes and his valuable collection of over four hundred paintings, which included portraits of Copernicus, Francis Bacon, Galileo, Boyle and Newton.

In 1755, Benedict XIV presented the Istituto with his personal library of twenty-five thousand volumes, including more than four hundred codices and manuscripts.

In 1756, in a solemn ceremony held in the Aula Magna, the Pope opened the library to the public. For this occasion, the portrait collections of Monti and Giovanni Nicolo Alidosi (who had earlier presented his collection of portraits of members of the greatest families of Bologna) were displayed on the walls of the library.

In 1758, Obstetrics became part of the curriculum of the Istituto delle Scienze e delle Arti. Despite the opposition of many university teachers, and with the determined support of the Pope, the chair was given, in 1757, to the natural philosopher and physician Giovanni Antonio Galli. The Pope paid 1,000 escudos (equivalent to 5,000 Bolognese lire) to acquire Galli's teaching museum for the Istituto.

This list refers only to Benedict XIV's activity in support of the study of natural philosophy in Bologna. He gave similar support in Ancona and, of course, in Rome, where he set up the Musei Capitolini (housing magnificent fine art collections that are still on show to the public today).

[7] A. Van Helden, 'The telescope in the seventeenth century', *ISIS*, 1974, **64**: 38–58; M. L. Bonelli and A. Van Helden, 'Divini and Campani: A forgotten chapter in the history of the Accademia del Cimento', *Ann. Ist. Mus. Stor. Sci. Firenze*, 1981, **6** (1): 1–176.

Benedict XIV's support for natural philosophy

What were the areas of the Pope's interest and competence in natural philosophy? The subject is rather complex, so we can do no more than consider a few possible influences. The contribution made by his formal education has already been mentioned. In addition, there are two figures who certainly influenced the future Benedict XIV.

The first was the Jesuit Roger Boscovich (1711–1787). He was an astronomer, mathematician and natural philosopher, born in Dalmatia, who worked for a long time in Rome, at the Collegio Romano, and then in Milan, where he founded the Brera astronomical observatory. Boscovich, whose work was in the Newtonian tradition, became one of the foremost natural philosophers in Italy. Benedict XIV sought his advice in connection with publication of the Paduan edition of Galileo's *Works* which we have already mentioned.

Benedict XIV's friendly inclination towards supporters of Newtonian ideas is further seen in his correspondence and his personal relationships with a number of prominent figures, among whom we may mention Voltaire (1694–1778) and Francesco Algarotti (1712–1764), both propagandists for Newton's work.

The second personality who influenced Benedict XIV was Luigi Ferdinando Marsigli, founder of the Istituto delle Scienze e delle Arti in Bologna. The most significant period in their relationship was in 1726.

The Istituto delle Scienze e delle Arti of Bologna

Marsigli had founded the Istituto delle Scienze e delle Arti under the patronage of Pope Clement XI, between 1711 and 1715. The Institute – sited in Palazzo Poggi (at that time Palazzo Cellesi) – included on the first floor the Accademia Clementina, the Academy of Painters, Sculptors and Architects (Accademia dei Pittori, Scultori e Architetti), on the second floor the Accademia delle Scienze dell'Istituto, and on the third floor the Specola, that is the astronomical observatory. It was officially opened to the public in 1714.[8] Marsigli established it specifically to carry out experimental research, in the Galilean–Newtonian tradition. Marsigli had shown great generosity in gifts of money and instruments to the Istituto and also in the expenditure of his energy on its behalf.

At the beginning of his diplomatic and military career in 1679, Marsigli was Secretary to the Venetian Ambassador in Constantinople. In 1683 he was at the Imperial Court in Vienna. On the Emperor's orders, he visited several European and Eastern countries. Following the foundation of the Istituto, Marsigli went abroad

[8] See 'Instrumentum Donationis favore Illustrissimi et Excelsi Senatus et Civitatis Bononiae in gratiam novae in eadem Scientiarum Institutionis' BUB MS Marsigli 146; legal act on 11 January 1712. These acts were printed by Stamperia di S. Tommaso d'Aquino, Bologna in 1727 (reprinted Bologna, 1981). In these papers it is possible to find the Charter (*Costituzioni*) of the Institute founded 1711. The Institute was solemnly inaugurated on 13 March 1714. See also ASB MS Assunteria, Diversorum, Busta 7, no 4, 1714 for the opening to the public.

again, pursuing his research interests and his desire to learn. He travelled to England, where he met Newton in 1722, and to Holland, where he met the publishers who were to print his *Histoire Physique de la Mer* (Amsterdam, 1725) and his *Danubius Pannonico-Mysicus* (Amsterdam, 1726). On his return to Bologna in 1723, Marsigli was very disappointed to find that the Istituto delle Scienze e delle Arti was not, in his terms, a success. He had intended it to be a centre for research rather than merely a place for intellectual entertainment.

Marsigli believed that the Istituto had been unsuccessful in several respects. For instance, he complained that the citizens of Bologna did not use the Istituto 'which they set foot in only to accompany visitors from abroad'.[9] That is, they did not attend lectures or demonstrations of experiments in the Istituto, but used it only as a place of entertainment, for instance in showing round foreign visitors. Moreover, in his Act of Donation – *Instrumentum Donationis* – Marsigli had included some requests that the Senate of the city had not respected. Marsigli wrote out a complete list of everything that had not worked as he, the founder, had hoped: the list of 'Disorders', *I Disordini*.[10] A serious conflict deveoped between Marsigli and the Senate of Bologna. The cause of all the problems was, in Marsigli's opinion, the lack of interest that the Senate had shown in the Istituto delle Scienze e delle Arti. In a document called 'Cures', *I Remedi*, Marsigli drew up a schedule of eighteen points which summarized the problems, and how they should be corrected.

Marsigli's most important points were:

Number 4. The need to have a librarian for the library of the Istituto.

Number 5. To allow people who were not citizens of Bologna to become lecturers at the Istituto. The restriction to Bolognese citizens was, Marsigli believed, one of the most important reasons why the Istituto's intellectual level was below that of comparable institutions elsewhere, such as the Royal Society of London, of which he was a Fellow.

Number 6. Salaries to lecturers must be improved. For this purpose, Marsigli promised to give the Istituto his Cabinet of Natural History of the Indies (that is from Dutch colonies both East and West).

Number 7. According to Marsigli, a new room was required so that the library could be better organized. Marsigli promised that, if the Senate accepted this point, he would undertake 'to give the total sum of the other notable books obtained in recompense for the Treatise on the Danube'. That is, to donate all the books he had been given in payment on the publication of his *Histoire Physique de la Mer* and *Danubius Pannonico-Mysicus*.

The *cahiers de doléances*, that is the list of things that did not work at the Istituto in Marsigli's opinion, ended with a request, addressed to Pope Benedict XIII (reigned 1724–1730), in case the Senate of Bologna did not accept Marsigli's conditions:

[9] 'dentro del quale pongono li piedi unicamente per accompagnare forestieri' in Marsigli to Ferrari quoted in Bertolotti, 'La Fondazione dell'Istituto delle Scienze', in Bolletti, reprint, *op. cit.* (note 6), 453.
[10] See ASB MS Assunteria, Diversorum, Busta 7, no 6.

> if appropriate remedies are not provided for the accumulation of so many disorders, which tend to the total demise of the Istituto, [the writer] intends to apply to the Holy Father to convert the sums donated, and the others accumulated since the said Donation, to other pious works.

That is, Marsigli asked the Pope for permission to use all his capital, including that already donated and that expected to accrue from the donation, for charitable purposes. He had become angry with the *Assunteria* of the Institute (that is, the committee of Senators who were responsible for running it), with the Senate of Bologna, city officials, some of his own family and teachers at the university (which at the time was still Aristotelian in outlook and disapproved of the Galilean experimental ideology of the Istituto delle Scienze e delle Arti). A letter Marsigli wrote to Benedict XIII on 30 May 1725 shows how matters stood. Marsigli wrote that the Istituto was unsuccessful in several respects:

> after having donated the major part of my fortune to create and found in the city of Bologna the Institute of all the Sciences, and Liberal Arts . . . and that, to run the Institute in an orderly manner, it was agreed between the Senate of Bologna and the writer that funds would be provided . . . [there were] various though groundless difficulties raised; I therefore have recourse to Your Holiness's Fatherly zeal to deign to order His Eminence Cardinal Paolucci . . . to provide, as Your Holiness's representative, for remedies for all the disorders that have come into being . . .[11]

Benedict XIII yielded to Marsigli's persistent requests and chose Monsignor Prospero Lambertini – the Bishop of Theodosia (since 1725) and a member of the Pope's staff in Rome – as his representative to visit Bologna to investigate the state of affairs, and bring about a reconciliation. To summarize, what Marsigli wanted was that there should be strict and effective regulations to compel the Senate to run the Istituto in a proper manner. Only on this condition was he willing to make to the Istituto a further donation, consisting of scientific material and books, most of which had been obtained from Holland. In the course of their meetings, Marsigli provided Prospero Lambertini with a detailed list of the main shortcomings and 'disorders' of the Istituto, together with a set of rules appropriate for running it satisfactorily.

Prospero Lambertini succeeded very skilfully in both calming Marsigli and making the city authorities accept his conditions. In July 1726 Lambertini wrote to the Secretary of State of the Vatican, Cardinal Lercari, to ask for the Pope's approval for his proposed solution to Marsigli's complaints.[12] The letter is a masterpiece of diplomacy, from which it is clear that Lambertini's intention was to persuade the Pope to issue a direct order to the governors (*Assunti*) of the Istituto in Bologna,

[11] 'dopo aver donato la maggior parte delle sue sostanze per erigere e fondare nella città di Bologna l'Instituto di tutte le Scienze, e Arti Liberali . . . e che a fine di ben regolare l'Instituto, furono concordati li capitoli da osservarsi fra il Senato di Bologna, e l'Oratore . . . [there were] varie, benche insussistenti difficoltà; ricorre perciò al Paterno zelo S.V., acciò si degni ordinare al Do. Emo Sig. Card. Paolucci . . . che proveda, come delegato della Santità Vostra a tutti quei disordini che sono nati . . .', ASB MS Assunteria, Diversorum, Busta 7, no 2, lettere varie del gen. Marsigli.

[12] ASB MS Assunteria, Diversorum, Busta 7.

commanding them to accept Marsigli's remedies. The reply from Lercari, dated 7 August 1726, was almost entirely favourable to Marsigli.[13] Marsigli was completely satisfied and at once decided to present the Istituto with the new materials he had promised.

Thus Marsigli transferred all the Dutch Indies material to the Istituto in 1727. In a letter of 7 August 1726 he gave his decision about the donation.

> In the name of God. Be it known by this document that I, the undersigned, recognize that the highest rectitude of Our Lord Benedict XIII wishes a speedy remedy for all disorders introduced into the Institute of Sciences and Arts of Bologna ... and obliges me to put myself in the position of immediately responding to his fatherly and sovereign clemency with these new donations to the said Institute, which donations have been suspended for two years ... [14]

The donations were made officially on 24 March 1727, and were as follows:

> These donations thus comprise: one, a very numerous collection of natural objects from the Indies, ... and the other many books, to the value of 10,000 Dutch florins, which I obtained in Amsterdam ...[15]

On 4 January 1727, Marsigli wrote to the Vatican that everything was ready for the donation:

> the Legate, Cardinal Ruffo, has already arranged everything so that my final official Donation can take place with the usual legal formalities ....[16]

Lambertini's diplomatic work was now complete. He had obtained Marsigli's confidence and had secured the donations for the Institute of Sciences. However, difficulties continued, probably due to too much red tape or to obstacles skilfully caused by Marsigli's enemies. These difficulties once more prevented the Istituto from operating satisfactorily.

Infuriated by the behaviour of the politicians and university teachers of Bologna, Marsigli eventually left the city. In a letter of 10 July 1728, addressed to the Pope, as usual through the Secretary of State to the Vatican, Cardinal Lercari, Marsigli, writing from Bologna, said

> Since, with the Paternal help of Your Holiness, my work of founding the Institute of Sciences and Arts is completed, work which took forty-six years of my life, I can without prejudice to the foundation follow my resolution to

[13] ASB MS Assunteria, Diversorum, Busta 7. In this letter it is possible to compare Marsigli's complaints and the Vatican's decision against the Bolognese Assunteria.

[14] 'Al nome del Sig. Iddio. E cosi sia la notizia, che tengo io sottoscrititto, che la somma rettitudine di Nostro Signore Benedetto XIII vogli il pronto rimedio a qualunque disordini introdotti nell'Instituto di Bologna delle Scienze, ed Arti, ... obbliga me di mettermi in stato di prontamente corrispondere alla di lui paterna e sovrana clemenza con queste nuove donazioni all'Instituto medesimo, che per due anni sono state sospese ...', ASB MS Assunteria, Diversorum, Busta 7.

[15] 'Tali donazioni dunque consistono: una in moltissime cose naturali delle Indie ... l'altra in molti libri per valore di 10 mila Fiorini d'Olanda da me tenuti in Amsterdam ...'. The second endowment was published in *Atti Legali per la Fondazione dell'Istituto delle Scienze ed Arti Liberali per Memoria degli Ordini Ecclesiastici e Secolari che compongono la Città di Bologna*, Bologna, 1728.

[16] 'gia che per l'Instituto questo Sr. Card.e Ruffo Legato ha ormai tutto disposto, perche possa seguire la mia finale Donazione autenticata con la consueta legalità ...', ASAV MS, Lettere di particolari, vol. 134, f. 10, p. 190.

> leave my homeland for ever . . . so that I may live in a little garden far from all society and given over entirely to my studies . . .[17]

Marsigli changed his name to Cavalier d'Aquino and went to live in Marseille. In a letter addressed to the Pope from Bologna, and dated 4 September 1728, Marsigli said

> I had strong motives for taking legal measures to set aside the coat of arms of my house, and the surname Marsigli in favour of that of d'Aquino . . .[18]

Some time later, Marsigli returned to Bologna, where he died on 1 November 1730.

Conclusions

At Marsigli's death, the future of the Institute he had founded was not yet assured. However, it seems that through their personal relationship Marsigli had communicated something of his scientific interests to Prospero Lambertini. Thus in 1731 when Lambertini became Cardinal of Bologna and later when he become Pope (as Benedict XIV, 1740), he used his great authority and power to further Marsigli's projects for the Istituto delle Scienze e delle Arti, eventually establishing the fruitful relationship chronicled in the earlier part of this essay. Elements of the local history of Bologna thus seem to throw light upon important acts of Papal patronage.

[17] 'Anche con le Paterne assistenze di V.ra Santità e terminata la mia Opera della fondazione dell'Instituto delle Scienze ed Arti, che mi costo quarantasei anni di vita, per poter senza pregiudizio a una tale Fondazione di eseguire la mia risoluzione di abbandonare per sempre la patria . . . perche vivero in un Giardinetto remoto da ogni commercio, e tutto dato ai miei studi . . .', ASAV MS, Lettere di particolari, vol. 137, p. 49.

[18] 'Sono stati forti li motivi per li quali già con Atti legali mi sono dimesso dell'Armi gentilizie della mia Casa, come del cognome Marsilli in quello d'Aquino . . .', ASAV MS, Lettere di particolari, vol. 137, pp. 253–4.

Afterword

Retrospection on the Scientific Revolution

A. RUPERT HALL

An advantage enjoyed by the writer of an Afterword is that he may presume his reader to have some acquaintance with the foregoing book. It is not for him to explain why Dr Who has chosen so strange a title, nor to reconcile the positions of well-known academic enemies. An Afterword may reflect on what precedes it without necessity for explanation or the enunciation of cosmic truths.

There is no particular magic in the word 'revolution' which rather possesses several demerits. It bears the suggestion of a completed cycle, therefore a return to a former state which, however, in its common political usage to mean 'revolt', 'overthrow of authority' is also specifically denied. A return to a previous political condition we confusingly call a 'counter-revolution' (a 'Thermidor'), even though time's arrow ensures the impossibility of the thing. Accordingly, 'revolution' is a rather bad emphasis of such weak words as 'change', 'alteration' which our minds reject as dull and unpicturesque, although we tend to accept other metaphorical nouns like 'rise' or 'evolution'. Would we not all expect two books, one entitled *The Rise of Quantitative Chemistry*, the other entitled *Lavoisier's Chemical Revolution* to cover much the same ground?

Similarly with the adjective 'scientific'. As everyone knows, far from conveying precise information this adjective is a trap, because the English language fails to distinguish between *Wissenschaft* and *Naturwissenschaft* in the manner of most European languages; we also have problems with *technique/technology* and the relation between these two concepts. Because of such linguistic problems we can no longer speak of 'moral sciences' and are pleased to introduce such a barbarous term as 'cognitive'. Imprecise authors use the word 'scientific' when they mean rather 'rational' or 'technical' or even 'industrial'. Again, many phrases are unchanged in meaning, perhaps even clarified, if a substitute for the word 'scientific' is employed: *The history of the Idea of Nature* is to be preferred as a title to *The history of Scientific Metaphysics* and it is surely better to discuss *Man and Nature* than *Man's Scientific Context*.

'Scientific Revolution' was thus not a particularly happy conjunction of terms. Yet as Bernard Cohen has shown, it has been sanctified by several centuries of use, though never free from ambiguity.[1] The notion that between non-specific dates – with the seventeenth century sharply in focus – humanity took a new view of external Nature (and to a less extent of itself) is one that can be encapsulated in various phrases with more or less precision and detail. 'The Scientific Revolution' has come to represent them all, but the historical concept – originally so bound up with the Battle of the Books between Ancients and Moderns and so fundamental to William Wotton's celebrated essay of 1694 – is neither defined nor limited by the expression. Herbert Butterfield's book (1949) which gave it wide currency was actually *entitled* by him *The Origins of Modern Science*, a less flamboyant metaphor; and it is curious that his discussion would have been open to fewer subsequent objections by critics if the final noun had been made plural. So complex and subtle are the nuances of language. When I chose the juxtaposition *Scientific Revolution* for the title of a book covering similar ground in 1954 I meant to pronounce upon no grand arcana of historiography, merely to find an expressive and striking title. Then, in 1962, Thomas S. Kuhn further muddied these far from limpid waters by writing about 'scientific revolutions' (now plural) divorced from any singular or quasi-singular historical context. Obviously Kuhn took up 'scientific revolution' as an historical 'term of art' (he had of course previously written about his 'Copernican revolution' himself) and modified it into a philosophical expression. But the older sense from Fontenelle to Tannery and beyond has remained valid.

Kuhn's *Structures of Scientific Revolutions* had another effect. *His* scientific revolution is typically a conceptual revolution, the displacement of one way of looking at things by another. Otherwise a 'paradigm shift'; of course Kuhn perceived that a revolutionized science – chemistry after Lavoisier – might embrace new ways of planning and performing experiments as well as new concepts like oxygenation and a *Nouvelle Nomenclature*. These were proper instrumental derivatives of Lavoisier's novel theory. And in this respect Kuhn did not differ from idealist historians of science of the twentieth century. However, the older historical tradition of a scientific revolution by which the natural philosophy and biological science of the Ancients had been overthrown did not see it this way at all. This tradition started from the notion that the revolution had at least as much to do with the methods and objectives chosen by investigators into Nature as with explanatory ideas; it envisaged a totality of change. Theories were the least valuable components of the 'new science' discussed by proto-Royal Society philosophers, who perceived its solid roots in new discoveries of fact (Jupiter's satellites, the lacteals), and in the establishment of such new facts by observation and experiment. Its paradigm was not heliocentricity or the circulation of the blood, but the notion that propositions about things in Nature should be demonstrable to the senses, not argued by syllogisms. It was a principle widely

[1] I. Bernard Cohen, *Revolution in Science* (Cambridge, MA, 1985).

regarded as properly Baconian that, in the words of Henry Oldenburg, a sound natural history must precede a new philosophy of Nature.

A number of writers in this volume have taken this point. Indeed, there has lately been something of a reaction against the homologization of scientific revolution and intellectual revolution, as though telescope and microscope, scalpel and balance, had been mere negligible tools of under-labourers in the great work. Several recent studies, on the contrary, have aimed to show how subtly the function of 'fact-gathering' with the aid of such instruments was mediated by the intellectual, social and religious context. Thereby they recognize the significance of 'fact-gathering'.

In what follows I shall briefly consider a number of points about the Scientific Revolution as an historical concept (and an *explicans*), some of them touched upon in the preceding papers.

Continuities, large and small

Continuities are affirmative and negative. The second I will explain in a moment, the first is obvious. The particulate theory of matter in the seventeenth century, and its larger context the mechanical philosophy of Nature, obviously derived from Greek atomism, effectively re-introduced to Europe in the fifteenth century. Though most historians would agree that the highly modified, kinematic version of this philosophy due to Descartes cut more ice than the 'hamous'[2] structural version favoured by Gassendi, both versions are as continuous with their Greek antecedents as geometry or Lovejoy's Great Chain of Being. It might not be altogether easy to explain why some moderns, like Descartes, boasted of the novelty of their clothes while others like Gassendi and Harvey were content to dress their innovations in antique garb. Copernicus we may be sure, and perhaps Newton also, was content to enhance his respectability in Pythagorean robes. We nowadays perceive some heroes of the Scientific Revolution as claiming an unreal continuity for their efforts: their supposed Greek precursors were mythical, back-formations giving rhetorical support to their own revolutionary positions. More interesting, in Copernicus's case, is the positive continuity evident in his geometrical machinery, since this provides a real historical puzzle. Where did he encounter the 'al-Ṭūsī couple' and the rest? We see that Copernicus's innovations have a wider context than renaissance Ptolemaic astronomy.

If the significance of continuity is sometimes pretty clear, as with the re-deployment of Greek mathematics by Commandino and others, it is not always so. The transition from medieval to renaissance optics is a case in point. In his *Opticks* Newton tackled the widely seen correlation between the density and the refractive index of transparent substances (Book II, Part III, Prop. X). He confirmed the correlation, save for the special case that 'unctuous and sulphureous Bodies refract more than others of the same density'. Unctuous bodies, besides the obvious vegetable

[2] The word is Charleton's, meaning 'hooked'.

oils, included (for Newton) camphor, amber and the diamond ('probably ... an unctuous Substance congelated') and to them, in Newton's chemical conspectus, 'ardent Spirits' might be related. The continuous historical background to this curious optical singularity of the unctuous body – whose transformation into *terra pinguis* was outside Newton's ken – has been traced in this volume by Gad Freudenthal. This particular continuity is of no particular significance to the Scientific Revolution, but it does confirm the unexpected tenacity of ancient ideas.

Continuity can be seen flowing through narrow meandering channels as well as broad ones, transmitting folk-myths as well as major systematic principles, details and large schemata. By negative continuity I mean such a union of opposites as is known to artisans as a male and female joint: the generation of a new thought from an old one both interlocked and antithetical. So Galilean physics is negatively continuous with Aristotelian physics; Copernican orbital geometry with that of Ptolemy; the concept of *horror vacui* with that of the experimental vacuum – which was far from being continuous with the Greek idea of the void. Negative continuity is entailed in the so-called *gestalt*-switch, whose relevance to the mirror-imagery of conceptual systems was examined by Russ Hanson long ago.[3] It is an historical mistake – of which Charles B. Schmitt has made us more keenly aware – to study only the 'modern' reaction against some thought or belief, without also thoroughly examining what is reacted against and the dependence of the former upon the latter. As the authors of *Father and Son* and *Sons and Lovers* recognized long ago, abreaction is just as much a relationship as is filiation.

The personalities of scientists

The qualities of intellect and character to be attributed to the various major figures of the Scientific Revolution have been much debated. A generation ago Arthur Koestler was accused by Giorgio di Santillana and Stillman Drake of depicting in *The Sleepwalkers* (1959) Copernicus as a 'moral dwarf', Galileo as having 'a cold sarcastic disposition' and treating Kepler unworthily, besides innumerable other errors of fact and interpretation.[4] In this instance, two established historians of science joined forces to attack, in a popular book, denigrations of character and imputations of irrationality, directed against two great astronomers. This was perhaps the last attempt to defend the conventionally sterling qualities once so universally attributed to the Great Men of Science. What has happened in the case of Newton is well known; every writer now feels certain that here was no moral and intellectual hero but a petty tyrant, a man of ungoverned passions intent upon pulling down his enemies, sexually abnormal and deeply confused about almost everything outside mathematics. The current trend is to seek the fount of Newton's majestic concepts of Nature in the arcana of alchemy or neo-Platonism, or maybe in some psychosis, and to depreciate

[3] N. R. Hanson, *Patterns of Discovery* (Cambridge, 1958), Chap. 1.

[4] *ISIS*, 1959, **50**: 255–60.

his experimental accuracy as stained by 'fudging'. Boyle is another British figure whose former cool pre-eminence as an impartial student of Nature by experiment has been shadowed by suggestions that he found out only those things which, for reasons having nothing to do with experiments, he wanted to find. Scientific ideas – at any rate in Britain – were in the Scientific Revolution (as some scholars now argue) little more than a shield for the defence of deeper and more momentous verities: those of religion and social structure.

As Hunter points out in this volume, this wider, contextualized version of the history of science has not always been buttressed by the firmest of scholarship and the most lucid of arguments. Nor is it obvious to me why its exercise has so far been confined to British 'heroes'. In my own view, Britain was pretty much of a backwater so far as scientific (or mathematical) achievement was concerned, until after the restoration of the Stuart monarchy. Its particular case-history is therefore not very illuminative of the Scientific Revolution as a European phenomenon, before 1660. Is the Mertonian argument strictly valid only for Britain (to which Robert K. Merton confined his discussion), and if so why? Similarly with the arguments of Simon Schaffer, the Jacobs and others. If we seek for general understanding of a European phenomenon, one in any event extremely various and complex in itself, we are hardly likely to find it by insisting upon the local peculiarities of the British scene, even if these have been correctly understood. Few, I suppose, would argue that without the special British contribution the Scientific Revolution would have become stultified, failed in the sands of Catholic obscurantism and North German mysticism. For my part, I defer to none in my sense of the power of Newton's mind and the splendour of his achievement, but I would certainly hesitate long before asserting that the state of scientific knowledge in the year 1900 would have been very different from what it actually was, if that special 'lad from Lincolnshire' had never been born.

And – to add a remark *en passant* – surely George Sarton was right to inveigh against the concept of national science, 'German science', 'French science' and so on (with political overtones obvious in his time). There are German, French, British scientists, he insisted, but the structure of science is international. Insistence on the minute particularities of this or that individual, or this or that group, tends to confuse this truth, which is not of recent validity only.

This is not to deny that it is an important and interesting part of our discipline to investigate the personal characteristics of individuals, where the material is rich enough to permit the exercise. To refer to Hunter once more, his paper on Robert Boyle illuminates a singular feature in the intellect and temperament of a major figure in the literature of both religion and science. Nothing more need be said about Newton's case. How far such dissection of individuals really contributes to the historical understanding of Western science as a cultural phenomenon I do not know. How to balance it against the apparently antithetical 'prosopological' analysis little practised in our field so far (Merton, Thackray, Westfall in this volume) I do not know either. If I may speak of *rational intellectual history*, meaning the analytical procedures practised on texts by such historians as A. E. Burtt, Hélène Metzger, Alexandre Koyré and Bernard Cohen (to name a few), now fallen from its former height of fashion, its strength seemed to be that it set aside the complex individual irrationalities unique to each person in order to attend only to thought-processes potentially accessible to all. That the entity responsible, engaged in these processes, without emotions, sex, religion, politics or social position was not a unique, complex, inconsistent human being is of course true.

Major and minor

In this volume a number of papers deal with lesser personae of the Scientific Revolution, each also making some general point or other of wider relevance. Hal Cook, for example, taking Swammerdam as a case in point, asks why the heroes of seventeenth-century biology seems less grandly heroic than those of mechanics. A similar question was posed by Charles Raven a generation ago.[5] It is a common, it may be an unjust, criticism of historians that they are always on the side of the larger, victorious battalions, and of the history of science that it is a saga of the deeds of great men. Should we not study significant *failures*, it is not infrequently asked, and some examples of this genre have proved either instructive or amusing. (No one proposes the study of obscure failures!) One has only to turn over the catalogues of antiquarian booksellers to see that, just as many books often quoted by historians never appear on the market, so it is flooded with barely recorded volumes to which historians never allude! A comprehensive history of astronomy would find a place for every astronomical book ever printed . . . or should it? If possible – and even Houzeau and Lancaster is not perfect, I believe – would the effort be worthwhile? Is *pointillisme* the best historiography?

To think that historical writing can by some process akin to holography recreate the totality of the past is an error. Clio, said Trevelyan quite rightly, is a muse. Academic history, and *a fortiori* that department which deals with natural science, is no longer a branch of literature: never will gentlemen's carriages queue outside a publisher's offices for the sake of the latest volume by Professors Dull and Dry. Nevertheless, history must be (even now) so far literature as to be read, if rarely for pure pleasure, and any pleasure there is in its reading comes from a certain magic in writing that transcends scholarship and 'cognition'. The art of history and its total depiction of an age are, even for a Braudel, incompatible. To create totality by plotting an infinite number of measured points upon the serial pages of a book is an illusion. The foundation of the art of history is the writer's ability to distill the essence of his mind, his sense of things in the past, into appropriate words. Distillation is a process that creates by omission.

I think therefore that Alan Gabbey is mistaken in suggesting elsewhere that a 'true'

[5] C. E. Raven, *Natural Religion and Christian Theology* (The Gifford Lectures 1951; First Series, *Science and Religion*), (Cambridge, 1953), Chap. VI.

history of mechanics would find room for every study of every problem.[6] True, that even a fairly summary exposition of all that lies behind that deceptively off-hand sentence penned by Newton: 'I judge that [this] proposition [Book II, Prop. 34, Scholium] will be of no little use in the construction of ships', about which Bernard Cohen wrote in 1974,[7] would require a fair-sized monograph alluding to at least fifty authors.[8] And such special studies as Gabbey asks for are indeed the bedrock of scholarship. But let us suppose that a general history of mechanics before Euler should give ten pages to this ship theme, five to the spinning-top, fifteen to the theory of structures (Galileo's other science, discussed at the Oxford Conference by Tom Settle), how many should be allowed to the physico-mechanical theory of the celestial motions? Five hundred, a thousand? Absurd questions, perhaps, but ones to be answered before large-scale books are written. I have much more sympathy with Gabbey's point in the present volume that the idea of a 'science of mechanics' or of 'rational mechanics' was obscure in its origins, carrying with it ambiguities whose neglect by the historian may be perilous.

In every walk of life, for reasons not to be analysed here, the greater subsumes the less. The private leaves a leg in the mud, his company-commander becomes a general. Prime Ministers and Presidents symbolize the labours of hundreds of speech-writers, researchers, diplomats, officials. Lavoisier subsumes Priestley, Faraday Henry, Darwin Wallace. To lament these orders of pre-eminence is as absurd as if a bacteriologist should lament that an amoeba is smaller than a louse. It is a natural and laudable act of critical scholarship to make the case that Smith is an under-rated composer, chemist, painter as compared with Jones. We may then hear a little more of Smith's music or see his pictures in a gallery. But these are marginal adjustments. Smith will not replace Matisse. True, Marcus Marci deserves a mention when Newton's optical investigations are canvassed, but only a very subtle or lengthy treatment will not buy comprehensiveness at the cost of confusion.

For inevitably too many details confuse the big picture; too many grey points make the whole image grey. I by no means deny that the big picture as we have it in 1992 has many grave defects, but these do not lie in treating Newton as a far more interesting and creative figure in the history of mechanics than Fabry or Pardies or Wallis. I believe too that we can examine the problems – *their* problems – more intimately in the documentary legacies of Newton, Lavoisier, Darwin, than by attending in great detail to those of whom we can know less. It may be unfair that posterity commonly gives more attention to those considerable masses of materials, but it is surely inevitable. If anyone should suggest that scholars like Nathan Reingold, D. T. Whiteside, the editors of Darwin and of Leibniz, are weighting the odds still more in favour of the Great Men by opening their volumes of rich material to scholarship, his

[6] Alan Gabbey, 'The case of mechanics: one revolution or many?' in David C. Lindberg and Robert S. Westman (eds.), *Reappraisals of the Scientific Revolution* (Cambridge, 1990), 493–528.
[7] I. Bernard Cohen 'Isaac Newton, the calculus of variations, and the design of ships' in R. S. Cohen *et al.*, *For Dirk Struik* (Boston Studies in the Philosophy of Science, 1974), 169–87.
[8] I myself made a trifling introduction to such a work in 'Architectura Navalis', *Trans. Newcomen Soc.*, 1979–80, **51**: 157–74 = *J. Naval Sci.*, 1984, **11**: 34–50.

plea must seem ridiculous. There *are* (in print) diaries of learned Pooters and boring they usually prove to be. Compare Samuel Pepys and Abraham de la Pryme!

Causes and queernesses

Richard S. Westfall remarks robustly in his paper that 'The concept of the Scientific Revolution rests on the radical reordering of the understanding of nature that did in fact take place in the sixteenth and seventeenth centuries. I am convinced that there is no way to understand the history of science without the recognition of this reality'. We could modify the label, of course, and (should we wish) precede this reality by a 'thirteenth-century scientific revolution' and follow it by 'nineteenth-century scientific revolutions' but juggling with labels does not diminish the reality of which Westfall speaks. Nearly all historians of science are, like him, convinced that this is so.

Whether this major historical phenomenon should be regarded as truly independent is, however, a question upon which there is less unanimity. A number of scholars make a bow to the idea that the Scientific Revolution derives from the Reformation of Christianity, and some (Professors Hooykaas and Stephen Mason) assert this derivation confidently. In my opinion those who would make Calvinism a *sine qua non* of modern science go a good bit beyond the Merton thesis. The rival Marxist assertion of the dependence of modern science upon Capitalism is not, perhaps, strongly enough asserted to merit further comment. On the other hand, a vague notion of science requiring certain 'religio-socio-economic structures' for its fruition is quite fashionable, though various analyses are far from agreeing one with another.

Such modern scholarly studies as I had in mind in composing the last paragraph make no claim, I think, to be studies of *causes*. Few would affirm that Henry VIII's lust for Anne Boleyn was the cause of *The Origine of Formes and Qualities* or Newton's *Principia mathematica*, however important they may judge Anglicanism to have been when those books were published. The search for 'causes of' rather than 'determining factors in' the Scientific Revolution attracts little support at present. Older generations of scholars tended to define as its chief cause the spiritual and technical liberation of Europe brought about by the renaissance of art and the revival of learning, the printing-press and the growth of literacy, and the voyages of exploration. Modern scholars, preferring to accentuate the positive, tend to deprecate such emphasis on discontinuity (or at best, negative continuity). Or, in Bernard Cohen's language, revolution is generalized into transformation. We no longer contemplate the start of modern history in 1453, or 1492, or 1543, as anything more than an arbitrary convention; if we may regret Koestler's dubbing Copernicus a 'timid canon', neither do we think of him as a Marx or a Marat. (The conspicuous iconoclasts, like Ramus and Bruno, were philosophers, not astronomers.) Since we no longer believe a new world to have been suddenly born at one of those sacred dates supposed to signalize the end of feudalism (or something of the sort) we no longer believe either that a new, modern science was suddenly fabricated by the inhabitants of this new world.

Yet because mathematicians and instrument-makers were rarely (if ever) imprisoned by the Inquisition, one should not forget that good work is still being done by those scholars pursuing an instrumental approach to the Scientific Revolution. That navigation, cartography etc. had a powerful stimulative influence upon applied mathematics and their various practical tools of wood, brass and ivory is a point upon which that splendid historian of early navigation, David Waters, has strongly insisted. But outside the applied mathematics–mechanics area (touched on by Frances Willmoth in this volume) other historians, philosophically inclined, have looked not to the European scene but inwardly to the individual psyche. How did people's emotions, beliefs, expectations change from the fifteenth to the seventeenth century? Can we measure changes in people's notions of magic and witchcraft – which assuredly were thought to operate in the natural world – and relate such changes to the rational investigation of that same natural world? What of neo-Platonism, astrology, alchemy? As I have noted previously in these remarks, the endeavour to set mathematics and natural science honestly in a total intellectual context produces strange bedfellows – cubic equations and demonology, gravitation and Diana's Doves, pneumatics and ghosts (in Britain, at any rate!).

No one can deplore this historical introspection, though I think that we are far from understanding its full significance. Few historians of fifty years ago would have been prepared mentally for the emergence of Thomas Hobbes and Thomas Sprat as major heroes of the Scientific Revolution in Britain, or for attitudes to millenarianism to be seen as major factors in its course, or to dismiss as a mythical perversion the once-commonplace bellicosity of science and religion. Let me mention Lynn Thorndike as an exception, as a man whose general attitudes to cultural history seem to have been closer to those now prevailing than were those of most of his older contemporaries.

Looking back to the first half of this century – a period through which the study of the history of science, philosophy and mathematics had already become established as academic subjects, starting to spread widely – one sees besides the provision of introductions, text-books, editions, hand-books, encyclopaedias and so on (George Sarton's massive *Introduction to the History of Science* may stand for all of these) many excellent articles, monographs, biographies and other synthetic or critical historial writings not even yet replaced by modern scholarship (let Clifford Dobell's *Antoni van Leeuwenhoek and his 'Little Animals'*, 1932, and Joseph Needham's *History of Embryology*, 1934, stand for all of these). I imagine that some of this 'ancient' work still helps to form the knowledge, if not the outlook, of younger scholars. Much of it came from the pens of men who (like Dobell) had worked long in the sciences; a few writers (like Thorndike) were professional historians. Others, like Sarton, Charles Singer, Paul Tannery, Joseph Needham, perhaps Sudhoff, had turned early from mathematics, medicine or science to historical studies (Tannery, employed in the *Régis des Tabacs*, was a part-time scholar). Few had missed training in the classical languages (as well as mathematics, etc.) and therefore a vast part of the ancient literature of science and learning lay open to them, in a manner rare now. (This apparent facility of access could dig pitfalls, on which I need not now insist.)

Among the many excellent pieces of scholarship and historical interpretation in the history of science produced in the first half of this century, the emphasis (it seems to me) was rather placed upon the eliciting of data from the sources, the construction of demonstrable chronologies and narratives, and the establishment of biography, than upon theoretical issues. Merton's thesis was exceptional. This characterization does not apply to the French school, however, represented by Emile Meyerson, Hélène Metzger and Alexandre Koyré; and A. E. Burtt was another exception for the same reason, that is, he approached history from a philosophical background. This French school of the 1920s and 1930s was not much admired where empirical-positivistic influences were strong, that is, particularly in the Anglo-Saxon world. Duhem's work also was less appreciated than it has been since; some detested him as 'an enemy of Galileo'.

George Sarton, in particular, represented a historiography very different from that of the French school. Possibly (as Charles Raven once remarked) the most learned man in the United States of America, immensely well equipped with scholarly skills, remorselessly industrious, Sarton excelled in clear, percipient summaries of other men's writings; his bibliographical knowledge was prodigious; and he had read enormously in printed original sources as well as in the secondary literature that he mastered so exhaustively. He never worked in a manuscript archive, and his study of texts was – dare I say it? – superficial. He tells us that in preparing his book on *Galen of Pergamon* he read through all Galen's writings in Kuhn's edition. Fine; and the bulk of Galen's numerous writings is, in fact, not so immense. But Sarton, in the time he could spare, must have read Kuhn as though it were a collected edition of Jules Verne. Galen, and the necessary studies ancillary to Galen, deserve the attention of a lifetime. As a result, wrote Ackerknecht, Sarton's book contained not one new thought . . . His monograph on Leonardo was no more successful for the same reason. The greatness of Sarton's scholarship, of his conception of scientific humanism and the liberation brought to the human spirit by the rational understanding of Nature, were restricted by his positivism, by his belief that the essence of history is the establishment of one accurate fact about the past. Was orach the spinach of antiquity, as an article in *ISIS* once asked? Was Maurolyco a priest or a monk, a point once settled by Edward Rosen? When one recalls the strange adventures of Michel Chasles (1793–1880), or Guglielmo Libri (1803–1869), both reputed as scholars in their day, and not so long before Sarton's time, one understands better how important the more basic and formally precise kinds of scholarship still were. And Sarton had strong prejudices: he detested Plato and despised the neo-Platonists, quite as much as Edward Gibbon had done.

But also in the United States, parallel to French historiography and in marked contrast to Sarton, should be set the 'history of ideas' movement founded by Arthur Lovejoy and George Boas, whose journal (like *ISIS*) flourishes still. But whereas *ISIS* had little enough to say about the general problems of the Scientific Revolution in the 1940s and 1950s, the *Journal of the History of Ideas* (founded in 1940) printed some of the best articles dealing with these questions from a variety of aspects. This journal, to one historian at least, was a helpful guide to recent thinking about the problems of the development of modern science. It certainly favoured increasingly those 'convictions about the importance of intellectual transformation and the relative insignificance of new experimental evidence in the development of early modern science' that David Lindberg attributes to the idealist tradition within the historiography of science.[9] It is my impression that this journal continues to this day to publish articles that contribute to the cutting-edge of modern treatments of the same problems.

As Lindberg goes on to point out, the historiography of science has become ever more complex and subtle during the last forty years. The effort to understand the scientific writings and correspondence of the seventeenth century in contemporary terms, to weigh all the influences affecting their content coming from non-scientific quarters, and (not least) the far greater profundity of investigations of the medieval European and Islamic heritage of early modern science, have all shifted the balance of interpretation and certainly made its equipoise more difficult to discover. Like Lindberg, I shall not pursue the inquiry further at this point.

[9] David C. Lindberg, 'Conceptions of the Scientific Revolution from Bacon to Butterfield: A preliminary sketch' in Lindberg and Westman *op. cit.* (note 6), 1–26, p. 18.

Bibliography

This list includes all works cited in the notes and in the text apart from manuscripts and official publications.

Adelmann, Howard B., *Marcello Malpighi and the Evolution of Embryology*, 5 vols., Ithaca, 1966.

Aiton, Eric J., *The Vortex Theory of Planetary Motion*, London, 1972.

Albertus Magnus, *Book of Minerals*, trans. Dorothy Wyckoff, Oxford, 1967.

Alexander, A. F. O'D., *The Planet Saturn*, London, 1962.

Alexander of Tralles, *Alexandri Tralliani medici lib. XII. Rhazae de pestilentia libellus . . .*, Paris, 1548.

Algarotti, Francesco, *Dialoghi sopra la luce e i colori*, Turin, 1977. (Also in *La Letteratura Italiana. Storia e Testi*, vol. XLVI, 2, Milan and Naples, 1969).

Algarotti, Francesco, *Opere*, 17 vols., Venice, 1784.

Allen, Woody, *Three Films of Woody Allen: Zelig, Broadway Danny Rose, The Purple Rose of Cairo*, London, 1990.

Alpers, Svetlana, *The Art of Describing: Dutch Art in the Seventeenth Century*, Chicago, 1983.

Andrê, François, *Chymical Disceptations: Or Discourses upon Acid and Alkali. Wherein are Examined the Objections of Mr. Boyle against these Principles*, trans. J. W., London, 1689.

André, François, *Entretiens sur l'acid et l'alcali, où sont examinées les objections de M. Boyle contre ces principes*, Paris, 1672.

Anon., *To the Memory Of my most Honoured Friend Sir Jonas Moore Knight, Late Surveyor General of His Majesties Ordnance and Armories* [London, *c.* 1679].

Arber, Agnes, *Herbals, Their Origins and Evolution: A Chapter in the History of Botany, 1470–1670*, Cambridge, 1912, 2nd edn, Cambridge, 1938.

Aretaeus, *Aretaei . . . De acutorum ac diuturnorum morborum causis & signis, lib . . . II . . .*, Paris, 1554.

Aristotle, *The Complete works of Aristotle. The Revised Oxford Translation*, ed. Jonathan Barnes, 2 vols., Princeton, 1984.

Armitage, Angus, 'The deviation of falling bodies', *Ann. Sci.*, 1947, **5**: 342–51.

Ashmole, Elias, *Elias Ashmole. His autobiographical and historical notes, his correspondence, and other contemporary sources relating to his life and work*, ed. C. H. Josten, 5 volumes, Oxford, 1966.

Ashworth, William B., Jr., 'Natural history and the emblematic world view' in Lindberg and Westman, 303–32.

Aubrey, John, '*Brief Lives*' *. . . set down by John Aubrey*, ed., Andrew Clark, 2 vols., Oxford, 1898.

Avicenna *see* Ibn Sīnā

Baader, Gerhard and Winau, Rolf (eds.), *Die hippokratischen Epidemien. Theorie – Praxis – Tradition*, *Sudhoffs Archiv, Beiheft*, **27**, Stuttgart, 1989.

Babson, Grace K., *A Descriptive catalogue of the Grace K. Babson collection of the Works of Sir Isaac Newton*, New York, 1950.

Bacon, Francis, *De augmentis scientiarum*, London, 1623.
Bacon, Francis, *De sapientia veterum*, London, 1609.
Bacon, Francis, *The Works of Francis Bacon*, ed., J. Spedding, R. L. Ellis and D. D. Heath, 14 vols., London, 1870, New York, 1968.
Badaloni, N., *Antonio Conti. Un abate libero pensatore tra Newton e Voltaire*, Milan, 1968.
Baily, Francis, *An Account of the Rev. John Flamsteed*, London, 1835, reprinted, 1966.
Barbaro, Daniele, *La Pratica della Perspettiva*, Venice, 1569.
Barozzi, G., called da Vignola, *Le Due regole della prospettiva pratica*, ed. E. Danti, Rome, 1583.
Barrow, Isaac, *Lectiones mathematicae XXIII*, London, 1683.
Barrow, Isaac, *The usefulness of mathematical learning explained and demonstrated: being mathematical lectures read in the publick schools at the University of Cambridge . . . To which is prefixed, the Oratorical Preface of our learned author, spoke before the University on his being elected Lucasian Professor of the Mathematics*, London, 1734.
Bauman, Richard, *Let your Words be Few*, Cambridge, 1983.
Baxter, Richard, *Reliquae Baxterianae*, ed. Matthew Sylvester, London, 1696.
Bechler, Z. (ed.), *Contemporary Newtonian Research*, Dordrecht, 1982.
Beltrami, Luca, *Vita di Aristotile da Bologna*, Bologna, 1912.
Benedict XIV, *Benedetto XIV (Prospero Lambertini)*, 3 vols., Ferrara, 1980.
Benedict XIV, *Diario Benedettino che contiene un'ampia serie di beneficenze fatte dalla Santità di N.S. Papa Benedetto Decimoquarto alla sua Patria*, Bologna, 1754.
Benedict XIV, *Lettere, brevi, chirografi, bolle ed apostoliche determinazione prese dalla Santità di Nostro Signore Papa Benedetto XIV nel suo pontificato per la Città di Bologna sua Patria*, Bologna, 1749.
Bennett, J. A., 'Robert Hooke as mechanic and natural philosopher', *Notes Rec. Roy. Soc. Lond.*, 1980, **35**: 33–48.
Bennett, J. A., 'The mechanics' philosophy and the mechanical philosophy', *Hist. Sci.*, 1986, **24**: 1–28.
Bertelli, Sergio, Rubinstein, Nicolai and Smyth, Craig Hugh (eds.), *Florence and Venice: Comparisons and Relations*, 2 vols., Florence, 1979–80.
Berthelot, Pierre Eugène Marcelin, *La Chimie au moyen âge*, 3 vols., Paris, 1893; Osnabrück and Amsterdam, 1967.
Biagi, M. L. Altieri and Basile, B. (eds.), *Scienziati del Settecento*, Milan and Naples, 1983.
Birch, Thomas, *The History of the Royal Scoiety of London*, 4 vols., London, 1756–7; Brussels, 1967–8.
Blasius, Gerard, *Ontleeding des Menschelyken Lichaems*, Amsterdam, 1675.
Boas, Marie, 'Hero's *Pneumatica*. A study of its transmission and influence', *ISIS*, 1949, **10**: 38–48.
Boas, Marie, 'Newton's chemical papers' in Cohen, 241–8.
Boas, Marie, *Robert Boyle and Seventeenth-century Chemistry*, Cambridge, 1958.
Boas, Marie, *see also* Hall, M. B.
Bolgar, R. R. (ed.), *Classical Influences on European Culture A.D. 1500–1700*, Cambridge, 1976.
Bolgar, R. R., *The Classical Heritage*, Cambridge, 1973.
Bolletti, G. G., *Dall'origine e dei Progressi dell'Istituto delle Scienze di Bologna*, Bologna, 1751, reprinted, Bologna, 1977.
Bonelli, M. L. and Van Helden, A., 'Divini and Campani: A forgotten chapter in the history of the Accademia del Cimento', *Ann. Ist. Mus. Stor. Sci. Firenze*, 1981, **6**(1): 1–176.
Boscovich, R. J., *De lumine dissertatio*, Rome, 1748.
Boscovich, R. J., 'Dialogi sull'aurora boreale', *Gior. Lett.*, 1748, 192–202, 264–75, 293–302, 329–36, 363–8. (Also in Biagi and Basile, 703–54).
Botfield, Beriah, *Prefaces to the First Editions of the Greek and Roman Classics and of the Sacred Scriptures*, London, 1861.
Bovio, Tommaso Zefiriele, *Flagello contro dei medici communi detti Rationali* in *Opere di Zefiriele Tomaso Bovio*, Venice, 1626.

Boxer, Charles R., *The Dutch Seaborne Empire 1600–1800*, London, 1965.
Boyle, Robert, *The Works of the Honourable Robert Boyle*, ed. Thomas Birch, 2nd edn, 6 vols, London, 1772.
Boyle, Robert, *see also* Fulton.
Brahe, Tycho, *Astronomiae instauratae mechanica*, Uraniborg, 1598.
Braudel, Fernand, *Civilisation matérielle, économie et capitalisme*, 3 vols., Paris, 1979.
Brewster, D., *Memoirs of the Life, Writings and Discoveries of Sir Isaac Newton*, 2 vols., Edinburgh, 1855.
Brewster, D., *The Martyrs of Science; or, the lives of Galileo, Tycho Brahe and Kepler*, London, 1841.
Briggs, John C., *Francis Bacon and the Rhetoric of Nature*, Cambridge, 1989.
Bronowski, Jacob, *Science and Human Values*, New York, 1965.
Brooke, John, *Science and Religion, Some Historical Perspectives*, Cambridge, 1992.
Brugmans, H., (ed.), *Gedenkboek van het Athenaeum en de Universiteit van Amsterdam 1632–1932*, Amsterdam, 1932.
Brunet, J., *L'Introduction des théories de Newton en France au 18ème siècle, tome 1 – Avant 1738*, Paris, 1931.
Buck, A. and Hartmann, K. (eds.), *Die Antike-Rezeption in den Wissenschaften während der Renaissance*, Weinheim, 1983.
Burke, John G. (ed.), *The Uses of Science in the Age of Newton*, Berkeley and Los Angeles, 1983.
Burtt, Edwin Arthur, *The Metaphysical Foundations of Modern Physical Science*, 2nd edn, Garden City, New York, 1954.
Busschof, Herman and Roonhuis, Hermann, *Two treatises, the One Medical, of the Gout, . . . the Other Partly Chirurgical, Partly Medical*, London, 1676.
Butterfield, H., *The Origins of Modern Science: 1300–1800*, Cambridge, 1949.
Campani, Giuseppe, *Ragguaglio di due Nuove Osservazioni*, Rome 1664.
Camporesi, Piero, 'Cultura popolare e cultura d'élite fra medioevo ed età moderna', *Stor. Ital.*, 1981, **4**: 79–157.
Camporesi, Piero, *The Incorruptible Flesh: Bodily Mutilation and Mortification in Religion and Folklore*, trans. Tania Croft-Murray, Cambridge, 1988.
Canny, Nicholas, *The Upstart Earl*, Cambridge, 1982.
Cantor, G. N., *Michael Faraday: Sandemanian and Scientist. A Study of Science and Religion in the Nineteenth Century*, London, 1991.
Carrara, Daniele Mugnai, *La biblioteca di Nicolò Leoniceno*, Florence, 1991.
Carter, B. A. R., 'A mathematical interpretation of Piero della Francesca's *Baptism of Christ*' in Lavin (1981), 149–63.
Casini, Paolo, *Newton e la coscienza europea*, Bologna, 1983.
Casini, Paolo, 'The Crudeli affair: inquisition and reason of state', in Gay, 133–52.
Cavazza, M., 'Bologna and the Royal Society in the seventeenth century', *Notes Rec. Roy. Soc. Lond.*, 1980, **35**: 105–23.
Céard, Jean, Fontaine, Marie Madeleine and Margolin, Jean-Claude (eds.), *Le corps à la Renaissance*, Paris, 1990.
Celsus, Aulus Cornelius, *De medicina*, ed. Friedrich Marx, Leipzig, 1915.
Chamberlayne, Edward, *The Present State of England*, 1st part, London 1669, 7th edn, 1673; 2nd part, London, 1671, 4th edn, 1673.
Charleton, Walter, *Physiologia Epicuro-Gassendo-Charltoniana: or A Fabrick of Science Natural, Upon the Hypothesis of Atoms*, London, 1654; New York and London, 1966.
Chastel, A. (ed.), *The Renaissance: Essays in Interpretation*, London, 1982, first published in Italian, 1979.
Chipman, R. A., 'An unpublished letter of Stephen Gray on electrical experiments, 1707–1708', *ISIS*, 1954, **45**: 33–40.

Christie, J. R. R. 'Sir David Brewster as an historian of science' in A. D. Morrison-Low and J. R. R. Christie, 53–6.
Christie, Richard Copley, *The Old Church and School Libraries of Lancashire*, Manchester, 1885.
Clagett, Marshall, *The Science of Mechanics in the Middle Ages*, Madison, 1959.
Clarke, Arthur C., *2001: A Space Odyssey*, London 1968, reissued London, 1991, includes a reprint of 'The Sentinel'.
Clarke, Arthur C., *The Sentinel*, London, 1991.
Classen, J., 'The first maps of the Moon', *Sky Telescope*, 1969, **37**: 82–3.
Cleidophorus Mystagogus, *Mercury's caducean rod: or, The great and wonderful office of the universal mercury, or God's Viceregent* . . ., London, 1702; 2nd edn, London, 1704.
Cleidophorus Mystagogus, *Trifertes sagani, or Immortal dissolvent. Being a brief . . . discourse of . . . preparing the Liquor Alkahest*, London, 1705.
Clough, Cecil H., 'Thomas Linacre, Cornelio Vitelli, and Humanistic Studies at Oxford', in Madison *et al.*, 1–23.
Coccaeus, Jacobus, *Brief, over de t'samenstellinghen des wereldt welcke in swangh gaen, ende over een nieuwe stellingh sekerder as de selve, een onghemeene bedenckinghe behelsende, door Jacobus Coccaeus: end evertaelt door I.K.V.W.*, Haarlem, 1660.
Coccaeus, Jacobus, *Epistola de mundi, quae circumferentur systematis et novo alio illis certiore dialogismum paradoxem complex; auctore Jacobo Coccaeo*, Amsterdam, 1660.
Cohen, I. Bernard (ed.), *Isaac Newton's Papers & Letters on Natural Philosophy*, Cambridge, 1958.
Cohen, I. Bernard, 'Isaac Newton, the calculus of variations, and the design of ships' in Cohen *et al.*, 169–87.
Cohen, I. Bernard, *The Newtonian Revolution*, Cambridge, 1980.
Cohen, I. Bernard, *Revolution in Science*, Cambridge, MA, 1985.
Cohen, I. B. and Taton, R. (eds.), *Mélanges Alexandre Koyré: Tome I, L'aventure de la science*, Paris, 1964.
Cohen, R. S., Stachel, J. J. and Wartofsky, M. W. (eds.), *For Dirk Struik, Scientific, Historical and Political Essays* (Boston Studies in the Philosophy of Science, **15**), Dordrecht and Boston, 1974.
Cole, F. J., *A History of Comparative Anatomy: From Aristotle to the Eighteenth Century*, London, 1944.
Collins, John, *Geometricall Dyalling*, London, 1659.
Cook, Harold J., *The Decline of the Old Medical Regime in Stuart London*, Ithaca, 1986.
Cook, Harold J., 'The new philosophy and medicine in seventeenth-century England', in Lindberg and Westman, 397–436.
Cook, Harold J., 'The Rose case reconsidered: physicians, apothecaries, and the law in Augustan England', *J. Hist. Med.*, 1990, **45**: 527–55.
Costa, G., *Saggi sul Settecento*, Naples, 1968.
Craig, John, *The Mint*, Cambridge, 1953.
Cremante, R. and Tega, W. (eds.), *Scienze e Letteratura nella Cultura Italiana del Settecento*, Bologna, 1984.
Cunningham, Andrew, 'How the *Principia* got its name; or, taking natural philosophy seriously', *Hist. Sci.*, 1991, **29**: 377–92.
Dal Fiume, Antonio, 'Un Medico astrologo a Verona nel '500: Tommaso Zefiriele Bovio', *Crit. Stor.*, 1983, **20**: 32–59.
Dall'Osso, Eugenio, 'Due lettere inedite di Leonardo Fioravanti', *Riv. Stor. Sci.*, 1956, **47**: 283–91.
Dary, Michael, *Art of Practical Gauging*, ed. John Newton, London, 1669.
Dary, Michael, *Interest Epitomized, Both Compound and Simple* with *A Short Appendix for the Solution of Adfected Equations in Numbers by Approachment*, London, 1677.
Dary, Michael, *Miscellanies*, London, 1669.
Daston, Lorraine, 'History of science in elegiac mode: E. A. Burtt's 'Metaphysical Foundations of Modern Science' revisited', *ISIS*, 1991, **82**: 522–31.

Dear, Peter, *Mersenne and the Learning of the Schools*, Ithaca, 1988.
De Backer, C. (ed.), *Festschrift for Willy Braekman*, Brussels, forthcoming.
De Beer, G. R., *Sir Hans Sloane and the British Museum*, Oxford, 1953.
Debus, Allen G., *Man and Nature in the Renaissance*, Cambridge, 1978.
Debus, Allen G., *The English Paracelsians*, New York, 1966.
De Dominis, Marco Antonio, *De radiis visus et lucis in perspectivis et iride*, Venice, 1611.
Demandt, Alexander, 'Was wäre Europa ohne die Antike', in Kneissl and Losemann, 113–129.
Dennis, Michael Aaron, 'Graphic understanding: instruments and interpretations in Robert Hooke's *Micrographia*', *Sci. Context*, 1989, **3**: 309–64.
Desaguliers, J. T., 'An account of an Optical Experiment made before the Royal Society, on Thursday, Dec. 6th, and repeated on the 13th, 1722', *Phil. Trans.*, 1722, **32**: 206–8.
Desaguliers, J. T., 'Optical Experiments made in the Beginning of August 1728, before the President and several Members of the Royal Society, and other Members of several Nations, upon the Occasion of Signor Rizzetti's Opticks, with an Account of the said Book', *Phil. Trans.*, 1728, **35**: 596–630.
Descartes, René, *Correspondance*, 8 vols., Paris, 1936–63.
Descartes, René, *Discours de la Méthode Pour bien conduire sa raison, & chercher la verité dans les sciences. Plus La Dioptrique. Les Meteores. et La Geometrie. Qui sont des essais de cete Methode*, Leiden, 1637.
Descartes, René, *Œuvres de Descartes*, ed., P. Costabel, J. Beaude and B. Rochot, 11 volumes, Paris, 1964–74.
Descartes, René, *Principes de la philosophie*, Paris, 1668.
Descartes, René, *Principia Philosophiae*, Amsterdam, 1664.
Descartes, René, *Principles of Philosophy*, trans. V. R. and R. P. Miller, Dordrecht, 1983.
Deursen, A. T. van, *Plain Lives in a Golden Age: Popular Culture, Religion and Society in Seventeenth-Century Holland*, trans. Maarten Ultee, Cambridge, 1991.
De Zan, M., 'La messa al Indice del "Newtonianismo per le dame" di Francesco Algarotti', in Cremante and Tega, 133–47.
Diels, H., *Anonymi Londinensis ex Aristotelis Iatricis Menoniis et aliis Medicis*, Berlin, 1893 (trans. W. H. S. Jones, *The Medical Writings of Anonymus Londinensis*, Cambridge, 1947).
Digby, Kenelm, *Two Treatises. In the One of Which, The Nature of Bodies, in the Other of Man's Soule; is Looked Into*, Paris, 1644.
Dijksterhuis, E. J., *Clio's Stiefkind*, ed. K. van Berkel, Amsterdam, 1990.
Dijksterhuis, E. J., *The Mechanization of the World Picture*, trans. C. Dikshoorn, Oxford, 1961. Original Dutch edition, Amsterdam, 1950.
Dionisotti, A. C., Grafton, Anthony and Kraye, Jill (eds.), *The Uses of Greek and Latin. Historical Essays*, London, 1988.
Dobbs, Betty Jo Teeter, 'Conceptual problems in Newton's early chemistry: a preliminary study' in Osler and Farber, 3–32.
Dobbs, Betty Jo Teeter, *Foundations of Newton's Alchemy, or 'The Hunting of the Greene Lyon'*, Cambridge, 1975.
Dobbs, Betty Jo Teeter, 'Newton's alchemy and his "active principle" of gravitation' in Scheurer and Debrock, 55–80.
Dobbs, Betty Jo Teeter, 'Newton's alchemy and his theory of matter', *ISIS*, 1982, **73**: 511–38.
Dobbs, Betty Jo Teeter, 'Newton's "Commentary" on the "Emerald Tablet" of Hermes Trismegistus: its scientific and theological significance' in Merkel and Debus, 182–91.
Dobbs, Betty Jo Teeter, 'Studies in the natural philosophy of Sir Kenelm Digby', *Ambix*, 1971, **18**: 1–25; 1973, **20**: 143–63; 1974, **21**: 1–28.
Douglas, Mary, *Natural Symbols: Explorations in Cosmology*, New York, 1973.
Douglas, Mary, *Purity and Danger*, London, 1966.
Dragoni, G., 'Vicende dimenticate del mecenatismo bolognese dell'ultimo '700: l'acquisto della collezione di strumentazione scientifiche di Lord Cowper', *Il Carrobbio*, 1985, **9**: 67–85.

Drake, Stillman, *Galileo Studies*, Ann Arbor, 1970.
Drake, S. and Drabkin, I. E. (eds.), *Mechanics in Sixteenth-Century Italy. Selections from Tartaglia, Benedetti, Guido Ubaldo, & Galileo*, Madison, 1969.
Drake, S. and O'Malley, C. D., *The Controversy of the Comets of 1618*, Philadelphia, 1960.
Dronke, P. (ed.), *A History of Twelfth-Century Western Philosophy*, Cambridge, 1988.
Duffy, Christopher, *Siege Warfare. The Fortress in the Early Modern World*, London, 1979.
Duffy, Eamon, 'Valentine Greatrakes, the Irish stroker: miracle, science and orthodoxy in Restoration England', *Stud. Church Hist.*, 1981, **17**: 251–73.
Dugas, René, *La Mécanique au XVIIème Siècle*, Paris, 1954.
Dürer, Albrecht, *Underweysung der Messung mit dem Zirkel und Richtscheyt*, Nuremberg, 1525.
Durling, Richard J., 'A chronological census of Renaissance editions and translations of Galen', *J. Warburg Courtauld Inst.*, 1961, **24**: 230–305.
Durling, Richard J., 'Linacre and medical humanism' in Maddison *et al.*, 84–103.
Duveen, Denis I., *Bibliotheca chemica et alchemica*, 2nd edn, London, 1965.
Eamon, William, 'Science and popular culture in sixteenth-century Italy: the "professors of secrets" and their books', *Sixteenth Cent. J.*, 1985, **16**: 471–85.
Eamon, William, 'Science as a *Venatio*' in De Backer.
Eamon, William, 'The *Secreti* of Alexis of Piedmont, 1555', *Res. Pub. Litt.*, 1979, **2**: 43–55.
Eastwood, Bruce, 'Plinian astronomy in the Middle Ages and Renaissance', in French and Greenaway, 197–235.
Ebels-Hoving, B. and Ebels, E. J., 'Erasmus and Galen', in Weiland and Frijhoff, 132–42.
Eckert, W. and Geyer-Kordesch, J. (eds.), *Heilberufe und Kranke im 17. und 18. Jahrhundert: Die Quellen- und Forschungssituation* (Münstersche Beitrage z. Geschichte u. Theorie d. Medizin, **18**), Münster, 1982.
Edgerton, Samuel Y., *The Heritage of Giotto's Geometry*, Ithaca, 1991.
Edgerton, Samuel Y., 'The Renaissance Artist as Quantifier' in Hagen, **1**: 179–212.
Elkins, J., 'Piero della Francesca and the Renaissance proof of linear perspective', *Art Bull.*, 1987, **69**(2): 220–30.
Elton, G. R., (ed.), *Renaissance and Reformation: 1300–1648*, London, 1963.
Euler, Leonhard, 'Recherches sur la connoissance mécanique des corps', *Opera Omnia*, vol. 2.8.
Evelyn, John, *The Diary of John Evelyn*, ed. E. S. de Beer, 6 volumes, Oxford, 1955.
Fantuzzi, G., *Memorie della vita del generale Co L. F. Marsigli*, Bologna, 1770.
Farrington, Benjamin, *Francis Bacon, philosopher of Industrial Science*, New York, 1949.
Fauvel, J., Flood, R., Shortland, M. and Wilson, R. (eds.) *Let Newton Be!*, Oxford, 1988.
Fauvel, J. and Gray, J. J., *A Source Book in the History of Mathematics*, London, 1987.
Fellmann, E. A., 'The *Principia* and continental mathematicians', *Notes Rec. Roy. Soc. Lond.*, 1988, **42**: 13–34.
Ferguson, John, *Bibliotheca chemica*, 2 volumes, Glasgow, 1906, reprinted, London, 1954.
Ferrone, V., *Scienze natura religione. Mondo newtoniano e cultura italiana nel primo settecento*, Naples, 1982.
Field, J. V., 'Giovanni Battista Benedetti on the mathematics of linear perspective', *J. Warburg Courtauld Inst.*, 1985, **48**: 71–99.
Field, J. V., 'Linear perspective and the projective geometry of Girard Desargues', *Nuncius*, 1987, **2**(2): 3–40.
Field, J. V., 'Piero della Francesca's treatment of edge distortion', *J. Warburg Courtauld Inst.*, 1986, **49**: 66–99.
Field, J. V., 'The natural philosopher as mathematician: Benedetti's mathematics and the tradition of *perspectiva*', in Manno, 247–70.
Field, J. V. and Gray, J. J., *The Geometrical Work of Girard Desargues*, New York, 1987.

Field, J. V., Lunardi, R. and Settle, T. B., 'The perspective scheme of Masaccio's *Trinity* fresco', *Nuncius*, 1988, **4**(2): 31–118.
Figala, Karin, ' "Die exakte Alchemie von Isaac Newton". Seine "gesetzmäßige" Interpretation der Alchemie – dargestellt am Beispiel einiger ihn beeinflussender Autoren', *Verhandl. Naturforsch. Ges. Basel*, 1984, **94**: 157–228.
Figala, Karin, 'Newton as alchemist', *Hist. Sci.*, 1977, **15**: 102–367.
Figala, Karin, 'Zwei Londoner Alchemisten um 1700: Sir Isaac Newton und Cleidophorus Mystagogus', *Physis*, 1976, **18**: 245–73.
Figala, Karin, Harrison, John and Petzold, Ulrich, '*De Scriptoribus Chemicis*: sources for the establishment of Isaac Newton's (al)chemical library' in Harman and Shapiro, 135–79.
Figala, Karin, and Petzold, Ulrich, 'Physics and poetry: Fatio de Duillier's "Ecloga" on Newton's "Principia" ', *Arch. Int. Hist. Sci.*, 1987, **37**: 316–49.
Findlen, Paula, 'Jokes of nature and jokes of knowledge: the playfulness of scientific discourse in early modern Europe', *Ren. Quart.*, 1990, **43**: 292–331.
Fioravanti, Leonardo, *Cappricci medicinali*, Venice, 1561, 1582.
Fioravanti, Leonardo, *Del compendio de i secreti medicinali*, Venice, 1564.
Fioravanti, Leonardo, *Del regimento della peste*, Venice, 1565.
Fioravanti, Leonardo, *Della fisica*, Venice, 1582.
Fioravanti, Leonardo, *Dello specchio de scientia universale*, Venice, 1572.
Fioravanti, Leonardo, *La cirurgia*, Venice, 1570.
Firoavanti, Leonardo, *Tesoro della vita humana*, Venice, 1570.
Firpo, Luigi, *Lo Stato ideale della Controriforma*, Bari, 1957.
Fontana, Francesco, *Novae coelestium, terrestriumque rerum observationes*, Naples, 1646.
Forbes, A., *A History of the Army Ordnance Services*, 3 vols., London, 1929.
Forbes, R. J., *Short History of the Art of Distillation*, Leiden, 1948.
Forster, W., *Janus Gruter's English years*, London, 1967.
Fouke, Daniel C., 'Mechanical and "organical" models in seventeenth-century explanations of biological reproduction', *Sci. Context*, 1989, **3**: 365–81.
Fournier, Marian, 'Jan Swammerdam en de 17e eeuwse microscopie', *Tijd. Gesch. Geneesk. Natuurw. Wisk. Tech.*, 1981, **4**: 74–86.
Fournier, Marian, *The Fabric of Life: The Rise and Decline of Seventeenth-century Microscopy*, Twente University Ph.D. dissertation, 1991.
Franci, R. and Rigatelli, L. Toti, 'Towards a history of algebra from Leonardo of Pisa to Luca Pacioli', *Janus*, 1985, **72**: 17–82.
Francissen, Frans P. M., 'Vroege nederlandese bijdragen tot de kennis van Ephemeroptera of Eendagsvliegen', *Tijd. Gesch. Geneesk. Natuurw. Wisk. Tech.*, 1984, **7**: 113–28.
Frank, Robert G., Jr., 'Institutional structure and scientific activity in the early Royal Society', *Proc. XIVth Int. Cong. Hist. Sci* (Tokyo, 1975), **4**: 82–101.
French, Roger, 'Berengario da Carpi and the use of commentary in anatomical teaching', in Wear *et al.*, 42–74.
French, Roger, 'Pliny and Renaissance medicine' in French and Greenaway, 252–81.
French, Roger and Greenaway, Frank (eds.), *Science in the Early Roman Empire: Pliny the Elder, his Sources and his Influence*, London and Sydney, 1986.
Freudenthal, Gad, '(Al)Chemical foundations for cosmological Ideas: Ibn Sīnā on the geology of an eternal world', in Unguru, 47–73.
Freudenthal, Gad, 'Die elektrische Anziehung im 17. Jahrhundert zwischen korpuskularer und alchemischer Deutung', in Meinel, 315–26.
Freudenthal, Gad, 'Early electricity between chemistry and physics: the simultaneous itineraries of Francis Hauksbee, Samuel Wall, and Pierre Polinière', *Hist. Stud. Phys. Sci.*, 1981, **11**: 203–29.

Freudenthal, Gad., 'The problem of cohesion between alchemy and natural philosophy: from unctuous moisture to phlogiston', in Martels, 107–16.
Fuchs, Leonhard *see* Hippocrates
Fulton, J. F., *A Bibliography of the Honourable Robert Boyle*, 2nd edn, Oxford, 1961.
Furfaro, Domenico, *La vita e l'opera de Leonardo Fioravanti*, Bologna, 1923.
Furth, Montgomery, 'Transtemporal stability in Aristotelian substances', *J. Phil.*, 1978, **75**: 624–46.
Gabbey, Alan, 'Descartes's physics and Descartes's mechanics: chicken and egg?' in Voss, 310–23.
Gabbey, Alan, 'Force and inertia in the seventeenth century: Descartes and Newton' in Gaukroger, 230–320.
Gabbey, Alan, 'Newton's *Mathematical Principles of Natural Philosophy*: a treatise on "mechanics"?' in Harman and Shapiro, 305–22.
Gabbey, Alan, 'The case of mechanics: one revolution or many?' in Lindberg and Westman, 493–528.
Gabbey, Alan, 'The mechanical philosophy and its problems: mechanical explanations, impenetrability, and perpetual motion' in Pitt, 9–84.
Gagnebin, Bernard, 'De la cause de la pesanteur', *Notes Rec. Roy. Soc. Lond.*, 1949, **6**: 105–60.
Galen, *Galeni . . . Exhortatio, etc*, trans. Erasmus, Basle, 1526.
Galen, *Galeni libri anatomici*, Bologna, 1529.
Galen, *Galeni librorum pars prima*, Venice, 1525.
Galen, [*Galeni Therapeuticorum lib. 14 . . .*] (in Greek), Venice, 1500.
Galen, *On Cohesive Causes*, (Corpus medicorum graecorum supplementum orientale **2**) ed. and trans. Malcolm Lyons, Berlin, 1969.
Galen, *See also* Hippocrates and Simon.
Galen, *pseudo*, *De philosopho historia* in Aristotle *Opera*, ed. A. Manutius, Venice, 1497.
Galen, *pseudo*, *Introductio & definitiones medicae*, ed., Sebastian Singkeler, Basle, 1537.
Galiani, C. and Grandi, G., *Carteggio 1714–1729*, ed. F. Palladino and L. Simonutti, Florence, 1989.
Galilei, Galileo, *Dialogo sopra i due massimi sistemi del mondo*, Florence, 1632; trans. S. Drake, Berkeley, 1953.
Galilei, Galileo, *Discoveries and Opinions of Galileo*, ed. and trans. Stillman Drake, New York, 1957.
Galilei, Galileo, *Il Saggiatore*, Rome, 1623.
Galilei, Galileo, *Sidereus Nuncius*, Venice, 1610; trans. Albert Van Helden, Chicago, 1989.
Galluzzi, Paolo and Torrini, Maurizio (eds.), *Le opere dei discepoli di Galileo Galilei. Il Carteggio 1642–1648*, Florence, 1978.
Gandt, François de, 'Les *Mécaniques* attribuées à Aristote et le renouveau de la science des machines au XVIe siècle, *Etudes Phil.*, 1986: 391–405.
Gardair, J. M., *Le 'Giornale de' Letterati' de Rome, 1668–1681*, Florence, 1984.
Gassendi, Pierre, *Mercurius in Sole visus et Venus invisa Parisiis mdcxxxi*, Paris, 1632.
Gauden, John, *A Discourse concerning Publick Oaths*, London, 1662.
Gaukroger, Stephen (ed.), *Descartes: Philosophy, Mathematics and Physics*, Brighton, 1980.
Gay, P. (ed.), *Enlightenment Essays Presented to Arthur M. Wilson*, Hanover NH, 1975.
Gentili, Giuseppe A., 'Leonardo Fioravanti Bolognese alla luce di ignorati documenti', *Riv. Stor. Sci.*, 1951, **42**: 16–41.
Gesner, Conrad, *The Treasure of Evonymus*, trans. Peter Morwyng, London, 1559.
Giannone, Pietro, *Vita scritta da lui medesimo*, Milan, 1960.
Giattini, Giovanni Battista, *Physica P. Io. Baptistae Giattini Panormitani Societatis Jesu in Collegio Romano, ter olim, Philosophiae nunc sacrae Theologiaae Professoris*, Rome, 1653.
Gill, Mary Louise, *Aristotle on Substance. The Paradox of Unity*, Princeton, 1989.
Ginzburg, Carlo, 'High and low: the Theme of forbidden knowledge in the sixteenth and seventeenth centuries', *Past Present*, 1976, **73**: 29–41.
Ginzburg, Carlo, 'The dovecote has opened its eyes: popular conspiracy in seventeenth-century Italy' in Henningsen and Tedeschi, 190–8.

Ginzburg, Carlo and Ferrari, Marco, 'La colombara ha aperto gli occhi', *Quad. Stor.*, 1978, **38**: 631–9.
Giordano, Davide, *Leonardo Fioravanti Bolognese*, Bologna, 1920.
Gjertsen, Derek, *The Newton Handbook*, London, 1986.
Glauber, Rudolph, *The Works . . .*, London, 1689.
Goldie, Mark, 'Sir Peter Pett, sceptical Toryism and the science of toleration in the 1680s', *Stud. Church Hist.*, 1984, **21**: 247–73.
Goldthwaite, R., *The Building of Renaissance Florence*, Baltimore, 1980.
Golinski, Jan, 'The secret life of an alchemist' in Fauvel *et al.*, 147–67.
Gombrich, E. H., *Art and Illusion: A Study in the Psychology of Pictorial Representation*, New York, 1960.
Goodman, David, *Power and Penury: Government, Technology and Science in Philip II's Spain*, Cambridge, 1988.
Gould, S. J., *The Mismeasurement of Man*, New York, 1981.
Grafton, Anthony, *Joseph Scaliger; a Study in the History of Classical Scholarship*, vol. 1, Oxford, 1983.
Grandi, G. *see* Galiani.
Greenberg, J. L., 'Isaac Newton et la théorie de la figure de la Terre', *Rev. Hist. Sci.*, 1987, **40**: 357–66.
Grendler, Paul F., *Critics of the Italian World, 1530–1560*, Madison, 1969.
Grössing, H., *Zur Geschichte der Wiener mathematischen Schülen des 15. und 16. Jahrhunderts*, Baden Baden, 1983.
Guerlac, Henry, 'Francis Hauksbee, expérimentateur au profit de Newton', *Arch. Int. Hist. Sci.*, 1963, **16**: 113–28.
Guerlac, Henry, *Newton on the Continent*, Ithaca and London, 1981.
Guerlac, Henry, 'Newton's optical aether: his draft of a proposed addition to his *Opticks*', *Notes Rec. Roy. Soc. Lond.*, 1967, **22**: 45–57.
Guerlac, Henry, 'Sir Isaac and the ingenious Mr. Hauksbee' in Cohen and Taton, 228–53.
Hacking, Ian, *The Emergence of Probability*, Cambridge, 1975.
Hagen, Margaret A. (ed.), *The Perception of Pictures*, 2 vols., New York, 1980.
Hale, John R., 'Terra ferma fortifications in the Cinquecento' in Bertelli *et al.*, **2**: 169–87.
Hale, John, R., 'The end of Florentine liberty: the Fortezza da Basso' in Rubinstein, 501–32.
Hale, John, R., 'The early develoment of the bastion: an Italian chronology *c.* 1450–*c.* 1534' in Hale *et al.*, 466–94.
Hale, J. R., Highfield, J. R. L. and Smalley, B. (eds.), *Europe in the Late Middle Ages*, Evanston, 1965.
Hall, A. Rupert, 'Architectura Navalis', *Trans. Newcomen Soc.*, 1979–80, **51**: 157–74. (Also in *J. Naval Sci.*, 1984, **11**: 34–50).
Hall, A. Rupert, *Ballistics in the Seventeenth Century*, Cambridge, 1952.
Hall, A. Rupert, 'Beyond the fringe: diffraction as seen by Grimaldi, Fabri, Hooke and Newton', *Notes Rec. Roy. Soc. Lond.*, 1990, **44**: 13–23.
Hall, A. Rupert, *From Galileo to Newton*, London, 1963; New York, 1981; London, 1982.
Hall, A. Rupert, 'Gunnery, science and the Royal Society', in Burke, 111–41.
Hall, A. Rupert, 'La matematica, Newton e la letteratura' in Cremante and Tega, 29–46.
Hall, A. Rupert, 'Newton in France: a new view', *Hist. Sci.*, 1975, **13**: 233–50.
Hall, A. Rupert, 'On Whiggism', *Hist. Sci.*, 1983, **21**: 45–59.
Hall, A. Rupert, 'Sir Isaac Newton's notebook, 1661–65', *Camb. Hist. J.*, 1948, **9**: 239–50.
Hall, A. Rupert, *The Revolution in Science 1500–1750*, London, 1983.
Hall, A. Rupert, *The Scientific Revolution, 1500–1800: The Formation of the Modern Scientific Attitude*, London, 1954, 2nd edn, London, 1962.
Hall, A. Rupert and Hall, Marie Boas, *The Correspondence of Henry Oldenburg*, 13 vols, Madison, London and Philadelphia, 1965–87.
Hall, A. Rupert and Hall, Marie Boas, *Unpublished Scientific Papers of Isaac Newton: A Selection from the Portsmouth Collection in the University Library, Cambridge*, Cambridge, 1962.
Hall, Marie Boas, *Robert Boyle on Natural Philosophy*, Bloomington, 1965.

Hall, Marie Boas, *The Scientific Renaissance, 1450–1630*, Cambridge and London, 1962; New York, 1966.
Hall, Marie Boas, *see also* Boas, M.
Halley, Edmond, *Correspondence and Papers of Edmond Halley*, ed. Eugene Fairfield McPike, London, 1932.
Hanson, N. R., *Patterns of Discovery*, Cambridge, 1958.
Harman, Peter and Shapiro, Alan (eds.), *The Investigation of Difficult Things: Essays on Newton and the History of the Exact Sciences*, Cambridge, 1992.
Harrison, John, *The Library of Isaac Newton*, Cambridge, 1978.
Harting, P., *Het microscoop, deszelfs gebruik, geschiedenis en tegenwoordige toestand*, 3 vols., Utrecht, 1848–50.
Harwood, J. T. (ed.), *The Early Essays and Ethics of Robert Boyle*, Edwardsville and Carbondale, 1991.
Hatfield, Gary, 'Metaphysics and the new science' in Lindberg and Westman, 93–166.
Hauksbee, Francis, 'An Account of Experiments concerning the proportion of the Power of the Load-Stone at different Distances', *Phil. Trans.*, 1710–12, **27**: 506–11.
Hawkins, R. L., 'The friendship of Joseph Scaliger and François Vertunien', *Romantic Rev.*, 1917, **8**: 117–44, 307–27.
Hay, C. (ed.), *Mathematics from Manuscript to Print*, Oxford, 1988.
Haydn, Hiram, *The Counter-Renaissance*, New York, 1950.
Haynes, R., *Philosopher King: the Humanist Pope Benedict XIV*, London, 1970.
Heckscher, W. S., *Rembrandt's Anatomy of Dr. Nicolaas Tulp: An Iconographical Study*, New York, 1958.
Heesakkers, Chris L. 'Foundation and early development of the Athenaeum Illustre at Amsterdam', *Lias*, 1982, **9**: 3–18.
Heilbron, John L., *Electricity in the Seventeenth and Eighteenth Centuries: a Study in Early Modern Physics*, Berkeley, 1979.
Heilbron, John L., *Elements of Early Modern Physics*, Berkeley, 1982.
Heilbron, John L., *Physics at the Royal Society during Newton's Presidency*, Los Angeles, 1983.
Heniger, Johann, 'Der wissenschaftliche Nachlass von Paul Hermann', *Wiss. Z. Univ. Halle*, 1969, **18**: 527–60.
Henning, B. D., *The House of Commons 1660–90*, 3 vols., London, 1983.
Henningsen, Gustav and Tedeschi, John (eds.), *The Inquisition in Early Modern Europe: Studies on Sources and Methods*, DeKalb, 1986.
Henry, John, 'A Cambridge Platonist's materialism: Henry More and the concept of soul', *J. Warburg Courtauld Inst.*, 1986, **49**: 172–95.
Henry, John, 'Occult qualities and the experimental philosophy: active principles in pre-Newtonian matter theory', *Hist. Sci.*, 1986, **24**: 355–81.
Henry, John, 'Robert Hooke, the incongruous mechanist', in Hunter and Schaffer, 149–80.
Hermannus, Paulus, *Paradisus Batavus*, Leiden, 1698.
Hevelius, Johannes, *Machinae Coelestis pars prior*, Gdańsk, 1673.
Hevelius, Johannes, *Selenographia: sive Lunae Descriptio*, Gdańsk, 1647.
Hill, Christopher, *Society and Puritanism in Pre-Revolutionary England*, London, 1964.
Hill, Christopher, 'William Harvey and the idea of monarchy', in Webster, 160–81. (First published in *Past Present* in 1964.)
Hill, Christopher, 'William Harvey (no Parliamentarian, no heretic) and the idea of monarchy', in Webster, 189–96.
Hippiatrica see *Veterinariae . . .*
Hippocrates, *Hippocrate. Des Chairs*, ed. and trans. R. Joly, in *Oeuvres*, **XIII**, Paris, 1978.
Hippocrates, *Hippocratis Epidemiarum liber sextus . . .* (ed. and trans. Leonhard Fuchs), Hagenau, 1532; 2nd edn, Basle, 1537.

Hippocrates, *Hippocratis et Galeni libri aliquot*, Lyon, 1532.
Hippocrates, *Hippokrates Über die Entstehung und Aufbau des menschlichen Körpers (Peri Sarkon)*, ed. and trans. K. Deichgräber, Berlin and Leipzig, 1935.
Hippocrates, *Omnia opera Hippocratis*, Venice, 1526.
Hippocrates, *see also* Baader and Winau.
Hirzgarter, Matthias, *Detectio Dioptrica*, Frankfurt, 1643.
Hobbes, Thomas, *Leviathan*, London, 1651; Oxford, 1909.
Holmes, George, *Florence, Rome and the Origins of the Renaissance*, Oxford, 1986.
Holmyard, E. J. and Mandeville, D. C. (eds. and trans.), *Avicennae De congelatione et conglutinatione lapidum, being sections of the Kitāb al-Shifā*, Paris, 1927.
Home, R. W., *Aepinus's Essay in the Theory of Electricity and Magnetism*, Princeton, 1979.
Home, R. W., 'Force, electricity, and the powers of living matter in Newton's mature philosophy of nature' in Osler and Farber, 95–117.
Home, R. W., 'Francis Hauksbee's theory of electricity', *Arch. Hist. Exact. Sci.*, 1967, **4**: 203–17.
Home, R. W., 'Newton on electricity and the aether' in Bechler, 191–213.
Home, R. W., '"Newtonianism" and the theory of the magnet', *Hist. Sci.*, 1977, **15**: 252–66.
[Hooke, Robert], 'A late Observation about Saturn', *Phil. Trans.*, 1666, **1**: 246–7.
Hooke, Robert, *The Diary of Robert Hooke for the years 1672–1680*, ed., Henry W. Robinson and Walter Adams, London, 1935.
Hooke, Robert, *Micrographia*, London, 1665; New York, 1961; Brussels, 1966.
Howard, Rio, 'Guy de la Brosse and the Jardin des Plantes in Paris' in Woolf, 195–224.
Howard, Rio, 'Guy de La Brosse: Botanique et chimie au début de la révolution scientifique', *Rev. Hist. Sci.*, 1978, **31**: 301–26.
Howard, Rio, *La bibliothèque et le laboratoire de Guy de la Brosse au Jardin des Plantes à Paris* (Ecole Pratique des Hautes Etudes, Histoire et civilisation du livre, n. 13), Geneva, 1983.
Howard, Rio, 'Medical politics and the founding of the Jardin des Plantes in Paris', *J. Soc. Bibliogr. Nat. Hist.*, 1980, **9**: 395–402.
Howse, D. (ed.), *Francis Place and the Early History of the Greenwich Observatory*, New York, 1975.
Howse, D., *Greenwich Observatory: Vol. 3, Buildings and Instruments*, London, 1975.
Hübner, J., *Die Theologie Johannes Keplers zwischen Orthodoxie und Naturwissenschaft*, Tübingen, 1975.
Hunter, Michael, 'Alchemy, magic and moralism in the thought of Robert Boyle', *Brit. J. Hist. Sci.*, 1990, **23**: 387–410.
Hunter, Michael, 'Casuistry in action: Robert Boyle's confessional interviews with Gilbert Burnet and Edward Stillingfleet, 1691', *J. Eccl. Hist.*, 1993, **44**: 80–98.
Hunter, Michael, *Establishing the New Science: The Experience of the Early Royal Society*, Woodbridge, 1989.
Hunter, Michael, *Letters and Papers of Robert Boyle: A Guide to the Manuscripts and Microfilm*, Bethseda, MD, 1992.
Hunter, Michael, 'Science and heterodoxy: an early modern problem reconsidered' in Lindberg and Westman, 437–60.
Hunter, Michael, *The Royal Society and its Fellows, 1660–1700*, Chalfont St Giles, 1982.
Hunter, Michael and Schaffer, Simon (eds.), *Robert Hooke: New Studies*, Woodbridge, 1989.
Hutchison, Keith, 'Dormitive virtues, scholastic qualities, and the new philosphies', *Hist. Sci.*, 1991, **29**: 245–78.
Hutchison, Keith, 'Idiosyncrasy, achromatic lenses, and early Romanticism', *Centaurus*, 1991, **34**: 125–71.
Hutchison, Keith, 'What happened to occult qualities in the Scientific Revolution?', *ISIS*, 1982, **73**: 233–53.
Huygens, Christiaan, *Discours de la Cause de la Pesanteur*, Leiden, 1690.

Huygens, Christiaan, *Horologium oscillatorium sive de motu pendulorum ad horologia* . . ., Paris, 1673. For an English translation *see* Yoder.
Huygens, Constantijn, *De Briefwisseling van Constantijn Huygens(1608–1687)*, ed. J. A. Worp, 6 vols., 's-Gravenhage, 1911–17.
Ibn Sīnā, *Kitāb al-Qānūn fī-l-Tibb* [Canon in Arabic], Rome, 1593.
Ikhwān al-Ṣafā [Encyclopedia of the Brothers of Purity], Rasā'īl, Beirut, 1957 (German translation in F. Dieterrici, *Die Naturanschauung und Naturphilosophie der Araber im zehnten Jahrhundert. Aus den Schriften der lautern Brüder*, Berlin, 1861).
Ingegno, Alfonso, 'Il Medico de' disperati e abbandonati: Tommaso Zeffiriele Bovio (1521–1609) tra Paracelso e l'alchimia del seicento' in *Cultura populare e cultura dotta nel seciento*, Milan, 1985, 164–74.
Instituto delle Scienze, Bologna, *I Materiali dell' Istituto delle Scienze*, Bologna, 1979.
Jacob, J. R., 'Boyle's atomism and the Restoration assault on pagan naturalism', *Soc. Stud. Sci.*, 1978, **8**: 211–33.
Jacob, J. R., *Robert Boyle and the English Revolution*, New York, 1977.
Jacob, J. R. and Jacob, M. C., 'The Anglican origins of modern science: the metaphysical foundations of the Whig constitution', *ISIS*, 1980, **71**: 251–67.
Jamnitzer, Wentzel, *Perspectiva corporum regularium*, Nuremberg, 1568.
Jardine, N., *The Birth of History and Philosophy of Science: Kepler's A Defence of Tycho against Ursus, with essays on its provenance and significance*, Cambridge, 1984.
Johnston, S., 'Mathematical practitioners and instruments in Elizabethan England', *Ann. Sci.*, 1991, **48**: 319–44.
Jorden, E., *A Discourse of Naturall Bathes, And Minerall Waters* . . ., 2nd edn, London, 1632.
Jurina, K., *Vom Quacksalber zum Doctor medicinae*, Cologne, 1985.
Kemp, M. J., *Leonardo, the Marvellous Works of Nature and Man*, London, 1981.
Kemp, M. J., *The Science of Art*, New Haven and London, 1990.
Kemp, M. J. and Massing, A., 'Paolo Uccello's "Hunt in the forest"', *Burlington Mag.*, 1991, **133**: 164–78.
Kepler, Johannes, *Astronomia nova* . . ., Heidelberg, 1609. (*KGW* **3**.)
Kepler, Johannes, *Dioptrice*, Augsburg, 1611. (*KGW* **4**.)
Kepler, Johannes, *Harmonices mundi libri V*, Linz, 1619. (*KGW* **6**.)
Kepler, Johannes, *Tabulae Rudolphinae*, Ulm, 1627. (*KGW* **10**.)
Kneissl, P. and Losemann, V. (eds.), *Alte Geschichte und Wissenschaftsgeschichte. Festschrift für Karl Christ*, Darmstadt, 1988.
Kopal, Zdeněk, *The Moon*, Dordrecht, 1969.
Kopal, Zdeněk and Carder, Robert W., *Mapping the Moon: Past and Present*, Dordrecht, 1974.
Koyré, Alexandre, *From the Closed World to the Infinite Universe*, Baltimore, 1957.
Koyré, Alexandre, *Newtonian Studies*, London, 1965; Chicago, 1968.
Kraus, Paul, *Jābir ibn Ḥayyān: Contribution à l'histoire des idées scientifiques dans l'Islam*. Vol. II: *Jābir et la science grecque* (Mémoires présentés à l'Institut d'Egypte, vol. 45), Cairo, 1945, reprinted Paris, 1986.
Kristeller, P. O. and Cranz, F. E. (eds.), *Catalogus translationum et commentariorum*, 4 vols., Washington, 1960–80.
Kubbinga, H. H., 'Newton's theory of matter', in Scheurer and Debrock, 321–41.
Kuhn, T. S., *The Copernican Revolution*, Cambridge MA, 1957.
Kuhn, T. S., *The Structure of Scientific Revolutions*, Chicago, 1962.
Kuijlen, J., Oldenburger-Ebbers, C. S. and Wijnands, D. O., *Paradisus batavus: Bibliografie van plantencatalogi van onderwijstuinen, particuliere tuinen en kwekerscollecties in de Nordelijke en Zuidelijke Nederlanden (1550–1839)*, Wageningen, 1983.
Kulischer, Josef, *Allgemeine Wirtschaft Geschichte des Mittelalters und der Neuzeit*, 2 vols., Berlin, 1954.

La Hire, Philippe de, *Traité de mécanique, ou l'on explique tout ce qui est nécessaire dans la pratique des arts, & les propriétés des corps pesants lesquelles ont un plus grand usage dans la physique*, Paris, 1695; reprinted in *Mém. Acad. Roy. Sci.*, 1730, **9**: 1–340.

Laird, W. R., 'Giuseppe Moletti's *Dialogue on Mechanics* (1576)', *Ren. Quart.*, 1987, **40**: 209–23.

Laird, W. R., 'The scope of Renaissance mechanics', *Osiris*, 1986, (2nd series), **2**: 43–68.

Lamy, B., *Traité de perspective* . . ., Paris, 1701.

Lange, Johannes, *Epistolae medicinales*, pref. Nicholas Reusner, Hanover, 1589.

Lapidge, Michael, 'Stoic cosmology' in Rist, 161–85.

Lapidge, Michael, 'The Stoic inheritance' in Dronke, 81–112.

Latronico, Nicola, 'Leonardo Fioravanti bolognese era un ciarlatano?', *Castalia*, 1965, **31**: 162–7.

Latronico, Nicola, 'Una disavventura milanese di Leonardo Fioravanti', *L'Ospedale magg.*, 1941, **29**: 481–2.

Lavin, M. A., *Piero della Francesca's 'Baptism of Christ'*, New Haven and London, 1981.

Lavin, M. A., *Piero della Francesca: 'The Flagellation'*, New York, 1972; 2nd edn, Chicago, 1990.

Leibniz, G. W., *Mathematische Schriften*, vol. 6, ed. C. I. Gerhard, Hildesheim, 1971.

Leites, Edmund (ed.), *Conscience and Casuistry in Early Modern Europe*, Cambridge, 1988.

Leeuwen, Henry van, *The Problem of Certainty in English Thought, 1630–90*, The Hague, 1963.

Levin, Harry, *The Myth of the Golden Age in the Renaissance*, New York, 1972.

Levine, Joseph M., *Doctor Woodward's Shield: History, Science and Satire in Augustan England*, Berkeley, 1977.

Levine, Joseph, M., 'Natural History and the History of the Scientific Revolution', *Clio*, 1983, **13**: 57–73.

Lieburg, J. van, 'Die medizinische Versorgung einer Stadtbevölkerung im 17. Jahrhundert: Die Quellen- und Forschungssituation für Rotterdam' in Eckert and Geyer-Kordesch, 29–48.

Lindberg, David C., 'Conceptions of the Scientific Revolution from Bacon to Butterfield: A preliminary sketch', in Lindberg and Westman, 1–26.

Lindberg, David C., *Theories of Vision from al-Kindi to Kepler*, Chicago, 1976.

Lindberg, David C., and Westman, Robert S. (eds.), *Reappraisals of the Scientific Revolution*, Cambridge, 1990.

Lindeboom, G. A., *Descartes and Medicine*, Amsterdam, 1979.

Lindeboom, G. A., 'Dog and Frog: Physiological Experiments', in Scheurleer and Meyjes, 279–93.

Lindeboom, G. A., *Het Cabinet van Jan Swammerdam (*1637–1680*)*, Amsterdam, 1980.

Lindeboom, G. A., 'Het Collegium Privatum Amstelodamense (1664–1673)', *Ned. Tijd. Geneesk.*, 1975, **119**: 1248–54.

Lindeboom, G. A., 'Jan Swammerdam als microscopist', *Tijd. Gesch. Geneesk. Natuurw. Wisk. Tech.*, 1981, **4**: 87–110.

Lindeboom, G. A., 'Jan Swammerdam (1637–1680) and his *Biblia Naturae*', *Clio Med.*, 1982, **17**: 113–31.

Lindeboom, G. A., *Ontmoeting met Jan Swammerdam*, Kampen, 1980.

Lindeboom, G. A., *The Letters of Jan Swammerdam to Melchisédec Thévenot*, Amsterdam, 1975.

Lloyd, G. E. R., *Magic, Reason and Experience*, Cambridge, 1979.

Lloyd, G. E. R., *The Revolutions of Wisdom: Studies in the Claims and Practice of Ancient Greek Science*, Berkeley, 1987.

Lonie, Iain M., 'The "Paris Hippocratics": teaching and research in Paris in the second half of the sixteenth century' in Wear *et al.*, 155–74, 318–26.

Luther, Martin, *Works*, vol. 1, ed., Jaroslav Pelikan, St Louis, 1958.

Lux, David S., *Patronage and Royal Science in Seventeenth-Century France: The Académie de Physique in Caen*, Ithaca, 1989.

Luyendijk-Elshout, Antonie M., 'Death enlightened: a study of Frederik Ruysch', *J. Am. Med. Ass.*, 1970, **212**(1): 121–6.

Lujendijk-Elshout, Antonie M., 'Le système lymphatique au dix-septième siècle: réalités et fantaisies', *Janus*, 1965, **52**: 283–8.
McAdoo, H. R., *The Structure of Caroline Moral Theology*, London, 1949.
McConnell, A., 'A profitable visit: Luigi Ferdinando Marsigli's studies, commerce and friendships in Holland, 1722–23' in C. S. Maffioli and L. C. Palm (eds.), 189–206.
MacGregor, Arthur, 'A magazin of all manner of inventions', *J. Hist. Coll.*, 1989, **1**: 207–12.
McGuire, J. E. 'Body and void in Newton's *De mundi systemate*: some new sources', *Arch. Hist. Exact Sci.*, 1966, **3**: 206–48.
McGuire, J. E., 'Force, active principles, and Newton's invisible realm', *Ambix*, 1968, **15**: 154–208.
McGuire, J. E. and Tamny, Martin (eds.), *Certain Philosophical Questions: Newton's Trinity Notebook*, Cambridge, 1983.
McKendrick, Neil, Brewer, John and Plumb, J. H. (eds.), *The Birth of a Consumer Society: The Commercialization of Eighteenth-century England*, Bloomington, 1982.
McMullin, Ernan, 'Conceptions of science in the Scientific Revolution' in Lindberg and Westman, 27–92.
McPhail, I. *et al.*, *Alchemy and the Occult. A Catalogue . . . of the Collection of Paul and Mary Mellon . . .*, 4 vols., New Haven, 1968–77.
McPike, Eugene Fairfield, *Hevelius, Flamsteed and Halley: Three Contemporary Astronomers and their Mutual Relations*, London, 1937.
Mach, Ernst, *The Science of Mechanics: A Critical and Historical Account of its Development*, (1st German edn, 1883) La Salle, 1974.
Maddison, F., Pelling, M. and Webster, C.(eds.), *Essays on the life and work of Thomas Linacre*, Oxford, 1977.
Maddison, R. E. W., *The Life of the Honourable Robert Boyle*, London, 1969.
Maffioli, C. S. and Palm, L. C. (eds.), *Italian Scientists in the Low Countries in the XVIIth and XVIIIth Centuries*, Amsterdam and Atlanta, 1989.
Magirus, Johannes, *Physiologiae Peripateticae libri sex cum commentaris*, Cambridge, 1642.
Majocchi, D., *Le vicende di alcuni antichi microscopi lasciati in dono al Museo di fisica della R. università di Bologna*, Bologna, 1928.
Marsigli, L. F., *Autobiografia di Luigi Ferdinando Marsigli*, ed. E. Lovarini, Bologna, 1930.
Manno, A. (ed.), *Cultura, scienze e techniche nella Venezia del cinquecento: Atti del Convegno internazionale di studio 'Giovan Battista Benedetti e il suo tempo'*, Venice: Istituto Veneto di Scienze, Lettere ed Arti, 1987.
Marquet, Yves, *La Philosophie des alchimistes et l'alchimie des philosophes. Jābir ibn Ḥayyān et les 'Frères de la Pureté'* (Islam d'hier et d'aujourd'hui, **31**), Paris, 1988.
Marquet, Yves, 'Quelles furent les relations entre "Jābir ibn Ḥayyān" et les Ihwān aṣ-Ṣafā?', *Stud. Islamica*, 1986, **64**: 39–51.
Martels, Z. R. W. M. von (ed.), *Alchemy Revisited. Proceedings of an International Congress at the University of Gronigen, 17–19 April 1989*, Leiden, 1990.
Martines, Lauro, *Power and Imagination: City-States in Renaissance Italy*, New York, 1979.
Martins, R. de A., 'Huygens e a gravitação Newtoniana', *Cad. Hist. Fil. Ci.*, 1989, **1**: 151–84.
Martins, R. de A., *Sobre o Papel dos Desiderata na Ciência*, Campinas, 1987.
Maurolyco, Francesco, *Photismi de lumine et umbra ad prospectivam radiorum incidentium facientes*, Venice, 1575.
Meinel, Christoph (ed.), *Die Alchemie in der europäischen Kultur- und Wissenschaftsgeschichte* (Wolfenbütteler Forschungen, Band 32), Wiesbaden, 1986.
Mémoires pour servir à l'histoire naturelle des animaux, Paris, 1671.
Mendelson, S. H., *The Mental World of Stuart Women*, Brighton, 1987.
Menghi, Girolamo, *Compendio dell'arte essorcistica*, Venice, 1576.

Mercuriale, Hieronymus, *De arte gymnastica*, intro. Christine Nutton, Stuttgart, 1978.
Mercurio, Scipione, *De gli errori populari d'Italia*, Venice, 1603.
Merkel, I. and Debus, A. G. (eds.), *Hermeticism and the Renaissance: Intellectual History and the Occult in Early Modern Europe*, Washington, 1988.
Merton, Robert K., 'Science, technology and society in seventeenth-century England', *Osiris*, 1938, **4**: 360–632. Reprinted in book form, New York, 1970.
Middleton, W. E. Knowles (ed. and trans.), *Lorenzo Magalotti at the Court of Charles II*, Waterloo, Ont., 1980.
Mignard, F., 'The theory of the figures of the Earth according to Newton and Huygens', *Vistas Astr.*, 1987, **30**: 291–311.
Molière, *Œuvres complètes*, ed. G. Couton, 2 vols., Paris, 1971.
Molière, *The Plays of Molière*, ed. A. R. Waller, Edinburgh, 1926.
Montague, M. F. Ashley, *Edward Tyson, M.D., F.R.S., 1650–1708, and the Rise of Human and Comparative Anatomy in England: A Study in the History of Science* (Memoirs of the American Philosophical Society, no. 20), Philadelphia, 1943.
Montalto, L., *Il Clementino (1595–1875)*, Rome, 1938.
Montucla, J. E., *Histoire des Mathématiques*, 2nd edn, 4 vols., Paris, 1799–1802.
More, Thomas, *Utopia*, London 1516; trans. Edward Surtz, New Haven, 1964.
Morelli, E., *Tre profili: Benedetto XIV, P.S. Mancini, P. Roselli*, Rome, 1955.
Moretti, Tommaso, *General Treatise of Artillery*, London, 1693.
Morrison-Low, A. D. and Christie, J. R. R. (eds.), *'Martyr of Science': Sir David Brewster 1781–1868*, Edinburgh, 1984.
Moschini, G., *Della Letteratura Veneziana*, Venice, 1806.
Mosse, G. L., *The Holy Pretence*, Oxford, 1957.
Mudry, Philippe, 'La médecine romaine: mythe et réalité', *Gesnerus*, 1990, **47**: 133–48.
Muratori, L. A., *Opere*, ed. G. Falco and F. Forti, 2 vols., Milan and Naples, 1964.
Nagel, Alexander, 'How German is it?', *J. Art*, December 1990, **4**: 50.
Nauert, C. G., 'Caius Plinius Secundus', in Kristeller and Cranz, **4**: 297–422.
Newman, William R., *The* Summa Perfectionis *of Pseudo-Geber, A Critical Edition, Translation and Study* (Collection de travaux de l'Académie internationale d'histoire des sciences, tome 35), Leiden, 1991.
Newton, Isaac, *I. Newtoni Optices ... accedunt ... Lectiones Opticae et Opuscula Omnia ad Lucem et Colores pertinentia*, Padua, 1749.
Newton, Isaac, *Isaac Newton's Philosophiae Naturalis principia mathematica*, eds. A. Koyré, I. B. Cohen, Anne Whitman, 2 vols., Cambridge, 1972.
Newton, Isaac, *Mathematical Principles of Natural Philosophy and his System of the World*, trans. Andrew Motte, ed. Florian Cajori, Berkeley and Los Angeles, 1934; 1947.
Newton, Isaac, *Opticks: or a Treatise of the Reflections, Refractions, Inflections and Colours of Light*, 1st edn, London, 1704; 4th edn, London, 1730, reprinted New York, 1952.
Newton, Isaac, *Opuscula Mathematica, Philosophica et Philologica. Collegit partimque Latine vertit ac recensuit Joh. Castilloneus*, 3 vols., Lausanne and Geneva, 1744.
Newton, Isaac, *Philosophiae naturalis principia mathematica*, London, 1687; 2nd edn, Cambridge, 1713; 3rd edn, London, 1726.
Newton, Isaac, *Philosophiae Naturalis Principia Mathematica perpetuis commentariis illustrata communi studio*, ed. T. Le Seur and F. Jacquier, 4 vols., Geneva, 1739–1742.
Newton, Isaac, *The Correspondence of Isaac Newton*, ed. H. W. Turnbull, J. F. Scott, A. R. Hall, L. Tilling, 7 vols., Cambridge, 1959–77.
Newton, Isaac, *Unpublished Scientific Papers of Isaac Newton. A Selection from the Portsmouth Collection in the University Library, Cambridge*, ed. A. Rupert Hall and Marie Boas Hall, Cambridge, 1962.
Niebyl, Peter H., 'Sennert, Van Helmont, and medical ontology', *Bull. Hist. Med.*, 1971, **45**: 115–37.

Noceti, C., *De iride et aurora boreali carmina . . . cum notis R. J. Boscovich*, Rome, 1747.
North, Roger, *Lives of the Norths*, ed. A.Jessopp, 3 vols., London, 1890.
Nutton, V., *From Democedes to Harvey: Studies in the History of Medicine*, London, 1988.
Nutton, V., 'Harvey, Goulston and Galen', *Koroth*, 1985, **8**: 112–22.
Nutton, V., 'Hippocrates in the Renaissance' in Baader and Winau, 420–39.
Nutton, V., 'John Caius and the Eton Galen: medical philology in the Renaissance', *Medizinhistorisches J.*, 1985, **20**: 227–52.
Nutton, V., *John Caius and the manuscripts of Galen*, Cambridge, 1987.
Nutton, V., 'Les exercises et la santé: Hieronymus Mercurialis et la gymnastique médicale' in Céard *et al.*, 295–308.
Nutton, V., 'Prisci dissectionum professores: Greek texts and Renaissance anatomists' in Dionisotti *et al.*, 111–26.
Nutton, V., 'The Legacy of Hippocrates: Greek medicine in the Library of the Medical Society of London', *Trans. Med. Soc. Lond.*, 1986–7, **103**: 30–1.
Nutton, V., 'The perils of patriotism: Pliny and Roman medicine', in French and Greenaway, 30–58.
Oestreich, G., 'Die antike Literatur als Vorbild der praktischen Wissenschaften im 16. und 17. Jahrhundert', in Bolgar, 315–24.
Olby, R. C., Cantor, G. N., Christie, J. R. R. and Hodge, M. J. S. (eds.), *Companion to the History of Modern Science*, London, 1990.
O'Neil, Mary, *Discerning Superstitions: Popular Errors and Orthodox Response in Late Sixteenth Century Italy*, Stanford University Ph.D. dissertation, 1981.
Oribasius, *Oribasii collectaneorum Artis Mecicae Libae . . .*, Paris, 1556.
Osler, Margaret J. and Farber, Paul Lawrence (edes.), *Religion, Science and Worldview: Essays in Honor of Richard S. Westfall*, Cambridge, 1985.
Pacioli, L., *De divina proportione*, Venice, 1509, partially reprinted Milan, 1956.
Pagel, Walter, *From Paracelsus to Van Helmont: Studies in Renaissance Medicine and Science*, ed. Marianne Winder, London, 1986.
Pagel, Walter, 'Harvey and Glisson on irritability with a note on Van Helmont', *Bull. Hist. Med.*, 1967, **47**: 497–514.
Pagel, Walter, *Paracelsus: An Introduction to Philosophical Medicine in the Era of the Renaissance*, Basle and New York, 1958.
Pagel, Walter, *Religion and Neoplatonism in Renaissance Medicine*, edited by M. Winder, London, 1985.
Pagel, Walter, 'Religious motives in the medical history of the seventeenth century', *Bull. Inst. Hist. Med.*, 1935, **3**: 97–128, 213–31, 265–312.
Pagel, Walter, 'The reaction to Aristotle in seventeenth-century biological thought: Campanella, Van Helmont, Glanvill, Charleton, Harvey, Glisson, Descartes', in Underwood, **1**: 489–509.
Panofsky, Erwin, *Renaissance and Renascences in Western Art*, London, 1960.
Pardies, Ignace-Gaston, *Discours du mouvement local. Avec des remarques sur le mouvement de la lumiere . . .*, 2nd edn, Paris, 1674.
Park, Katherine and Daston, Lorraine J., 'Unnatural conceptions: the study of monsters in 16th and 17th-Century France and England', *Past Present*, 1981, **92**: 20–54.
Pastor, L. von, *Storia dei Papi*, Rome, 1933.
Patey, D. L., *Probability and Literary Form*, Cambridge, 1984.
Pearce, E. H., *Annals of Christ's Hospital*, London, 1908.
Pepper, Simon and Adams, Nicholas, *Firearms and Fortifications. Military Architecture and Siege Warfare in Sixteenth-Century Sienna*, Chicago, 1986.
Pepys, Samuel, *The Tangier Papers of Samuel Pepys*, ed. Edwin Chappell, Navy Records Society, **73**, 1935.
Pérez-Ramos, Antonio, *Francis Bacon's Idea of Science and the Maker's Knowledge Tradition*, Oxford, 1988.
Perrault, Claude, *Description anatomique d'un cameleon, d'un castor, d'un dromadaire, d'un ours, et d'une gazelle*, Paris, 1669.

Peters, F. E., *Greek Philosophical Terms: A Historical Lexicon*, New York and London, 1967.
[Pett, Peter], *A Discourse Concerning Liberty of Conscience*, London, 1661.
Pett, Peter, *The Happy Future State of England*, London, 1688.
Piemontese, Alessio, *Secreti del reverendo donno Alessio Piemontese*, Venice, 1555.
Piero della Francesca, *De prospectiva pingendi*, ed. G. Nicco Fasola, Florence, 1942; 2nd edn, ed. E. Battisti, *et al.*, Florence, 1984.
Piero della Francesca, 'L'Opera "De corporibus regularibus" di Pietro Franceschi detto della Francesca, usurpata da Fra' Luca Pacioli', ed G. Mancini, *Mem. R. Accad. Lincei*, 1916, series 5, **14**, fasc. 8B: 441–580.
Piero della Francesca, *Trattato d'abaco*, ed. G. Arrighi, Pisa, 1970.
Pirenne, M. H., *Optics, Painting and Photography*, Cambridge, 1970.
Pitt, J. C. (ed.), *Change and Progress in Modern Science: Papers Related to and Arising from the 4th International Congress of History & Philosophy of Science, Blacksburg, Virginia, November 1982*, Dordrecht, 1985.
Plumb, J. H., 'The Acceptance of Modernity', in McKendrick *et al.*, 316–34.
Pöhlmann, Olga, *Jan Swammerdam: Natuuronderzoeker en Medicus*, trans. H. W. J. Schaap, Amsterdam, 1944.
Polanyi, Michael, *Science, Faith and Society*, London, 1946, Chicago, 1964.
Pope-Hennessy, J., *The Piero della Francesca Trail*, London, 1991.
Pope-Hennessy, J., 'Whose *Flagellation*?', *Apollo*, 1986, **124**: 162–5.
Porter, Roy, *Health for Sale. Quackery in England 1660–1850*, Manchester, 1989.
Posthumus, N. W., *Inquiry into the History of Prices in Holland*, Leiden, 1946–64.
Pratensis, Jason (van de Velde), *De pariente et partu*, Antwerp, 1527.
Prest, John, *The Garden of Eden: the Botanic Garden and the Re-creation of Paradise*, New Haven, 1981.
Ptolemy, Claudius, *Megale Syntaxis [Almagest]* (in Greek) ed. Grynaeus, Basle, 1538.
Ptolemy, Claudius, *Geographia*, ed. Philipp Melanchthon, Basle, 1553.
Ptolemy, Claudius, *Tetrabiblos*, Nuremberg, 1535.
Quincy, L. D. C. H. D., *Mémoires sur la vie de M. le Comte de Marsigli*, Zurich, 1741.
Rattansi, P. M., 'Recovering the Paracelsian milieu', in Shea, 1–26.
Raven, Charles E., *English Naturalists from Neckham to Ray: A Study in the Making of the Modern World*, Cambridge, 1947.
Raven, Charles, E., *Natural Religion and Christian Theology* (The Gifford Lectures 1951), Cambridge, 1953.
Redondi, P., *Galileo eretico*, Turin, 1983.
Redondi, P., *Galileo Heretic*, Princeton, 1987.
Reeds, Karen Meier, 'Renaissance humanism and botany', *Ann. Sci.*, 1976, **33**: 519–42.
Renazzi, F. M., *Storia dell'Università di Roma*, 4 vols., Rome, 1803–6, reprinted Bologna, 1971.
Reusner, Nicholas *see* Lange.
Révérend, Dominique, *La Physique des anciens . . .*, Paris, 1701.
Rex, F., *Zur Theorie der Naturprozesse in der früharabischen Wissenschaft. Das 'Kitāb al-Ikhrāǧ', übersetzt und erklärt. Ein Beitrag zum alchemistischen Weltbild der Gābir-Schriften*, (Collection de travaux de l'Académie internationale d'histoire des sciences, tome 22), Wiesbaden, 1975.
Rhazes *see* Alexander of Tralles
Rhijne, Willem Ten, *Dissertatio de arthritide*, London, 1683.
Riccioli, Giovanni Baptista, *Almagestum novum*, 2 vols., Bologna, 1651.
Richter, Friedrich, 'De Systemate Opticae Newtonianae', *Acta Eruditorum*, supp. VIII, 1724: 227–46.
Riddle, John, 'Dioscorides', in Kristeller and Cranz, **4**: 1–143.
Rigaud, Stephen J. (ed.), *Correspondence of Scientific Men of the Seventeenth Century*, 2 vols., Oxford, 1891.

Risner, Frederick, *Opticae thesaurus*, Basle, 1572.
Rist, John M. (ed.), *The Stoics*, Berkeley, 1978.
Rizzetti, J., *De luminis affectionibus specimen physico-mathematicum*, Treviso and Venice, 1727.
[Rizzetti, J.], *Saggio dell'anti-newtonianismo sopra le leggi del moto e dei colori*, Venice, 1741.
Robertus, Gaudentius (ed.), *Miscellanea Italica Physico-Mathematica*, Bologna, 1692.
Rocchi, Enrico, *Le Fonti storiche dell'architettura militare*, Rome, 1908.
Rose, Paul Lawrence, *The Italian Renaissance of Mathematics*, Geneva, 1975.
Rose, Paul Lawrence and Drake, Stillman, 'The Pseudo-Aristotelian *Questions of Mechanics* in Renaissance culture', *Stud. Ren.*, 1971, **18**: 65–104.
Rosen, George, 'Sir William Temple and the therapeutic use of moxa for gout in England', *Bull. Hist. Med.*, 1970, **44**: 31–9.
Rossi, Paolo, *Francesco Bacone: Dalla magia alla scienza*, Bari, 1957.
Rossi, Paolo, *Francis Bacon: From Magic to Science*, trans. S. Rabinovitch, Chicago and London, 1968.
Rostino, Pietro and Rostino, Ludovico, *Compendio di tutta la cirurgia*, ed. Leonardo Fioravanti, Venice, 1588.
Royal Society, *The Record of the Royal Society of London*, 4th edn, London, 1940.
Royal Society, *The Royal Society Tercentenary Celebrations 15–19 July 1946*, Cambridge, 1947.
Rubinstein, Nicolai (ed.), *Florentine Studies: Politics and Society in Renaissance Florence*, Evanston, 1968.
Ruestow, Edward G., 'The rise of the doctrine of vascular secretion in the Netherlands', *J. Hist. Med.*, 1980, **35**: 265–87, 272.
Rufus of Ephesus, *Ruffi de Vesicae renumque morbis . . . Sorani de utero & muliebri pudendo*, Paris, 1554.
Rupp, Jan C. C., 'Matters of life and death: the social and cultural conditions of the rise of anatomical theatres, with special reference to seventeenth-century Holland', *Hist. Sci.*, 1990, **28**: 263–87.
Ruska, J., *Al-Rāzī's Buch Geheimnis der Geheimnisse* (Quellen und Studien zur Geschichte der Naturwissenschaften und der Medizin, 7), Berlin, 1937.
Ruska, J., 'Der Salmiak in der Geschichte der Alchemie', *Z. Angew. Chem.*, 1928, **41**: 1321–4.
Saccardino, Costantino, *Il Libro nominato la verità de diverse cose, quale minutamente tratta di molte salutifere operationi spagiriche et chimiche*, Bologna, 1621.
Salmon, William, *Medica practica . . .*, London, 1707.
Sambursky, Samuel, *The Physical World of the Greeks*, trans. Merton Dagut, London, 1963.
Sambursky, Samuel, *The Physics of the Stoics*, London, 1959.
Sampson, Margaret, 'Laxity and liberty in seventeenth-century English political thought' in Leites, 72–118.
Sanderson, Robert, *Bishop Sanderson's Lectures on Conscience and Human Law*, ed. C. Wordsworth, Lincoln and London, 1877.
Sanderson, Robert, *De juramento*, London, 1647; English trans., London, 1655.
Sanderson, Robert, *Works*, ed. Jacobson, Oxford, 1854.
Sargent, Rose-Mary, 'Scientific experiment and legal expertise: the way of experience in seventeenth-century England', *Stud. Hist. Phil. Sci.*, 1989, **20**: 19–45.
Sarton, George, *Sarton on the History of Science. Essays by George Sarton*, selected and ed. D. Stimson, Cambridge, MA, 1962.
Sarton, George, *The Appreciation of Ancient and Medieval Science During the Renaissance (1450–1600)*, Philadelphia, 1955.
Sarton, George, *The History of Science and the New Humanism*, Cambridge MA, 1937.
Schaffer, Simon, 'Godly men and mechanical philosophers: souls and spirits in Restoration natural philosophy', *Sci. Context*, 1987, **1**: 55–85.
Scheiner, Christoph, *Rosa Ursina*, Bracciano, 1630.
Scheurer, P. B. and Debrock, G. (eds.), *Newton's Scientific and Philosphical Legacy*, Dordrecht, 1988.
Scheurleer, T. H. Lunsingh, 'Un amphitheâtre d'anatomie moralisée' in Scheurleer and Meyjes, 217–77.

Scheurleer, T. H. Lunsingh and Meyjes, G. H. M. Posthumus (eds.), *Leiden University in the Seventeenth Century: An Exchange of Learning*, Leiden, 1975.
Schmitt, Charles B., 'Theophrastus', in Kristeller and Cranz, **2**: 239–322.
Schmitt, Charles B., 'Thomas Linacre and Italy', in Madison *et al.*, 36–75.
Schmitt, Charles B., Skinner, Quentin, Kessler, Eckhard and Kraye, Jill (eds.), *The Cambridge History of Renaissance Philosophy*, Cambridge and New York, 1988.
Schupbach, W., *The Paradox of Rembrandt's Anatomy of Dr. Tulp*, London, 1982.
Schyrlaeus de Rheita, A. M., *Novem stellae circum Iovem, circa Saturnum sex, circa Martem nonnullae, a Antonio Rheita detectae & satellitibus adiudicatae. De primus (& si mavelis de universis) Petri Gassendi Iudicium. Ioannis Caramuel Lobkowitz eiusdem Iudicij censura*, Louvain,1643.
Screech, M. A. and Rawles, S., *A New Rabelais Bibliography*, Geneva, 1987.
Severinus, Petrus, *Idea medicinae philosophicae*, The Hague, 1660.
Shanahan, Timothy, 'God and nature in the thought of Robert Boyle', *J. Hist. Phil.*, 1988, **26**: 547–69.
Shapin, Steven, 'Pump and circumstance: Robert Boyle's literary technology', *Soc. Stud. Sci.*, 1984, **14**: 481–520.
Shapin, Steven, 'The house of experiment in seventeenth-century England', *ISIS*, 1988, **79**: 373–404.
Shapin, Steven, 'The invisible technician', *Am. Scientist*, 1989, **77**: 554–63.
Shapin, Steven, 'Robert Boyle and mathematics: reality, representation, and experimental practice', *Sci. Context*, 1988, **2**: 23–58.
Shapin, Steven, 'Who was Robert Hooke' in Hunter and Schaffer, 253–85.
Shapin, Steven and Schaffer, Simon, *Leviathan and the Air Pump: Hobbes, Boyle, and the Experimental Life*, Princeton, 1985.
Shapiro, Barbara, *Probability and Certainty in Seventeenth-century England*, Princeton, 1983.
Shea, William R. (ed.), *Revolutions in Science: Their Meaning and Relevance*, Canton MA, 1988.
Sherborn, Charles Davies, *A History of the Family of Sherborn*, London, 1901.
Sherburne, Edward, *The Poems and Translations of Sir Edward Sherburne*, ed. F. J. Van Beeck, Assen, 1959.
Sherburne, Edward, *The Sphere of Marcus Manilius*, London, 1675.
Sherrington, Charles S., *The Endeavour of Jean Fernel*, Cambridge, 1946.
Simon, M., *Sieben Bücher Anatomie des Galen*, Leipzig, 1906, (trans. W. H. L. Duckworth, M. C. Lyons, B. Towers, as *Galen, On Anatomical Procedures. The Later Books*, Cambridge, 1962 and I. Garofalo, *Galeno, Procedimenti anatomici*, 3 vols., Milan, 1991.
Singer, Charles, *A Short History of Anatomy and Physiology from the Greeks to Harvey*, New York, 1957.
Singer, Charles, *The Evolution of Anatomy*, London and New York, 1925.
Singkeler *see* Galen, *Pseud*
Siraisi, Nancy G., 'Some current trends in the study of Renaissance medicine', *Ren. Quart.*, 1984, **37**: 585–90.
Sirtori, Girolamo, *Telescopium: sive ars perficiendi*, Frankfurt, 1618.
Slights, Camille, 'Ingenious piety: Anglican causistry of the seventeenth century', *Harvard Theo. Rev.*, 1970, **63**: 409–32.
Sloan, Phillip R., 'Natural history, 1670–1802' in Olby *et al.*, 295–313.
Sloane, Hans, *A Voyage To the Islands Madera, Barbados, Nieves, S. Christophers and Jamaica, with the Natural History of the Herbs and Tress, Four-footed Beasts, Fishes, Birds, Insects, Reptiles, &c.*, London, 1707.
Smith, Wesley D., *The Hippocratic Tradition*, Ithaca, 1979.
Solerti, Angelo, *Ferrara e la corte estense nella seconda metà del secolo decimosesto*, Città di Castello, 1900.
Soppelsa, M. L., *Leibniz e Newton in Italia. Il dibattito padovano, 1687–1790*, Trieste, 1989.
Sorabji, Richard, *Matter, Space, and Motion*, Ithaca, 1988.
Soranus *see* Rufus of Ephesus
Sotheby, *Catalogue of the Newton Papers sold by order of the Viscount Lymington*, London, 1936.

Spini, G., 'Giovan Francesco Salvemini "de Castillon" tra illuminismo e protestantismo', in *I Valdesi e l'Europa*, Torre Pellice, 1982, 319–49.

Stapleton, H. E., Azo, R. F. and Husain, Hidāyat, 'Chemistry in Irāq and Persia in the tenth century A.D.', *Mem. Asiatic Soc. Bengal*, 1927, **8**: 317–418.

Starkey, George, *A true light of alchymy*, London, 1709.

Stearns, R. P., 'The relations between science and society in the later seventeenth century' in *The Restoration of the Stuarts: Blessing or Disaster?*, Washington, 1960, 67–75.

Stendhal, *La Chartreuse de Parme*, Paris, 1839.

Steneck, N. H. and Kaplan, Barbara, 'Greatrakes the stroker', *ISIS*, 1982, **73**: 161–85.

Stroup, Alice, *A Company of Scientists: Botany, Patronage, and Community at the Seventeenth-century Parisian Royal Academy of Sciences*, Berkeley, 1990.

Sudduth, W. H., 'Eighteenth-century identifications of electricity with phlogiston', *Ambix*, 1978, **25**: 131–47.

Swammerdam, Jan, *Ephemeri vita; or, the natural history and anatomy of the Ephemeron*, London, 1681.

Sydenham, Thomas, *Tractatus De Podagra et Hydrope*, London, 1683.

Symonds, John Addington, *Renaissance in Italy*, 7 vols., London, 1875–86.

Targioni-Tozzetti, G., *Notizie della vita e delle opera di Pier' Antonio Micheli botanico fiorentino*, Florence, 1858.

Taylor, Brook, 'An Account of an Experiment made by Dr. Brook Taylor assisted by Mr. Hauksbee, in order to discover the Law of Magnetical Attraction', *Phil. Trans.*, 1714–16, **29**: 294–5.

Taylor, Brook, 'Some experiments relating to magnetism', *Phil. Trans.*, 1720–21, **31**: 204–8.

Taylor, E. G. R., *Mathematical Practitioners in Tudor and Stuart England*, Cambridge, 1954.

Taylor, W. B., 'Kinetic theories of gravitation', *Ann. Rep. Smithsonian Inst.*, 1876: 205–82.

Tega, W. (ed.), *Anatomie Accademiche*, 2 vols., Bologna, 1986–7.

Thackray, Arnold, *Atoms and Powers. An Essay on Newtonian Matter-Theory and the Development of Chemistry*, Cambridge, MA, 1970.

Thomas, Keith, *Man and the Natural World: Changing Attitudes in England 1500–1800*, London, 1983.

Thomas, Keith, *Religion and the Decline of Magic*, London, 1971.

Thoren, V. E., *The Lord of Uraniborg: A Biography of Tycho Brahe*, Cambridge, 1990.

Thorndike, Lynn, *A History of Magic and Experimental Science*, vol. 1, New York, 1923.

Tillotson, John, *The Lawfulness, and Obligation of Oaths*, London, 1681.

Todhunter, I., *A History of the Mathematical Theories of Attraction and the Figure of the Earth – from the time of Newton to that of Laplace*, 2 vols., London, 1873; New York, 1962.

Tomlinson, H. C., *Guns and Government*, London, 1979.

Turnbull, H. W., *James Gregory Tercentenary Memorial Volume*, London, 1939.

Tyson, Edward, *Orang-outang, sive homo silvestris: or, the anatomy of a pygmie compared with that of a monkey, an ape, and a man To which is added a philological essay concerning the pygmies, the cynocephali, the satyrs, and sphinges of the ancients. Wherein it will appear that they were either apes or monkeys, and not men, as formerly pretended*, London, 1699.

Underwood, E. A. (ed.), *Science, Medicine and History: Essays on the Evolution of Scientific Thought and Medical Practice written in Honour of Charles Singer*, 2 vols., Oxford, 1953.

Unguru, Sabetai (ed.), *Physics, Cosmology and Astronomy, 1300–1700: Tension and Accommodation* (Boston Studies in the Philosophy of Science, vol. 126), Dordrecht, Boston and London, 1991.

Urbach, Peter, *Francis Bacon's Philosophy of Science: An Account and Reappraisal*, La Salle, 1987.

Urceus Codrus, Antonius, *A.C.V.U. Sermo primus . . .*, Paris, 1515.

Van Helden, A., 'The telescope in the seventeenth century', *ISIS*, 1974, **64**: 38–58.

Van Helden, A., *see also* Winkler.

Varignon, Pierre, *Nouvelle mécanique, ou statique, dont le project fut donné en M.D.LXXXVII. Ouvrage posthume . . .*, 2 vols., Paris, 1725.

Vasari, Giorgio, *Le Vite dei piu eccellenti architetti, pittori e scultori italiani da Cimabue a tempi nostri*, Florence, 1550, 1568, ed. P. Barocchi and R. Bettarini, Florence, 1971.
Vermij, Rienk H., *Secularisering en natuurwetenschap in de zeventiende en achttiende eeuw: Bernard Nieuwentijt*, Amsterdam, 1991.
Veterinariae medicinae libri duo, Basle, 1537.
Vickers, Brian (ed.), *Occult and Scientific Mentalities in the Renaissance*, Cambridge, 1984.
Vignola *see* Barozzi
Visser, R. P. W., 'Theorie en Praktijk van Swammerdams wetenschappellijke methode in zijn entomologie', *Tijd. Gesch. Geneesk. Natuurw. Wisk. Tech.*, 1981, **4**: 63–73.
Voltaire, *Correspondence, definitive*, ed. T. Besterman, 107 vols., Geneva and Oxford, 1968–77.
Voss, Stephen (ed.), *Essays on the Philosophy and Science of René Descartes*, Oxford, 1993.
Walker, D. P., *Studies in Musical Science in the Late Renaissance*, London, 1978.
Wallace, William A., *Galileo and his Sources: The Heritage of the Collegio Romano in Galileo's Science*, Princeton, 1984.
Wallace, Wiliam A., 'Traditional natural philosophy' in Schmitt *et al.*, 201–35.
Wallis, John, *Mechanica: sive, de motu, tractatus geometricus*, Oxford, 1670–1.
Wallis, John, *Opera mathematica*, 3 vols., Oxford, 1693–99.
Wallis, P. J. and Wallis, R. V. *et al.*, *Eighteenth Century Medics*, 2nd edn, Newcastle, 1988.
Walton, Isaac, *Lives*, ed. George Saintsbury, London, 1927.
Warner, Deborah Jean, 'What is a scientific instrument, when did it become one, and why?', *Brit. J. Hist. Sci.*, 1990, **23**: 83–93.
Wear, A., French, R. K. and Lonie, I.M. (eds.), *The Medical Renaissance of the Sixteenth Century*, Cambridge, 1985.
Webster, Charles, *The Great Instauration: Science, Medicine and Reform 1626–1660*, New York, 1975; Harmondsworth, 1976.
Webster, Charles, (ed.), *The Intellectual Revolution of the Seventeenth Century*, London, 1974.
Weiland, J. Sperna and Frijhoff, W. (eds.), *Erasmus of Rotterdam*, Leiden, 1988.
Westfall, Richard S., 'Alchemy in Newton's library', *Ambix*, 1984, **31**: 97–101.
Westfall, Richard S., *Force in Newton's Physics: The Science of Dynamics in the Seventeenth Century*, London and New York, 1971.
Westfall, Richard S., *Never at Rest: A Biography of Isaac Newton*, Cambridge, 1980.
Westfall, Richard S., 'Newton and alchemy' in Vickers, 315–35.
Westfall, Richard S., *Science and Religion in Seventeenth-century England*, New Haven, 1958.
Westfall, Richard S., *The Construction of Modern Science: Mechanisms and Mechanics*, New York, 1971; Cambridge, 1977.
Westfall, Richard S., 'The foundations of Newton's philosophy of nature', *Brit. J. Hist. Sci.*, 1962, **1**: 171–82.
Whewell, W., *History of the Inductive Sciences from the Earliest to the Present Time*, 3 vols., London, 1837.
Whitaker, Ewen A., 'Galileo's lunar observations and the dating of the composition of "Sidereus Nuncius" ', *J. Hist. Ast.*, 1978, **9**: 155–69.
Whitehead, Alfred North, *Science and the Modern World*, Cambridge 1926, New York, 1967.
Whitteridge, G., 'William Harvey: a Royalist and no Parliamentarian', in Webster, 182–8.
Wilamowitz, U. von, *History of Classical Scholarship*, London, 1982.
Willmoth, Frances, *Sir Jonas Moore: practical mathematics and Restoration science*, Woodbridge, 1993.
Willmoth, Frances, 'Sir Jonas Moore (1617–79): practical mathematician and patron of science', University of Cambridge Ph.D. thesis, 1990.
Wilson, Catherine, 'Visual surface and visual symbol: the microscope and the occult in Early Modern Europe', *J. Hist. Ideas*, 1988, **49**: 85–108.

Winkler, Mary G. and Van Helden, Albert, 'Representing the Heavens: Galileo and Visual Astronomy', *ISIS*, 1992, **83**: 195–217.

Witelo *see* Risner

Wittkower, R. and Carter, B. A. R., 'The perspective of Piero della Francesca's "Flagellation" ', *J. Warburg Courtauld Inst.*, 1953, **16**: 292–302.

Wolf, C., 'Memoires sur le pendule' in Société Française de Physique, *Collection de Mémoires relatifs à la Physique*, 5 vols., Paris, 1884–91, vol. 4.

Wood, Thomas, *English Casuistical Divinity during the Seventeenth Century with Special Reference to Jeremy Taylor*, London, 1952.

Woolf, Harry (ed.), *The Analytic Spirit*, Ithaca, 1981.

Worcester, Marquis of, *A Century of the Names and Scantlings of such Inventions, As at present I can call to mind to have tried and perfected* . . ., London, 1663.

Yates, Frances A., *The Theatre of the World*, London, 1969.

Yoder, J., *Unrolling Time, Christiaan Huygens and the Mathematization of Nature*, Cambridge, 1984.

Yworth, Theophrastus, *A Brief . . . Account of the Vertue, . . . of Certain . . . Medicines, Faithfully prepared as in my Father's Days*, n.p., n.d.

Yworth, W., *A new art of making wines, brandy and other spirits compliant to the late Act of Parliament* . . ., London, 1691.

Yworth, W., *A new treatise of artificial wines, or A Bacchean Magazine, in three parts* . . ., London, 1690.

Yworth, W., *Cerevisarii comes: or, The new and true art of brewing* . . ., London, 1692.

Yworth, W., *Chymicus rationalis: or, The fundamental grounds of the chymical art* . . ., London, 1692.

Yworth, W., *Introitus apertus ad artem distillationis; or The Whole art of distillation practically stated* . . ., London, 1692.

Yworth, W., *Pharmacopœa spagyrica nova: or, An Helmontian course* . . ., London, 1705.

Yworth, W., *The Britannian Magazine: or, A new art* . . ., London, 1694.

Yworth, W., *The compleat distiller* . . ., London, 1705.

Zambarelli, L., *Il nobile pontificio Collegio Clementino*, Rome, 1936.

Zanier, Giancarlo, 'La Medicina paracelsiana in Italia: aspetti di un'accoglienza particolare', *Riv. Stor. Fil.*, 1985, **4**: 627–53.

Zannini, G. L. Masetti, 'Prospero Lambertini e la sua educazione al Collegio Clementino (1689–1692)' in Benedict XIV, *Benedetto.*, 141–60.

Zeuthen, H. G., *Geschichte der Mathematik im XVI. und XVII. Jahrhundert*, ed. R. Meyer, Leipzig, 1903.

Index